Conjectures and Refutations

'The central thesis of the essays and lectures gathered together in this stimulating volume is that our knowledge, and especially our scientific knowledge, progresses by unjustified (and unjustifiable) anticipations, by guesses, by tentative solutions to our problems, in a word by conjectures. Professor Popper puts forward his views with a refreshing self-confidence.'

The Times Literary Supplement

'Professor Popper holds that truth is not manifest, but extremely elusive, he believes that men need above all things, open-mindedness, imagination, and a constant willingness to be corrected. In summarizing his views in this way, I have done scant justice to the subtlety and importance of his argument. His own presentation of his case is luminously clear.'

Maurice Cranston, The Listener

Karl

Popper

Conjectures and Refutations

The Growth of Scientific Knowledge

London and New York

First published 1963 by Routledge & Kegan Paul

First published in Routledge Classics 2002
by Routledge
2 Park Square, Milton Park, Abingdon, OX14 4RN
711 Third Ave, New York, NY10017

Routledge is an imprint of the Taylor & Francis Group

Typeset in Joanna by RefineCatch Limited, Bungay, Suffolk

British Library Cataloguing in Publication Data
A catalogue record for this book is available from the British Library

Library of Congress Cataloging in Publication Data
A catalog record for this book has been requested

ISBN 10: 0-415-28593-3 (hbk)
ISBN 10: 0-415-28594-1 (pbk)

ISBN 13: 978-0-415-28593-3 (hbk)
ISBN 13: 978-0-415-28594-0 (pbk)

Printed and bound in Great Britain by
TJ International Ltd, Padstow, Cornwall

TO

F. A. von HAYEK

Experience is the name every one gives to their mistakes.

OSCAR WILDE

Our whole problem is to make the mistakes as fast as possible . . .

JOHN ARCHIBALD WHEELER

CONTENTS

REFUTATIONS

Preface

The essays and lectures of which this book is composed are variations upon one very simple theme—the thesis that *we can learn from our mistakes*. They develop a theory of knowledge and of its growth. It is a theory of reason that assigns to rational arguments the modest and yet important role of criticizing our often mistaken attempts to solve our problems. And it is a theory of experience that assigns to our observations the equally modest and almost equally important role of tests which may help us in the discovery of our mistakes. Though it stresses our fallibility it does not resign itself to scepticism, for it also stresses the fact that knowledge can grow, and that science can progress—just because we can learn from our mistakes.

The way in which knowledge progresses, and especially our scientific knowledge, is by unjustified (and unjustifiable) anticipations, by guesses, by tentative solutions to our problems, by *conjectures*. These conjectures are controlled by criticism; that is, by attempted *refutations*, which include severely critical tests. They may survive these tests; but they can never be positively justified: they can be established neither as certainly true nor even as 'probable' (in the sense of the probability calculus). Criticism of our conjectures is of decisive importance: by bringing out our mistakes it makes us understand the difficulties of the problem which we are trying to solve. This is how we become better acquainted with our problem, and able to propose more mature solutions: the very refutation of a theory—that is, of any serious tentative solution to our problem—is always a step forward that takes us nearer to the truth. And this is how we can learn from our mistakes.

As we learn from our mistakes our knowledge grows, even though we may never know—that is, know for certain. Since our knowledge can grow, there can be no reason here for despair of reason. And since we can never know for certain, there can be no authority here for any claim to authority, for conceit over our knowledge, or for smugness.

Those among our theories which turn out to be highly resistant to criticism, and which appear to us at a certain moment of time to be better approximations to truth than other known theories, may be described, together with the reports of their tests, as 'the science' of that time. Since none of them can be positively justified, it is essentially their critical and progressive character—the fact that we can *argue* about their claim to solve our problems better than their competitors—which constitutes the rationality of science.

This, in a nutshell, is the fundamental thesis developed in this book and applied to many topics, ranging from problems of the philosophy and history of the physical sciences and of the social sciences to historical and political problems.

I have relied upon my central thesis to give unity to the book, and upon the diversity of my topics to make acceptable the marginal overlapping of some of the chapters. I have revised, augmented, and re-written most of them, but I have refrained from changing the distinctive character of the lectures and broadcast addresses. It would have been easy to get rid of the tell-tale style of the lecturer, but I thought that my readers would rather make allowances for that style than feel that they had not been taken into the author's confidence. I have let a few repetitions stand so that every chapter of the book remains self-contained.

As a hint to prospective reviewers I have also included a review—a severely critical one; it forms the last chapter of the book, and contains an essential part of my argument which is not stated elsewhere in it. I have excluded all those papers which presuppose acquaintance on the part of the reader with technicalities in the field of logic, probability theory, etc. But in the *Addenda* I have put together a few technical notes which may be useful to those who happen to be interested in these things. The *Addenda* and four of the chapters are published here for the first time.

To avoid misunderstandings I wish to make it quite clear that I use the terms 'liberal', 'liberalism', etc., always in a sense in which they are still generally used in England (though perhaps not in America): by a liberal I do not mean a sympathizer with any one political party but simply a man who values individual freedom and who is alive to the dangers inherent in all forms of power and authority.

Berkeley, California, Spring 1962 K. R. P.

Acknowledgements

The place and date of the first publication of the papers here collected are mentioned in each case at the bottom of the first page of each chapter. I wish to thank the editors of the various periodicals for giving me permission to include these papers in the present book.

I have been helped in various ways with the revision of the text, the reading of the proofs, and the preparation of the index, by Richard Gombrich, Lan Freed and Dr. Julius Freed, J. W. N. Watkins, Dr. William W. Bartley, Dr. Ian Jarvie, Bryan Magee, and A. E. Musgrave. I am greatly indebted to all of them for their help. My greatest indebtedness is to my wife. She has worked on the book even harder than I, and her acute criticism has led to innumerable improvements.

K. R. P.

Preface to the Second Edition

This new edition contains, apart from a general revision of the text, a considerable amount of historical material which accumulated since the first edition was printed. As far as possible I have tried to leave the pagination unchanged so that references to the first edition will in almost all cases agree with the second edition. There is also an addition to the end of chapter 5, and a new Addendum (6) at the end of the book. Alan Musgrave has completely revised the indexes, and has also given me much help with improvements to the body of the book.

Having tried, in my first Preface, to sum up my thesis in one sentence—that we can learn from our mistakes—I may perhaps add to it a word or two here. It is part of my thesis that *all* our knowledge grows *only* through the correcting of our mistakes. For example, what is called today 'negative feed back' is only an application of the general method of learning from our mistakes—the method of trial and error.

Now it appears that in order to apply this method we must already have *some aim*: we err if we stray from this aim. (A feedback thermostat depends on *some aim*—some definite temperature—which must be selected in advance.) Yet though in this way some aim must precede any particular instance of the trial and error method, this does not mean that our aims are not in their turn subject to this method. Any particular aim can be changed by trial and error, and many are so changed. (We can change the setting on our thermostat, selecting by trial and error one that better satisfies some aim—an aim of a different level.) And our system of aims not only *changes*, but it can also *grow* in a way closely similar to the way in which our knowledge grows.

Penn, Buckinghamshire, January 1965 K. R. P.

Preface to the Third Edition

Apart from a great number of minor revisions several additions have been made to the text, among them a clearer statement of my views on Tarski's theory of truth (pp. 303 f.). There are also some new *Addenda*.

Penn, Buckinghamshire, April 1968 K. R. P.

Introduction

But I shall let the little I have learnt go forth into the day in order that someone better than I may guess the truth, and in his work may prove and rebuke my error. At this I shall rejoice that I was yet a cause whereby such truth has come to light.

ALBRECHT DÜRER

I can now rejoice even in the falsification of a cherished theory, because even this is a scientific success.

JOHN CAREW ECCLES

ON THE SOURCES OF KNOWLEDGE AND OF IGNORANCE

It follows, therefore, that truth manifests itself . . .

BENEDICTUS DE SPINOZA

Every man carries about him a touchstone . . . to distinguish . . . truth from appearances.

JOHN LOCKE

. . . it is impossible for us to *think* of any thing, which we have not antecedently *felt*, either by our external or internal senses.

DAVID HUME

The title of this lecture is likely, I fear, to offend some critical ears. For although 'Sources of Knowledge' is in order, and 'Sources of Error' would have been in order too, the phrase 'Sources of Ignorance' is another matter. 'Ignorance is something negative: it is the absence of knowledge. But how on earth can the absence of anything have

Annual Philosophical Lecture read before the British Academy on January 20th, 1960. First published in the Proceedings of the British Academy, **46**, 1960, *and separately by Oxford University Press*, 1961.

sources?'[1] This question was put to me by a friend when I confided to him the title I had chosen for this lecture. Hard pressed for a reply I found myself improvising a rationalization, and explaining to my friend that the curious linguistic effect of the title was actually intended. I told him that I hoped to direct attention, through the phrasing of this title, to a number of unrecorded philosophical doctrines and among them (apart from the doctrine that *truth is manifest*) especially to the *conspiracy theory of ignorance* which interprets ignorance not as a mere lack of knowledge but as the work of some sinister power, the source of impure and evil influences which pervert and poison our minds and instil in us the habit of resistance to knowledge.

I am not quite sure whether this explanation allayed my friend's misgivings, but it did silence him. Your case is different since you are silenced by the rules of the present transactions. So I can only hope that I have allayed your misgivings sufficiently, for the time being, to allow me to begin my story at the other end—with the sources of knowledge rather than with the sources of ignorance. However, I shall presently come back to the sources of ignorance, and also to the conspiracy theory of these sources.

I

The problem which I wish to examine afresh in this lecture, and which I hope not only to examine but to solve, may perhaps be described as an aspect of the old quarrel between the British and the Continental schools of philosophy—the quarrel between the classical empiricism of Bacon, Locke, Berkeley, Hume, and Mill, and the classical rationalism or intellectualism of Descartes, Spinoza, and Leibniz. In this quarrel the British school insisted that the ultimate source of all knowledge was

[1] Descartes and Spinoza went even further, and asserted that not only ignorance but also error is 'something negative'—a 'privation' of knowledge, and even of the proper use of our freedom. (See Descartes' *Principles*, Part I, 33–42, and the Third and Fourth *Meditations*; also Spinoza's *Ethics*, Part II, *propos.* 35 and *schol.*; and his *Principles of Descartes' Philosophy*, Part I, *propos.* 15 and *schol.*) Nevertheless, they speak (e.g. *Ethics*, Part II, *propos.* 41) also of the 'cause' of falsity (or error), as does Aristotle, *Met.* 1046a30–35; see also 1008b35; 1009a6; 1052a1; *Top.* 147b29; *An. Post.* 79b23; and *Cat.* 12a26–13a35.

observation, while the Continental school insisted that it was the intellectual intuition of clear and distinct ideas.

Most of these issues are still very much alive. Not only has empiricism, still the ruling doctrine in England, conquered the United States, but it is now widely accepted even on the European Continent as the true theory of *scientific* knowledge. Cartesian intellectualism, alas, has been only too often distorted into one or another of the various forms of modern irrationalism.

In this lecture I shall try to show of the two schools of empiricism and rationalism that their differences are much smaller than their similarities, and that both are mistaken. I hold that they are mistaken although I am myself an empiricist and a rationalist of sorts. But I believe that, though observation and reason have each an important role to play, these roles hardly resemble those which their classical defenders attributed to them. More especially, I shall try to show that neither observation nor reason can be described as a source of knowledge, in the sense in which they have been claimed to be sources of knowledge, down to the present day.

II

Our problem belongs to the theory of knowledge, or to epistemology, reputed to be the most abstract and remote and altogether irrelevant region of pure philosophy. Hume, for example, one of the greatest thinkers in the field, predicted that, because of the remoteness and abstractness and practical irrelevance of some of his results, none of his readers would believe in them for more than an hour.

Kant's attitude was different. He thought that the problem 'What can I know?' was one of the three most important questions a man could ask. Bertrand Russell, in spite of being closer to Hume in philosophic temperament, seems to side in this matter with Kant. And I think Russell is right when he attributes to epistemology practical consequences for science, ethics, and even politics. For he says that epistemological relativism, or the idea that there is no such thing as objective truth, and epistemological pragmatism, or the idea that truth is the same as usefulness, are closely linked with authoritarian and totalitarian ideas. (Cf. *Let the People Think*, 1941, pp. 77 ff.)

Russell's views are of course disputed. Some recent philosophers have developed a doctrine of the essential impotence and practical irrelevance of all genuine philosophy, and thus, one can assume, of epistemology. Philosophy, they say, cannot by its very nature have any significant consequences, and so it can influence neither science nor politics. But I think that ideas are dangerous and powerful things, and that even philosophers have sometimes produced ideas. Indeed, I do not doubt that this new doctrine of the impotence of all philosophy is amply refuted by the facts.

The situation is really very simple. The belief of a liberal—the belief in the possibility of a rule of law, of equal justice, of fundamental rights, and a free society—can easily survive the recognition that judges are not omniscient and may make mistakes about facts and that, in practice, absolute justice is never fully realized in any particular legal case. But the belief in the possibility of a rule of law, of justice, and of freedom, can hardly survive the acceptance of an epistemology which teaches that there are no objective facts; not merely in this particular case, but in any other case; and that the judge cannot have made a factual mistake because he can no more be wrong about the facts than he can be right.

III

The great movement of liberation which started in the Renaissance and led through the many vicissitudes of the reformation and the religious and revolutionary wars to the free societies in which the English-speaking peoples are privileged to live, this movement was inspired throughout by an unparalleled epistemological optimism: by a most optimistic view of man's power to discern truth and to acquire knowledge.

At the heart of this new optimistic view of the possibility of knowledge lies the doctrine that *truth is manifest*. Truth may perhaps be veiled. But it may reveal itself.[2] And if it does not reveal itself, it may be

[2] See my mottoes: Spinoza, *Of God, Man, and Human Happiness*, ch. 15 (parallel passages are: *Ethics*, ii, *scholium to propos.* 43: 'Indeed, as light manifests itself and darkness, so with truth: it is its own standard, and that of falsity.' *De intell. em.*, 35, 36; letter 76 [74], end of para. 5 [7]); Locke, *Cond. Underst.*, 3. (Cp. also *Romans*, i, 19, and see ch. 17, below.)

revealed by us. Removing the veil may not be easy. But once the naked truth stands revealed before our eyes, we have the power to see it, to distinguish it from falsehood, and to know that it *is* truth.

The birth of modern science and modern technology was inspired by this optimistic epistemology whose main spokesmen were Bacon and Descartes. They taught that there was no need for any man to appeal to authority in matters of truth because each man carried the sources of knowledge in himself; either in his power of sense-perception which he may use for the careful observation of nature, or in his power of intellectual intuition which he may use to distinguish truth from falsehood by refusing to accept any idea which is not clearly and distinctly perceived by the intellect.

Man can know: thus he can be free. This is the formula which explains the link between epistemological optimism and the ideas of liberalism.

This link is paralleled by the opposite link. Disbelief in the power of human reason, in man's power to discern the truth, is almost invariably linked with distrust of man. Thus epistemological pessimism is linked, historically, with a doctrine of human depravity, and it tends to lead to the demand for the establishment of powerful traditions and the entrenchment of a powerful authority which would save man from his folly and his wickedness. (There is a striking sketch of this theory of authoritarianism, and a picture of the burden carried by those in authority, in the story of *The Grand Inquisitor* in Dostoievsky's *The Brothers Karamazov.*)

The contrast between epistemological pessimism and optimism may be said to be fundamentally the same as that between epistemological traditionalism and rationalism. (I am using the latter term in its wider sense in which it is opposed to irrationalism, and in which it covers not only Cartesian intellectualism but empiricism also.) For we can interpret traditionalism as the belief that, in the absence of an objective and discernible truth, we are faced with the choice between accepting the authority of tradition, and chaos; while rationalism has, of course, always claimed the right of reason and of empirical science to criticize, and to reject, any tradition, and any authority, as being based on sheer unreason or prejudice or accident.

IV

It is a disturbing fact that even an abstract study like pure epistemology is not as pure as one might think (and as Aristotle believed) but that its ideas may, to a large extent, be motivated and unconsciously inspired by political hopes and by Utopian dreams. This should be a warning to the epistemologist. What can he do about it? As an epistemologist I have only one interest—to find out the truth about the problems of epistemology, whether or not this truth fits in with my political ideas. But am I not liable to be influenced, unconsciously, by my political hopes and beliefs?

It so happens that I am not only an empiricist and a rationalist of sorts but also a liberal (in the English sense of this term); but just because I am a liberal, I feel that few things are more important for a liberal than to submit the various theories of liberalism to a searching critical examination.

While I was engaged in a critical examination of this kind I discovered the part played by certain epistemological theories in the development of liberal ideas; and especially by the various forms of epistemological optimism. And I found that, as an epistemologist, I had to reject these epistemological theories as untenable. This experience of mine may illustrate the point that our dreams and our hopes need not necessarily control our results, and that, in searching for the truth, it may be our best plan to start by criticizing our most cherished beliefs. This may seem to some a perverse plan. But it will not seem so to those who want to find the truth and are not afraid of it.

V

In examining the optimistic epistemology inherent in certain ideas of liberalism, I found a cluster of doctrines which, although often accepted implicitly, have not, to my knowledge, been explicitly discussed or even noticed by philosophers or historians. The most fundamental of them is one which I have already mentioned—the doctrine that truth is manifest. The strangest of them is the conspiracy theory of ignorance, which is a curious outgrowth from the doctrine of manifest truth.

By the doctrine that truth is manifest I mean, you will recall, the optimistic view that truth, if put before us naked, is always recognizable as truth. Thus truth, if it does not reveal itself, has only to be unveiled, or dis-covered. Once this is done, there is no need for further argument. We have been given eyes to see the truth, and the 'natural light' of reason to see it by.

This doctrine is at the heart of the teaching of both Descartes and Bacon. Descartes based his optimistic epistemology on the important theory of the *veracitas dei*. What we clearly and distinctly see to be true must indeed be true; for otherwise God would be deceiving us. Thus the truthfulness of God must make truth manifest.

In Bacon we have a similar doctrine. It might be described as the doctrine of the *veracitas naturae*, the truthfulness of Nature. Nature is an open book. He who reads it with a pure mind cannot misread it. Only if his mind is poisoned by prejudice can he fall into error.

This last remark shows that the doctrine that truth is manifest creates the need to explain falsehood. Knowledge, the possession of truth, need not be explained. But how can we ever fall into error if truth is manifest? The answer is: through our own sinful refusal to see the manifest truth; or because our minds harbour prejudices inculcated by education and tradition, or other evil influences which have perverted our originally pure and innocent minds. Ignorance may be the work of powers conspiring to keep us in ignorance, to poison our minds by filling them with falsehood, and to blind our eyes so that they cannot see the manifest truth. Such prejudices and such powers, then, are sources of ignorance.

The conspiracy theory of ignorance is fairly well known in its Marxian form as the conspiracy of a capitalist press that perverts and suppresses truth and fills the workers' minds with false ideologies. Prominent among these, of course, are the doctrines of religion. It is surprising to find how unoriginal this Marxist theory is. The wicked and fraudulent priest who keeps the people in ignorance was a stock figure of the eighteenth century and, I am afraid, one of the inspirations of liberalism. It can be traced back to the protestant belief in the conspiracy of the Roman Church, and also to the beliefs of those dissenters who held similar views about the Established Church. (Elsewhere I have traced the pre-history of this belief back to Plato's uncle Critias; see chapter 8, section ii, of my *Open Society*.)

This curious belief in a conspiracy is the almost inevitable consequence of the optimistic belief that truth, and therefore goodness, must prevail if only truth is given a fair chance. 'Let her and falsehood grapple; who ever knew Truth put to the worse, in a free and open encounter?' (*Areopagitica*. Compare the French proverb, *La vérité triomphe toujours*.) So when Milton's Truth was put to the worse, the necessary inference was that the encounter had not been free and open: if the manifest truth does not prevail, it must have been maliciously suppressed. One can see that an attitude of tolerance which is based upon an optimistic faith in the victory of truth may easily be shaken. (See J. W. N. Watkins on Milton in *The Listener*, 22nd January 1959.) For it is liable to turn into a conspiracy theory which would be hard to reconcile with an attitude of tolerance.

I do not assert that there was never a grain of truth in this conspiracy theory. But in the main it was a myth, just as the theory of manifest truth from which it grew was a myth.

For the simple truth is that truth is often hard to come by, and that once found it may easily be lost again. Erroneous beliefs may have an astonishing power to survive, for thousands of years, in defiance of experience, with or without the aid of any conspiracy. The history of science, and especially of medicine, could furnish us with a number of good examples. One example is, indeed, the general conspiracy theory itself. I mean the erroneous view that whenever something evil happens it must be due to the evil will of an evil power. Various forms of this view have survived down to our own day.

Thus the optimistic epistemology of Bacon and of Descartes cannot be true. Yet perhaps the strangest thing in this story is that this false epistemology was the major inspiration of an intellectual and moral revolution without parallel in history. It encouraged men to think for themselves. It gave them hope that through knowledge they might free themselves and others from servitude and misery. It made modern science possible. It became the basis of the fight against censorship and the suppression of free thought. It became the basis of the nonconformist conscience, of individualism, and of a new sense of man's dignity; of a demand for universal education, and of a new dream of a free society. It made men feel responsible for themselves and for others, and eager to improve not only their own condition but also that of their fellow men. It is a case of a bad idea inspiring many good ones.

VI

This false epistemology, however, has also led to disastrous consequences. The theory that truth is manifest—that it is there for everyone to see, if only he wants to see it—this theory is the basis of almost every kind of fanaticism. For only the most depraved wickedness can refuse to see the manifest truth; only those who have reason to fear truth conspire to suppress it.

Yet the theory that truth is manifest not only breeds fanatics—men possessed by the conviction that all those who do not see the manifest truth must be possessed by the devil—but it may also lead, though perhaps less directly than does a pessimistic epistemology, to authoritarianism. This is so, simply, because truth is not manifest, as a rule. The allegedly manifest truth is therefore in constant need, not only of interpretation and affirmation, but also of re-interpretation and re-affirmation. An authority is required to pronounce upon, and lay down, almost from day to day, what is to be the manifest truth, and it may learn to do so arbitrarily and cynically. And many disappointed epistemologists will turn away from their own former optimism and erect a resplendent authoritarian theory on the basis of a pessimistic epistemology. It seems to me that the greatest epistemologist of all, Plato, exemplifies this tragic development.

VII

Plato plays a decisive part in the pre-history of Descartes' doctrine of the *veracitas dei*—the doctrine that our intellectual intuition does not deceive us because God is truthful and will not deceive us; or in other words, the doctrine that our intellect is a source of knowledge because God is a source of knowledge. This doctrine has a long history which can easily be traced back at least to Homer and Hesiod.

To us, the habit of referring to one's sources would seem natural in a scholar or an historian, and it is perhaps a little surprising to find that this habit stems from the poets; but it does. The Greek poets refer to the sources of their knowledge. The sources are divine. They are the Muses. '. . . the Greek bards', Gilbert Murray observes (*The Rise of the Greek Epic*, 3rd edn., 1924, p. 96), 'always owe, not only what we should call their inspiration, but their actual knowledge of facts to the Muses. The Muses "are present and know all things" . . . Hesiod . . . always explains that he is dependent on the Muses for his knowledge. Other sources of knowledge are indeed recognized. . . . But most often he consults the Muses. . . . So does Homer for such subjects as the Catalogue of the Greek army.'

As this quotation shows, the poets were in the habit of claiming not only divine sources of inspiration, but also divine sources of knowledge—divine guarantors of the truth of their stories.

Precisely the same two claims were raised by the philosophers Heraclitus and Parmenides. Heraclitus, it seems, sees himself as a prophet who 'talks with raving mouth, . . . possessed by the god'—by Zeus, the source of all wisdom (DK,[3] B 92, 32; cf. 93, 41, 64, 50). And Parmenides, one could almost say, forms the missing link between Homer or Hesiod on the one side and Descartes on the other. His guiding star and inspiration is the goddess Dikē, described by Heraclitus (DK, B 28) as the guardian of truth. Parmenides describes her as the guardian and keeper of the keys of truth, and as the source of all his knowledge. But Parmenides and Descartes have more in common than the doctrine of divine veracity. For example, Parmenides is told by his divine guarantor of truth that in order to distinguish between truth and falsehood, he must rely upon the intellect alone, to the exclusion of the senses of sight, hearing, and taste. (Cf. Heraclitus, B 54, 123; 88 and 126 hint at *unobservable* changes yielding observable opposites.) And even the principle of his physical theory which, like Descartes, he founds upon his intellectualist theory of knowledge, is the same as that adopted by Descartes: it is the impossibility of a void, the necessary fullness of the world.

In Plato's *Ion* a sharp distinction is made between divine inspiration—the divine frenzy of the poet—and the divine sources or origins of true knowledge. (The topic is further developed in the *Phaedrus*, especially from 259e on; and in 275b–c Plato even insists, as

[3] DK = Diels-Kranz, *Fragmente der Vorsokratiker*.

Harold Cherniss pointed out to me, on the distinction between questions of origin and of truth.) Plato grants that the poets are inspired, but he denies to them any divine authority for their alleged knowledge of facts. Nevertheless, the doctrine of the divine source of our knowledge plays a decisive part in Plato's famous theory of *anamnēsis* which in some measure grants to each man the possession of divine sources of knowledge. (The knowledge considered in this theory is knowledge of the *essence* or *nature* of a thing rather than of a particular historical fact.) According to Plato's *Meno* (81b–d) there is nothing which our immortal soul does not know, prior to our birth. For as all natures are kindred and akin, our soul must be akin to all natures. Accordingly it knows them all: it knows all things. (On kinship and knowledge see also *Phaedo*, 79d; *Republic*, 611d; *Laws*, 899d.) In being born we forget; but we may recover our memory and our knowledge, though only partially: only if we see the truth again shall we recognize it. All knowledge is therefore re-cognition—recalling or remembering the essence or true nature that we once knew. (Cp. *Phaedo*, 72e ff.; 75e.)

This theory implies that our soul is in a divine state of omniscience as long as it dwells, and participates, in a divine world of ideas or essences or natures, prior to being born. The birth of a man is his fall from grace; it is his fall from a natural or divine state of knowledge; and it is thus the origin and cause of his ignorance. (Here may be the seed of the idea that ignorance is sin, or at least related to sin; cp. *Phaedo*, 76d.)

It is clear that there is a close link between this theory of *anamnēsis* and the doctrine of the divine origin or source of our knowledge. At the same time, there is also a close link between the theory of *anamnēsis* and the doctrine of manifest truth: if, even in our depraved state of forgetfulness, we see the truth, we cannot but recognize it as the truth. So, as the result of *anamnēsis*, truth is restored to the status of that which is not forgotten and not concealed (*alēthēs*): it is that which is manifest.

Socrates demonstrates this in a beautiful passage of the *Meno* by helping an uneducated young slave to 'recall' the proof of a special case of the theorem of Pythagoras. Here indeed is an optimistic epistemology, and the root of Cartesianism. It seems that, in the *Meno*, Plato was conscious of the highly optimistic character of his theory, for he describes it as a doctrine which makes men eager to learn, to search, and to discover.

Yet disappointment must have come to Plato; for in the *Republic* (and also in the *Phaedrus*) we find the beginnings of a pessimistic epistemology. In the famous story of the prisoners in the cave (514 ff.) he shows that the world of our experience is only a shadow, a reflection, of the real world. And he shows that even if one of the prisoners should escape from the cave and face the real world, he would have almost insuperable difficulties in seeing and understanding it—to say nothing of his difficulties in trying to make those understand who stayed behind. The difficulties in the way of an understanding of the real world are all but super-human, and only the very few, if anybody at all, can attain to the divine state of understanding the real world—the divine state of true knowledge, of *epistēmē*.

This is a pessimistic theory with regard to almost all men, though not with regard to all. (For it teaches that truth may be attained by a few—the elect. With regard to these it is, one might say, more wildly optimistic than even the doctrine that truth is manifest.) The authoritarian and traditionalist consequences of this pessimistic theory are fully elaborated in the *Laws*.

Thus we find in Plato the first transition from an optimistic to a pessimistic epistemology. Each of these forms a basis for one of two diametrically opposed philosophies of the state and of society: on the one hand an anti-traditionalist, anti-authoritarian, revolutionary and Utopian rationalism of the Cartesian kind, and on the other hand an authoritarian traditionalism.

This development may well be connected with the fact that the idea of an epistemological fall of man can be interpreted not only in the sense of the optimistic doctrine of *anamnēsis*, but also in a pessimistic sense.

In this latter interpretation, the fall of man condemns all mortals—or almost all—to ignorance. I think one can discern in the story of the cave (and perhaps also in the story of the fall of the city, when the Muses and their divine teaching are neglected; see *Republic*, 546d) an echo of an interesting older form of this idea. I have in mind Parmenides' doctrine that the opinions of mortals are delusions, and the result of a misguided convention. (This may stem from Xenophanes' doctrine that all human knowledge is guesswork, and that his own theories are, at best, merely *similar to the truth*.[4]) The misguided convention is a linguistic one: it consists in *giving names* to what is non-existent. The idea of an epistemological fall of man can perhaps be found, as Karl Reinhardt suggested, in those words of the goddess that mark the transition from the way of truth to the way of delusive opinion.[5]

> But you also shall learn how it was that delusive opinion,
> Bound to be taken for real, was forcing its way through all things . . .
>
> Now of this world thus arranged to seem wholly like truth I shall tell you;
> Then you will be nevermore led astray by the notions of mortals.

Thus though the fall affects all men, the truth may be revealed to the elect by an act of grace—even the truth about the unreal world of the delusions, opinions, conventional notions and decisions, of mortals: the unreal world of appearance, destined to be accepted, and to be approved of, as real.[6]

The revelation received by Parmenides, and his conviction that a few may reach certainty about both the unchanging world of eternal reality and the unreal and changing world of verisimilitude and deception, were two of the main inspirations of Plato's philosophy. It was a theme to which he was for ever returning, oscillating between hope, despair, and resignation.

[4] Xenophanes' fragment here alluded to is DK, B 35, quoted here in ch. 5, section xii, below. For the idea of *truthlikeness*—of a doctrine that partly corresponds to the facts (and so may be '*taken for real*', as Parmenides has it here)—see especially pp. 320 f., below, where *verisimilitude* is contrasted with *probability*, and the *Addenda* 3, 4, 6, and 7.

[5] See Karl Reinhardt, *Parmenides*, 2nd ed., p. 26; see also pp. 5–11 for the text of Parmenides, DK, B 1: 31–32, which are the first two lines here quoted. My third line is Parmenides, DK, B 8: 60, cf. Xenophanes, B 35. My fourth line is Parmenides, DK, B 8: 61.

[6] It is interesting to contrast this pessimistic view of the necessity of error with the optimism of Descartes, or of Spinoza who scorns (letter 76[74], paragraph 5[7]) those 'who dream of an impure spirit inspiring us with false ideas which are similar to true ones (*veris similes*)'; see also ch. 10, section xiv, and *Addendum* 6, below.

VIII

Yet what interests us here is Plato's optimistic epistemology, the theory of *anamnēsis* in the *Meno*. It contains, I believe, not only the germs of Descartes' intellectualism, but also the germs of Aristotle's and especially of Bacon's theories of induction.

For Meno's slave is helped by Socrates' judicious questions to remember or recapture the forgotten knowledge which his soul possessed in its pre-natal state of omniscience. It is, I believe, this famous Socratic method, called in the *Theaetetus* the art of midwifery or *maieutic*, to which Aristotle alluded when he said (*Metaphysics*, 1078b17–33; see also 987b1) that Socrates was the inventor of the method of induction.

Aristotle, and also Bacon, I wish to suggest, meant by 'induction' not so much the inferring of universal laws from particular observed instances as a method by which we are guided to the point whence we can intuit or perceive the essence or the true nature of a thing.[7] But this, as we have seen, is precisely the aim of Socrates' *maieutic*: its aim is to help or lead us to *anamnēsis*; and *anamnēsis* is the power of seeing the true nature or essence of a thing, the nature or essence with which we were acquainted before birth, before our fall from grace. Thus the aims of the two, *maieutic* and induction, are the same. (Incidentally, Aristotle taught that the result of an induction—the intuition of the essence—was to be expressed by a definition of that essence.)

Now let us look more closely at the two procedures. The *maieutic* art of Socrates consists, essentially, in asking questions designed to destroy

[7] Aristotle meant by 'induction' (*epagōgē*) at least two different things which he sometimes links together. One is a method by which we are 'led to intuit the general principle' (*Anal. Pr.* 67a 22 f., on *anamnēsis* in the *Meno*; *An. Post.*, 71a 7). The other (*Topics* 105a 13, 156a 4; 157a 34; *Anal. Posteriora* 78a 35; 81b 5 ff.) is a method of *adducing* (particular) *evidence*—*positive* evidence rather than *critical* evidence or counter-examples. The first method seems to me the older one, and the one which can be better connected with Socrates and his *maieutic* method of criticism and counter-examples. The second method seems to originate in the attempt to systematize induction logically or, as Aristotle (*Anal. Priora*, 68b 15 ff.) puts it, to construct a valid 'syllogism which springs out of induction'; this, to be valid, must of course be a syllogism of perfect or complete induction (complete enumeration of instances); and ordinary induction in the sense of the second method here mentioned is just a weakened (and invalid) form of this valid syllogism. (Cp. my *Open Society*, note 33 to ch. 11.)

prejudices; false beliefs which are often traditional or fashionable beliefs; false answers, given in the spirit of ignorant cocksureness. Socrates himself does not pretend to know. His attitude is described by Aristotle in the words, 'Socrates raised questions but gave no answers; for he confessed that he did not know.' (*Sophist. El.*, 183b7; cp. *Theaetetus*, 150c–d, 157c, 161b.) Thus Socrates' *maieutic* is not an art that aims at teaching any belief, but one that aims at purging or cleansing (cf. the allusion to the *Amphidromia* in *Theaetetus* 160e; cp. *Phaedo* 67b, 69b/c) the soul of its false beliefs, its seeming knowledge, its prejudices. It achieves this by teaching us to doubt our own convictions.

Fundamentally the same procedure is part of Bacon's induction.

IX

The framework of Bacon's theory of induction is this. He distinguishes in the *Novum Organum* between a true method and a false method. His name for the true method, '*interpretatio naturae*', is ordinarily translated by the phrase 'interpretation of nature', and his name for the false method, '*anticipatio mentis*', by 'anticipation of the mind'. Obvious as these translations may seem, they are misleading. What Bacon means by '*interpretatio naturae*' is, I suggest, the reading of, or better still, *the spelling out of, the book of Nature*. (Galileo, in a famous passage of his *Il saggiatore*, section 6, of which Mario Bunge has kindly reminded me, speaks of 'that great book which lies before our eyes—I mean the universe'; cf. Descartes' *Discourse*, section 1.)

The term 'interpretation' has in modern English a decidedly subjectivistic or relativistic tinge. When we speak of Rudolf Serkin's interpretation of the *Emperor Concerto*, we imply that there are different interpretations, but that this one is Serkin's. We do not of course wish to imply that Serkin's is not the best, the truest, the nearest to Beethoven's intentions. But although we may be unable to imagine that there is a better one, by using the term 'interpretation' we imply that there are other interpretations or readings, leaving the question open whether some of these other readings may, or may not, be equally true.

I have here used the word 'reading' as a synonym for 'interpretation', not only because the two meanings are so similar but also because 'reading' and 'to read' have suffered a modification analogous to that of 'interpretation' and 'to interpret'; except that in the case of 'reading' both meanings are still in full use. In the phrase 'I have read John's letter', we have the ordinary, non-subjectivist meaning. But 'I read this passage of John's letter quite differently' or perhaps 'My reading of this passage is very different' may illustrate a later, a subjectivistic or relativistic, meaning of the word 'reading'.

I assert that the meaning of 'interpret' (though not in the sense of 'translate') has changed in exactly the same way, except that the original meaning—perhaps 'reading aloud for those who cannot read themselves'—has been practically lost. Today even the phrase 'the judge must interpret the law' means that he has a certain latitude in interpreting it; while in Bacon's time it would have meant that the judge had the duty to read the law as it stood, and to expound it and to apply it in the one and only right way. *Interpretatio juris* (or *legis*) means either this or else the expounding of the law to the layman. (Cp. Bacon, *De Augmentis* VI, xlvi; and T. Manley, *The Interpreter: . . . Obscure Words and Terms used in the Lawes of this Realm*, 1672.) It leaves the legal interpreter no latitude; at any rate no more than would be allowed to a sworn interpreter translating a legal document.

Thus the translation 'the interpretation of nature' is misleading; it should be replaced by something like 'the (true) reading of nature'; analogous to 'the (true) reading of the law'. And I suggest that 'reading the book of Nature as it is' or better still 'spelling out the book of Nature' is what Bacon meant. The point is that the phrase should suggest the avoidance of all interpretation in the modern sense, and that it should *not* contain, more especially, any suggestion of an attempt to interpret what is manifest in nature in the light of non-manifest causes or of hypotheses; for all this would be an *anticipatio mentis*, in Bacon's sense. (It is a mistake, I think, to ascribe to Bacon the teaching that hypotheses—or conjectures—may result from his method of induction; for Baconian induction results in *certain* knowledge rather than in conjecture.)

As to the meaning of '*anticipatio mentis*' we have only to quote Locke: 'men give themselves up to the first anticipations of their minds' (*Conduct Underst.*, 26). This is, practically, a translation from Bacon; and it makes it amply clear that '*anticipatio*' means 'prejudice' or even 'superstition'. We can also refer to the phrase '*anticipatio deorum*' which means harbouring naïve or primitive or superstitious views about the gods. But to make matters still more obvious: 'prejudice' (cp. Descartes, *Princ.* I, 50) derives from a legal term, and according to the *Oxford English Dictionary* it was Bacon who introduced the verb 'to prejudge' into the English language, in the sense of 'to judge adversely in advance'—that is, in violation of the judge's duty.

Thus the two methods are (1) 'the spelling out of the open book of Nature', leading to knowledge or *epistēmē*, and (2) 'the prejudice of the mind that wrongly prejudges, and perhaps misjudges, Nature', leading to *doxa*, or mere guesswork, and to the misreading of the book of Nature. This latter method, rejected by Bacon, is in fact a method of interpretation, in the modern sense of the word. It is the *method of conjecture or hypothesis* (a method of which, incidentally, I happen to be a convinced advocate).

How can we prepare ourselves to read the book of Nature properly or truly? Bacon's answer is: by purging our minds of all anticipations or conjectures or guesses or prejudices (*Nov. Org.* i, 68, 69 end). There are various things to be done in order so to purge our minds. We have to get rid of all sorts of 'idols', or generally held false beliefs; for these distort our observations (*Nov. Org.* i, 97). But we have also, like Socrates, to look out for all sorts of counter-instances by which to destroy our prejudices concerning the kind of thing whose true essence or nature we wish to ascertain. Like Socrates, we must, by purifying our intellects, prepare our souls to face the eternal light of essences or natures (cf. St Augustine, *Civ. Dei*, VIII, 3): our impure prejudices must be exorcized by the invocation of counter-instances (*Nov. Org.* ii, 16 ff.).

Only after our souls have been cleansed in this way may we begin the work of spelling out diligently the open book of Nature, the manifest truth.

In view of all this I suggest that Baconian (and also Aristotelian) induction is the same, fundamentally, as Socratic *maieutic*; that is to say, the preparation of the mind by cleansing it of prejudices, in order to enable it to recognize the manifest truth, or to read the open book of Nature.

Descartes' method of systematic doubt is also fundamentally the same: it is a method of destroying all false prejudices of the mind, in order to arrive at the unshakeable basis of self-evident truth.

We can now see more clearly how, in this optimistic epistemology, the state of knowledge is the natural or the pure state of man, the state of the innocent eye which can see the truth, while the state of ignorance has its source in the injury suffered by the innocent eye in man's fall from grace; an injury which can be partially healed by a course of purification. And we can see more clearly why this epistemology, not only in Descartes' but also in Bacon's form, remains essentially a religious doctrine in which the source of all knowledge is divine authority.

One might say that, encouraged by the divine 'essences' or divine 'natures' of Plato, and by the traditional Greek opposition between the truthfulness of nature and the deceitfulness of man-made convention, Bacon substitutes, in his epistemology, 'Nature' for 'God'. This may be the reason why we have to purify ourselves before we may approach the goddess *Natura*: when we have purified our minds, even our sometimes unreliable senses (held by Plato to be hopelessly impure) will be pure. The sources of knowledge must be kept pure, because any impurity may become a source of ignorance.

X

In spite of the religious character of their epistemologies, Bacon's and Descartes' attacks upon prejudice, and upon traditional beliefs which we carelessly or recklessly harbour, are clearly anti-authoritarian and anti-traditionalist. For they require us to shed all beliefs except those whose truth we have perceived ourselves. And their attacks were certainly intended to be attacks upon authority and tradition. They were part of the war against authority which it was the fashion of the time to wage, the war against the authority of Aristotle and the tradition of the schools. Men do not need such authorities if they can perceive the truth themselves.

But I do not think that Bacon and Descartes succeeded in freeing their epistemologies from authority; not so much because they appealed to religious authority—to Nature or to God—but for an even deeper reason.

In spite of their individualistic tendencies, they did not dare to appeal to our critical judgment—to your judgment, or to mine;

perhaps because they felt that this might lead to subjectivism and to arbitrariness. Yet whatever the reason may have been, they certainly were unable to give up thinking in terms of authority, much as they wanted to do so. They could only replace one authority—that of Aristotle and the Bible—by another. Each of them appealed to a new authority; the one to *the authority of the senses*, and the other to *the authority of the intellect*.

This means that they failed to solve the great problem: How can we admit that our knowledge is a human—an all too human—affair, without at the same time implying that it is all individual whim and arbitrariness?

Yet this problem had been seen and solved long before; first, it appears, by Xenophanes, and then by Democritus, and by Socrates (the Socrates of the *Apology* rather than of the *Meno*). The solution lies in the realization that all of us may and often do err, singly and collectively, but that this very idea of error and human fallibility involves another one—the idea of *objective truth*: the standard which we may fall short of. Thus the doctrine of fallibility should not be regarded as part of a pessimistic epistemology. This doctrine implies that we may seek for truth, for objective truth, though more often than not we may miss it by a wide margin. And it implies that if we respect truth, we must search for it by persistently searching for our errors: by indefatigable rational criticism, and self-criticism.

Erasmus of Rotterdam attempted to revive this Socratic doctrine—the important though unobtrusive doctrine, 'Know thyself, and thus admit to thyself how little thou knowest!' Yet this doctrine was swept away by the belief that truth is manifest, and by the new self-assurance exemplified and taught in different ways by Luther and Calvin, by Bacon and Descartes.

It is important to realize, in this connection, the difference between Cartesian doubt and the doubt of Socrates, or Erasmus, or Montaigne. While Socrates doubts human knowledge or wisdom, and remains firm in his rejection of any pretension to knowledge or wisdom, Descartes doubts everything—but only to end up with the possession of *absolutely certain* knowledge; for he finds that his universal doubt would lead him to doubt the truthfulness of God, which is absurd. Having proved that universal doubt is absurd, he concludes that we *can* know securely, that we *can* be wise—by distinguishing, in the natural light of reason, between clear and distinct ideas whose source is God, and all other ideas whose source is our own impure imagination. Cartesian doubt, we see, is merely a *maieutic* instrument for establishing a criterion of truth and, with it, a way to secure knowledge and wisdom. Yet for the Socrates of the *Apology*, wisdom consisted in the awareness of our limitations; in knowing how little we know, every one of us.

It was this doctrine of an essential human fallibility which Nicolas of Cusa and Erasmus of Rotterdam (who refers to Socrates) revived; and it was this 'humanist' doctrine (in contradistinction to the optimistic doctrine on which Milton relied, the doctrine that truth will prevail) which Nicolas and Erasmus, Montaigne and Locke and Voltaire, followed by John Stuart Mill and Bertrand Russell, made the basis of the doctrine of tolerance. 'What is tolerance?' asks Voltaire in his *Philosophical Dictionary*; and he answers: 'It is a necessary consequence of our humanity. We are all fallible, and prone to error; let us then pardon each other's follies. This is the first principle of natural right.' (More recently the doctrine of fallibility has been made the basis of a theory of political freedom; that is, freedom from coercion. See F. A. Hayek, *The Constitution of Liberty*, especially pp. 22 and 29.)

XI

Bacon and Descartes set up observation and reason as new authorities, and they set them up within each individual man. But in doing so they split man into two parts, into a higher part which had authority with respect to truth—Bacon's observations, Descartes' intellect—and a lower part. It is this lower part which constitutes our ordinary selves, the old Adam in us. For it is always 'we ourselves' who are alone responsible for error, if truth is manifest. It is we, with *our* prejudices, *our* negligence, *our* pigheadedness, who are to blame; it is we ourselves who are the sources of our ignorance.

Thus we are split into a human part, we ourselves, the part which is the source of our fallible opinions (*doxa*), of our errors, and of our ignorance; and a super-human part, such as the senses or the intellect, the sources of real knowledge (*epistēmē*), whose authority over us is almost divine.

But this will not do. For we know that Descartes' physics, admirable as it was in many ways, was mistaken; yet it was based only upon ideas which, he thought, were clear and distinct, and which therefore should have been true. And as to the authority of the senses as sources of knowledge, the fact that the senses are not reliable was known to the ancients even before Parmenides, for example to Xenophanes and Heraclitus; and of course to Democritus and to Plato. (Cp. pp. 222 f., below.)

It is strange that this teaching of antiquity could be almost ignored by modern empiricists, including phenomenalists and positivists; yet it is ignored in most of the problems posed by positivists and phenomenalists, and in the solutions they offer. The reason is this: they believe that it is not our senses that err, but that it is always 'we ourselves' who err in *our interpretation* of what is 'given' to us by our senses. Our senses tell the truth, but we may err, for example, when we try to put into *language—conventional, man-made, imperfect language*—what they tell us. It is our linguistic description which is faulty because it may be tinged with prejudice.

(So our man-made language was at fault. But then it was discovered that our language too was 'given' to us, in an important sense: that it embodied the wisdom and experience of many generations, and that it should not be blamed if we misused it. So language too became a truthful authority that could never deceive us. If we fall into temptation and use language in vain, then it is we who are to blame for the trouble that ensues. For Language is a jealous God and will not hold him guiltless that taketh His words in vain, but will throw him into darkness and confusion.)

By blaming *us*, and our language (or misuse of Language), it is possible to uphold the divine authority of the senses (and even of Language). But it is possible only at the cost of widening the gap between this authority and ourselves: between the pure sources from which we can obtain an authoritative knowledge of the truthful goddess Nature, and our impure and guilty selves: between God and man. As indicated before, this idea of the truthfulness of Nature which, I believe, can be discerned in Bacon, derives from the Greeks; for it is part of the classical opposition between *nature* and human *convention* which, according to Plato, is due to Pindar; which may be discerned in

Parmenides; and which is identified by him, and by some Sophists (for example, by Hippias) and partly also by Plato himself, with the opposition between divine truth and human error, or even falsehood. After Bacon, and under his influence, the idea that nature is divine and truthful, and that all error or falsehood is due to the deceitfulness of our own human conventions, continued to play a major role not only in the history of philosophy, of science, and of politics, but also in that of the visual arts. This may be seen, for example, from Constable's most interesting theories on nature, veracity, prejudice, and convention, quoted in E. H. Gombrich's *Art and Illusion*. It also played a role in the history of literature, and even in that of music.

XII

Can the strange view that the truth of a statement may be decided upon by inquiring into its sources—that is to say its *origin*—be explained as due to some logical mistake which might be cleared up? Or can we do no better than explain it in terms of religious beliefs, or in psychological terms—referring perhaps to parental authority? I think that it is indeed possible to discern here a logical mistake which is connected with the close analogy between the *meaning* of our words, or terms, or concepts, and the *truth* of our statements or propositions. (See the table opposite.)

It is easy to see that the meaning of our words does have some connection with their history or their origin. A word is, logically considered, a conventional sign; psychologically considered, it is a sign whose meaning is established by usage or custom or association. Logically considered, its meaning is indeed established by an initial decision—something like a primary definition or convention, a kind of original social contract; and psychologically considered, its meaning was established when we originally learned to use it, when we first formed our linguistic habits and associations. Thus there is a point in the complaint of the schoolboy about the unnecessary artificiality of French in which 'pain' means bread, while English, he feels, is so much more natural and straightforward in calling pain 'pain' and bread 'bread'. He may understand the conventionality of the usage perfectly well, but he gives expression to the feeling that there is no

	IDEAS	
	that is	
DESIGNATIONS or TERMS or CONCEPTS		STATEMENTS or PROPOSITIONS or THEORIES
	may be formulated in	
WORDS		ASSERTIONS
	which may be	
MEANINGFUL		TRUE
	and their	
MEANING		TRUTH
	may be reduced, by way of	
DEFINITIONS		DERIVATIONS
	to that of	
UNDEFINED CONCEPTS		PRIMITIVE PROPOSITIONS
	the attempt to establish (rather than reduce) by these means their	
MEANING		TRUTH
	leads to an infinite regress	

reason why the original conventions—original for him—should not be binding. So his mistake may consist merely in forgetting that there can be several equally binding original conventions. But who has not made, implicitly, the same mistake? Most of us have caught ourselves in a feeling of surprise when we find that in France even little children speak French fluently. Of course, we smile about our own naïvety; but we do not smile about the policeman who discovers that *the real name* of the man called 'Samuel Jones' was 'John Smith'—though here is, no doubt, a last vestige of the magical belief that we gain power over a man or a god or a spirit by gaining knowledge of his *real* name: by pronouncing it we can summon or cite him.

Thus there is indeed a familiar as well as a logically defensible sense in which the 'true' or 'proper' meaning of a term is its original meaning; so that if we understand it, we do so because we learned it correctly—from a true authority, from one who knew the language.

This shows that the problem of the meaning of a word is indeed linked to the problem of the authoritative source, or the origin, of our usage.

It is different with the problem of the truth of a statement of fact, a proposition. For anybody can make a factual mistake—even in matters on which he should be an authority, such as his own age or the colour of a thing which he has just this moment clearly and distinctly perceived. And as to origins, a statement may easily have been false when it was first made, and first properly understood. A word, on the other hand, must have had a proper meaning as soon as it was ever understood.

If we thus reflect upon the difference between the ways in which the meaning of words and the truth of statements is related to their origins, we are hardly tempted to think that the question of origin can have much bearing on the question of knowledge or of truth. There is, however, a deep analogy between meaning and truth; and there is a philosophical view—I have called it 'essentialism'—which tries to link meaning and truth so closely that the temptation to treat both in the same way becomes almost irresistible.

In order to explain this briefly, we may once more contemplate our table of Ideas, noting the relation between its two sides.

How are the two sides of this table connected? If we look at the left side of the table, we find there the word '*Definitions*'. But a definition is a kind of *statement* or *theory* or *proposition*, and therefore one of those things which stand on the right side of our table. (This fact, incidentally, does not spoil the symmetry of our table of Ideas; for derivations also transcend the kind of thing—statements, etc.—which stand on the side where the word 'derivation' occurs: just as a definition is formulated by a special kind of *sequence of words* rather than by a word, so a derivation is formulated by a special kind of *sequence of statements* rather than by a statement.) The fact that definitions, which occur on the left side of our table, are nevertheless statements suggests that somehow they may form a link between the left and the right side of the table.

That they do this is, indeed, part of that philosophic doctrine to which I have given the name 'essentialism'. According to essentialism (especially Aristotle's version of it) a definition is a statement of the inherent essence or nature of a thing. At the same time, it states the meaning of a word—of the name that designates the essence. (For example, Descartes, and also Kant, hold that the word 'body' designates something that is, essentially, extended.)

Moreover, Aristotle and all other essentialists held that *definitions are 'principles'*; that is to say, they yield primitive propositions (example: 'All bodies are extended') which cannot be derived from other propositions, and which form the basis, or are part of the basis, of every demonstration. They thus form the basis of every science. (Cf. my *Open Society*, especially notes 27 to 33 to chapter 11.) It should be noted that this particular tenet, though an important part of the essentialist creed, is free of any reference to 'essences'. This explains why it was accepted by some nominalistic opponents of essentialism such as Hobbes or, say, Schlick. (See the latter's *Erkenntnislehre*, 2nd edition, 1925, p. 62.)

I think we have now the means at our disposal by which we can explain the logic of the view that questions of origin may decide questions of factual truth. For if origins can determine the *true meaning* of a term or word, then they can determine the *true definition* of an important idea, and therefore some at least of the basic 'principles' which are descriptions of the essences or natures of things and which underlie our demonstrations and consequently our scientific knowledge. *So it will then appear that there are authoritative sources of our knowledge.*

Yet we must realize that essentialism is mistaken in suggesting that definitions can add to our *knowledge of facts* (although *qua* decisions about conventions they may be influenced by our knowledge of facts, and although they create instruments which may in their turn influence the formation of our theories and thereby the evolution of our knowledge of facts). Once we see that definitions never give any factual knowledge about 'nature', or about 'the nature of things', we also see the break in the logical link between the problem of origin and that of factual truth which some essentialist philosophers tried to forge.

XIII

I will now leave all these largely historical reflections aside, and turn to the problems themselves, and to their solution.

This part of my lecture might be described as an attack on *empiricism*, as formulated for example in the following classical statement of Hume's: 'If I ask you why you believe any particular matter of fact . . . , you must tell me some reason; and this reason will be some other fact, connected with it. But as you cannot proceed after this manner, *ad infinitum*, you must at last terminate in some fact, which is present to your memory or senses; or must allow that your belief is entirely without foundation.' (*Enquiry Concerning Human Understanding*, Section v, Part I; Selby-Bigge, p. 46; see also my motto, taken from Section vii, Part I; p. 62.)

The problem of the validity of empiricism may be roughly put as follows: is observation the ultimate source of our knowledge of nature? And if not, what are the sources of our knowledge?

These questions remain, whatever I may have said about Bacon, and even if I should have managed to make those parts of his philosophy on which I have commented somewhat unattractive for Baconians and for other empiricists.

The problem of the source of our knowledge has recently been restated as follows. If we make an assertion, we must justify it; but this means that we must be able to answer the following questions.

'*How do you know? What are the sources of your assertion?*'

This, the empiricist holds, amounts in its turn to the question,

'*What observations* (or memories of observations) *underlie your assertion?*' I find this string of questions quite unsatisfactory.

First of all, most of our assertions are not based upon observations, but upon all kinds of other sources. 'I read it in *The Times*' or perhaps 'I read it in the *Encyclopaedia Britannica*' is a more likely and a more definite answer to the question 'How do you know?' than 'I have observed it' or 'I know it from an observation I made last year'.

'But', the empiricist will reply, 'how do you think that *The Times* or the *Encyclopaedia Britannica* got their information? Surely, if you only carry on your inquiry long enough, you will end up with *reports of the observations of eyewitnesses* (sometimes called "protocol sentences" or—by yourself—"basic statements"). Admittedly', the empiricist will continue, 'books are largely made from other books. Admittedly, a historian, for example, will work from documents. But ultimately, in the last analysis, these other books, or these documents, must have been based upon observations. Otherwise they would have to be described as poetry, or invention, or lies, but not as testimony. It is in this sense that we empiricists assert that observation must be the ultimate source of our knowledge.'

Here we have the empiricist's case, as it is still put by some of my positivist friends.

I shall try to show that this case is as little valid as Bacon's; that the answer to the question of the sources of knowledge goes against the empiricist; and, finally, that this whole question of ultimate sources—sources to which one may appeal, as one might to a higher court or a higher authority—must be rejected as based upon a mistake.

First I want to show that if you actually went on questioning *The Times* and its correspondents about the sources of their knowledge, you would in fact never arrive at all those observations by eyewitnesses in the existence of which the empiricist believes. You would find, rather, that with every single step you take, the need for further steps increases in snowball-like fashion.

Take as an example the sort of assertion for which reasonable people might simply accept as sufficient the answer 'I read it in *The Times*'; let us say the assertion 'The Prime Minister has decided to return to London several days ahead of schedule'. Now assume for a moment that somebody doubts this assertion, or feels the need to investigate its truth. What shall he do? If he has a friend in the Prime Minister's office, the simplest and most direct way would be to ring him up; and if this friend corroborates the message, then that is that.

In other words, the investigator will, if possible, try to check, or to examine, *the asserted fact itself*, rather than trace the source of the information. But according to the empiricist theory, the assertion 'I have read it in *The Times*' is merely a first step in a justification procedure consisting in tracing the ultimate source. What is the next step?

There are at least two next steps. One would be to reflect that 'I have read it in *The Times*' is also an assertion, and that we might ask 'What is the source of your knowledge that you read it in *The Times* and not, say, in a paper looking very similar to *The Times?*' The other is to ask *The Times* for the sources of its knowledge. The answer to the first question may be 'But we have only *The Times* on order and we always get it in the morning' which gives rise to a host of further questions about sources which we shall not pursue. The second question may elicit from the editor of *The Times* the answer: 'We had a telephone call from the Prime Minister's Office.' Now according to the empiricist procedure, we should at this stage ask next: 'Who is the gentleman who received the telephone call?'

and then get his observation report; but we should also have to ask that gentleman: 'What is the source of your knowledge that the voice you heard came from an official in the Prime Minister's office', and so on.

There is a simple reason why this tedious sequence of questions never comes to a satisfactory conclusion. It is this. Every witness must always make ample use, in his report, of his knowledge of persons, places, things, linguistic usages, social conventions, and so on. He cannot rely merely upon his eyes or ears, especially if his report is to be of use in justifying any assertion worth justifying. But this fact must of course always raise new questions as to the sources of those elements of his knowledge which are not immediately observational.

This is why the programme of tracing back all knowledge to its ultimate source in observation is logically impossible to carry through: it leads to an infinite regress. (The doctrine that truth is manifest cuts off the regress. This is interesting because it may help to explain the attractiveness of that doctrine.)

I wish to mention, in parenthesis, that this argument is closely related to another—that all observation involves interpretation in the light of our theoretical knowledge,[8] or that pure observational knowledge, unadulterated by theory, would, if at all possible, be utterly barren and futile.

The most striking thing about the observationalist programme of asking for sources—apart from its tediousness—is its stark violation of common sense. For if we are doubtful about an assertion, then the normal procedure is to test it, rather than to ask for its sources; and if we find independent corroboration, then we shall often accept the assertion without bothering at all about sources.

Of course there are cases in which the situation is different. Testing an *historical* assertion always means going back to sources; but not, as a rule, to the reports of eyewitnesses.

Clearly, no historian will accept the evidence of documents uncritically. There are problems of genuineness, there are problems of bias, and there are also such problems as the reconstruction of earlier sources. There are, of course, also problems such as: was the writer

[8] See my *Logic of Scientific Discovery*, last paragraph of section 25, and new appendix *x, (2). For an anticipation by Mark Twain of my *Times* argument, see p. 557 below.

present when these events happened? But this is not one of the characteristic problems of the historian. He may worry about the reliability of a report, but he will rarely worry about whether or not the writer of a document was an eyewitness of the event in question, even assuming that this event was of the nature of an observable event. A letter saying 'I changed my mind yesterday on this question' may be most valuable historical evidence, even though changes of mind are unobservable (and even though we may conjecture, in view of other evidence, that the writer was lying).

As to eyewitnesses, they are important almost exclusively in a court of law where they can be cross-examined. As most lawyers know, eyewitnesses often err. This has been experimentally investigated, with the most striking results. Witnesses most anxious to describe an event as it happened are liable to make scores of mistakes, especially if some exciting things happen in a hurry; and if an event suggests some tempting interpretation, then this interpretation, more often than not, is allowed to distort what has actually been seen.

Hume's view of historical knowledge was different: '. . . we believe', he writes in the *Treatise* (Book I, Part III, Section iv; Selby-Bigge, p. 83), 'that Caesar was kill'd in the Senate-house on the *ides of March* . . . because this fact is establish'd on the unanimous testimony of historians, who agree to assign this precise time and place to that event. Here are certain characters and letters present either to our memory or senses; which characters we likewise remember to have been us'd as the signs of certain ideas; and these ideas were either in the minds of such as were immediately present at that action, and receiv'd the ideas directly from its existence; or they were deriv'd from the testimony of others, and that again from another testimony . . . 'till we arrive at those who were eye-witnesses and spectators of the event.' (See also *Enquiry*, Section x; Selby-Bigge, pp. 111 ff.)

It seems to me that this view must lead to the infinite regress described above. For the problem is, of course, whether 'the unanimous testimony of historians' is to be accepted, or whether it is, perhaps, to be rejected as the result of their reliance on a common yet spurious source. The appeal to 'letters present to our memory or our senses' cannot have any bearing on this or on any other relevant problem of historiography.

XIV

But what, then, are the sources of our knowledge?

The answer, I think, is this: there are all kinds of sources of our knowledge; but *none has authority*.

We may say that *The Times* can be a source of knowledge, or the *Encyclopaedia Britannica*. We may say that certain papers in the *Physical Review* about a problem in physics have more authority, and are more of the character of a source, than an article about the same problem in *The Times* or the *Encyclopaedia*. But it would be quite wrong to say that the source of the article in the *Physical Review* must have been wholly, or even partly, observation. The source may well be the discovery of an inconsistency in another paper or, say, the discovery of the fact that a hypothesis proposed in another paper could be tested by such and such an experiment; all these non-observational discoveries are 'sources' in the sense that they all add to our knowledge.

I do not, of course, deny that an experiment may also add to our knowledge, and in a most important manner. But it is not a source in any ultimate sense. It has always to be checked: as in the example of the news in *The Times* we do not, as a rule, question the eyewitness of an experiment, but, if we doubt the result, we may repeat the experiment, or ask somebody else to repeat it.

The fundamental mistake made by the philosophical theory of the ultimate sources of our knowledge is that it does not distinguish clearly enough between questions of origin and questions of validity. Admittedly, in the case of historiography, these two questions may sometimes coincide. The question of the validity of an historical assertion may be testable only, or mainly, in the light of the origin of certain sources. But in general the two questions are different; and in general we do not test the validity of an assertion or information by tracing its sources or its origin, but we test it, much more directly, by a critical examination of what has been asserted—of the asserted facts themselves.

Thus the empiricist's questions 'How do you know? What is the source of your assertion?' are wrongly put. They are not formulated in an inexact or slovenly manner, but *they are entirely misconceived*: they are questions that beg for an authoritarian answer.

XV

The traditional systems of epistemology may be said to result from yes-answers or no-answers to questions about the sources of our knowledge. *They never challenge these questions, or dispute their legitimacy*; the questions are taken as perfectly natural, and nobody seems to see any harm in them.

This is quite interesting, for these questions are clearly authoritarian in spirit. They can be compared with that traditional question of political theory, 'Who should rule?', which begs for an authoritarian answer such as 'the best', or 'the wisest', or 'the people', or 'the majority'. (It suggests, incidentally, such silly alternatives as 'Who should be our rulers: the capitalists or the workers?', analogous to 'What is the ultimate source of knowledge: the intellect or the senses?') This political question is wrongly put and the answers which it elicits are paradoxical (as I have tried to show in chapter 7 of my *Open Society*). It should be replaced by a completely different question such as '*How can we organize our political institutions so that bad or incompetent rulers* (whom we should try not to get, but whom we so easily might get all the same) *cannot do too much damage?*' I believe that only by changing our question in this way can we hope to proceed towards a reasonable theory of political institutions.

The question about the sources of our knowledge can be replaced in a similar way. It has always been asked in the spirit of: 'What are the best sources of our knowledge—the most reliable ones, those which will not lead us into error, and those to which we can and must turn, in case of doubt, as the last court of appeal?' I propose to assume, instead, that no such ideal sources exist—no more than ideal rulers—and that *all* 'sources' are liable to lead us into error at times. And I propose to replace, therefore, the question of the sources of our knowledge by the entirely different question: '*How can we hope to detect and eliminate error?*'

The question of the sources of our knowledge like so many authoritarian questions, is a *genetic* one. It asks for the origin of our knowledge, in the belief that knowledge may legitimize itself by its pedigree. The nobility of the racially pure knowledge, the untainted knowledge, the knowledge which derives from the highest authority, if possible from God: these are the (often unconscious) metaphysical ideas behind the question. My modified question, 'How can we hope to detect error?' may be said to derive from the view that such pure, untainted and certain sources do not exist, and that questions of origin or of purity should not be confounded with questions of validity, or of truth. This view may be said to be as old as Xenophanes. Xenophanes knew that our knowledge is guesswork, opinion—*doxa* rather than *epistēmē*—as shown by his verses (DK, B, 18 and 34):

> The gods did not reveal, from the beginning,
> All things to us; but in the course of time,
> Through seeking we may learn, and know things better.
>
> But as for certain truth, no man has known it,
> Nor will he know it; neither of the gods,
> Nor yet of all the things of which I speak.
> And even if by chance he were to utter
> The perfect truth, he would himself not know it;
> For all is but a woven web of guesses.

Yet the traditional question of the authoritative sources of knowledge is repeated even today—and very often by positivists and by other philosophers who believe themselves to be in revolt against authority.

The proper answer to my question 'How can we hope to detect and eliminate error?' is, I believe, 'By *criticizing* the theories or guesses of others and—if we can train ourselves to do so—by *criticizing* our own theories or guesses.' (The latter point is highly desirable, but not indispensable; for if we fail to criticize our own theories, there may be others to do it for us.) This answer sums up a position which I propose to call 'critical rationalism'. It is a view, an attitude, and a tradition, which we owe to the Greeks. It is very different from the 'rationalism' or 'intellectualism' of Descartes and his school, and very different even from the epistemology of Kant. Yet in the field of ethics, of moral knowledge, it was approached by Kant with his *principle of autonomy*. This principle expresses his realization that we must not accept the command of an authority, however exalted, as the basis of ethics. For whenever we are faced with a command by an authority, it is for us to judge, critically, whether it is moral or immoral to obey. The authority may have power to enforce its commands, and we may be powerless to resist. But if we have the physical power of choice, then the ultimate responsibility remains with us. It is our own critical decision whether to obey a command; whether to submit to an authority.

Kant boldly carried this idea into the field of religion: '. . . in whatever way', he writes, 'the Deity should be made known to you, and even . . . if He should reveal Himself to you: it is you . . . who must judge whether you are permitted to believe in Him, and to worship Him.'[9]

In view of this bold statement, it seems strange that in his philosophy of science Kant did not adopt the same attitude of critical rationalism, of the critical search for error. I feel certain that it was only his acceptance of the authority of Newton's cosmology—a result of its almost unbelievable success in passing the most severe tests—which prevented Kant from doing so. If this interpretation of Kant is correct, then the critical rationalism (and also the critical empiricism) which I advocate merely puts the finishing touch to Kant's own critical philosophy. And this was made possible by Einstein, who taught us that Newton's theory may well be mistaken in spite of its overwhelming success.

So my answer to the questions 'How do you know? What is the source or the basis of your assertion? What observations have led you to it?' would be: 'I do *not* know: my assertion was merely a guess. Never mind the source, or the sources, from which it may spring—there are many possible sources, and I may not be aware of half of them; and origins or pedigrees have in any case little bearing upon truth. But if you are interested in the problem which I tried to solve by my tentative assertion, you may help me by criticizing it as severely as you can; and if you can design some experimental test which you think might refute my assertion, I shall gladly, and to the best of my powers, help you to refute it.'

This answer[10] applies, strictly speaking, only if the question is asked about some scientific assertion as distinct from an historical one. If my

[9] See Immanuel Kant, *Religion Within the Limits of Pure Reason*, 2nd edition (1794), Fourth Chapter, Part II, § 1, the first footnote. The passage (*not* in the 1st edition, 1793) is quoted more fully in ch. 7 of the present volume, text to note 22.

[10] This answer, and almost the whole of the contents of the present section XV, are taken with only minor changes from a paper of mine which was first published in *The Indian Journal of Philosophy*, 1, No. 1, 1959.

conjecture was an historical one, sources (in the non-ultimate sense) will of course come into the critical discussion of its validity. Yet fundamentally, my answer will be the same, as we have seen.

XVI

It is high time now, I think, to formulate the epistemological results of this discussion. I will put them in the form of ten theses.

1. There are no ultimate sources of knowledge. Every source, every suggestion, is welcome; and every source, every suggestion, is open to critical examination. Except in history, we usually examine the facts themselves rather than the sources of our information.

2. The proper epistemological question is not one about sources; rather, we ask whether the assertion made is true—that is to say, whether it agrees with the facts. (That we may operate, without getting involved in antinomies, with the idea of objective truth in the sense of correspondence to the facts, has been shown by the work of Alfred Tarski.) And we try to find this out, as well as we can, by examining or testing the assertion itself; either in a direct way, or by examining or testing its consequences.

3. In connection with this examination, all kinds of arguments may be relevant. A typical procedure is to examine whether our theories are consistent with our observations. But we may also examine, for example, whether our historical sources are mutually and internally consistent.

4. Quantitatively and qualitatively by far the most important source of our knowledge—apart from inborn knowledge—is tradition. Most things we know we have learnt by example, by being told, by reading books, by learning how to criticize, how to take and to accept criticism, how to respect truth.

5. The fact that most of the sources of our knowledge are traditional condemns anti-traditionalism as futile. But this fact must not be held to support a traditionalist attitude: every bit of our traditional knowledge (and even our inborn knowledge) is open to critical examination and may be overthrown. Nevertheless, without tradition, knowledge would be impossible.

6. Knowledge cannot start from nothing—from a *tabula rasa*—nor

yet from observation. The advance of knowledge consists, mainly, in the modification of earlier knowledge. Although we may sometimes, for example in archaeology, advance through a chance observation, the significance of the discovery will usually depend upon its power to modify our earlier theories.

7. Pessimistic and optimistic epistemologies are about equally mistaken. The pessimistic cave story of Plato is the true one, and not his optimistic story of *anamnēsis* (even though we should admit that all men, like all other animals, and even all plants, possess inborn knowledge). But although the world of appearances is indeed a world of mere shadows on the walls of our cave, we all constantly reach out beyond it; and although, as Democritus said, the truth is hidden in the deep, we can probe into the deep. There is no criterion of truth at our disposal, and this fact supports pessimism. But we do possess criteria which, *if we are lucky*, may allow us to recognize error and falsity. Clarity and distinctness are not criteria of truth, but such things as obscurity or confusion *may* indicate error. Similarly coherence cannot establish truth, but incoherence and inconsistency do establish falsehood. And, when they are recognized, our own errors provide the dim red lights which help us in groping our way out of the darkness of our cave.

8. Neither observation nor reason is an authority. Intellectual intuition and imagination are most important, but they are not reliable: they may show us things very clearly, and yet they may mislead us. They are indispensable as the main sources of our theories; but most of our theories are false anyway. The most important function of observation and reasoning, and even of intuition and imagination, is to help us in the critical examination of those bold conjectures which are the means by which we probe into the unknown.

9. Although clarity is valuable in itself, exactness or precision is not: there can be no point in trying to be more precise than our problem demands. Linguistic precision is a phantom, and problems connected with the meaning or definition of words are unimportant. Thus our table of Ideas (on p. 25), in spite of its symmetry, has an important and an unimportant side: while the left-hand side (words and their meanings) is unimportant, the right-hand side (theories and the problems connected with their truth) is all-important. Words are significant only

as instruments for the formulation of theories, and verbal problems are tiresome: they should be avoided at all cost.

10. Every solution of a problem raises new unsolved problems; the more so the deeper the original problem and the bolder its solution. The more we learn about the world, and the deeper our learning, the more conscious, specific, and articulate will be our knowledge of what we do not know, our knowledge of our ignorance. For this, indeed, is the main source of our ignorance—the fact that our knowledge can be only finite, while our ignorance must necessarily be infinite.

We may get a glimpse of the vastness of our ignorance when we contemplate the vastness of the heavens: though the mere size of the universe is not the deepest cause of our ignorance, it is one of its causes. 'Where I seem to differ from some of my friends', F. P. Ramsey wrote in a charming passage of his *Foundations of Mathematics* (p. 291), 'is in attaching little importance to physical size, I don't feel in the least humble before the vastness of the heavens. The stars may be large but they cannot think or love; and these are qualities which impress me far more than size does. I take no credit for weighing nearly seventeen stone.' I suspect that Ramsey's friends would have agreed with him about the insignificance of sheer physical size; and I suspect that if they felt humble before the vastness of the heavens, this was because they saw in it a symbol of their ignorance.

I believe that it would be worth trying to learn something about the world even if in trying to do so we should merely learn that we do not know much. This state of learned ignorance might be a help in many of our troubles. It might be well for all of us to remember that, while differing widely in the various little bits we know, in our infinite ignorance we are all equal.

XVII

There is a last question I wish to raise.

If only we look for it we can often find a true idea, worthy of being preserved, in a philosophical theory which must be rejected as false. Can we find an idea like this in one of the theories of the ultimate sources of our knowledge?

I believe we can; and I suggest that it is one of the two main ideas which underlie the doctrine that the source of all our knowledge is super-natural. The first of these ideas is false, I believe, while the second is true.

The first, the false idea, is that we must *justify* our knowledge, or our theories, by *positive* reasons, that is, by reasons capable of establishing them, or at least of making them highly probable; at any rate, by better reasons than that they have so far withstood criticism. This idea implies, I suggest, that we must appeal to some ultimate or authoritative source of true knowledge; which still leaves open the character of that authority—whether it is human, like observation or reason, or super-human (and therefore super-natural).

The second idea—whose vital importance has been stressed by Russell—is that no man's authority can establish truth by decree; that we should submit to truth; that *truth is above human authority*.

Taken together these two ideas almost immediately yield the conclusion that the sources from which our knowledge derives must be super-human; a conclusion which tends to encourage self-righteousness and the use of force against those who refuse to see the divine truth.

Some who rightly reject this conclusion do not, unhappily, reject the first idea—the belief in the existence of ultimate sources of knowledge. Instead they reject the second idea—the thesis that truth is above human authority. They thereby endanger the idea of the objectivity of knowledge, and of common standards of criticism or rationality.

What we should do, I suggest, is to give up the idea of ultimate sources of knowledge, and admit that all knowledge is human; that it is mixed with our errors, our prejudices, our dreams, and our hopes; that all we can do is to grope for truth even though it be beyond our reach. We may admit that our groping is often inspired, but we must be on our guard against the belief, however deeply felt, that our inspiration carries any authority, divine or otherwise. If we thus admit that there is no authority beyond the reach of criticism to be found within the whole province of our knowledge, however far it may have penetrated into the unknown, then we can retain, without danger, the idea that truth is beyond human authority. And we must retain it. For without this idea there can be no objective standards of inquiry; no criticism of our conjectures; no groping for the unknown; no quest for knowledge.

Conjectures

There could be no fairer destiny for any . . . theory than that it should point the way to a more comprehensive theory in which it lives on, as a limiting case.

ALBERT EINSTEIN

1

SCIENCE: CONJECTURES AND REFUTATIONS

> Mr. Turnbull had predicted evil consequences, . . . and was now doing the best in his power to bring about the verification of his own prophecies.
>
> ANTHONY TROLLOPE

I

When I received the list of participants in this course and realized that I had been asked to speak to philosophical colleagues I thought, after some hesitation and consultation, that you would probably prefer me to speak about those problems which interest me most, and about those developments with which I am most intimately acquainted. I therefore decided to do what I have never done before: to give you a report on my own work in the philosophy of science, since the autumn of 1919 when I first began to grapple with the problem, '*When should a*

A lecture given at Peterhouse, Cambridge, Summer 1953, *as part of a course on developments and trends in contemporary* British *philosophy, organized by the British Council; originally published under the title 'Philosophy of Science: a Personal Report'* in British Philosophy in Mid-Century, *ed. C. A. Mace,* 1957.

theory be ranked as scientific?' or '*Is there a criterion for the scientific character or status of a theory?*'

The problem which troubled me at the time was neither, 'When is a theory true?' nor, 'When is a theory acceptable?' My problem was different. I *wished to distinguish between science and pseudo-science*; knowing very well that science often errs, and that pseudo-science may happen to stumble on the truth.

I knew, of course, the most widely accepted answer to my problem: that science is distinguished from pseudo-science—or from 'metaphysics'—by its *empirical method*, which is essentially *inductive*, proceeding from observation or experiment. But this did not satisfy me. On the contrary, I often formulated my problem as one of distinguishing between a genuinely empirical method and a non-empirical or even a pseudo-empirical method—that is to say, a method which, although it appeals to observation and experiment, nevertheless does not come up to scientific standards. The latter method may be exemplified by astrology, with its stupendous mass of empirical evidence based on observation—on horoscopes and on biographies.

But as it was not the example of astrology which led me to my problem I should perhaps briefly describe the atmosphere in which my problem arose and the examples by which it was stimulated. After the collapse of the Austrian Empire there had been a revolution in Austria: the air was full of revolutionary slogans and ideas, and new and often wild theories. Among the theories which interested me Einstein's theory of relativity was no doubt by far the most important. Three others were Marx's theory of history, Freud's psycho-analysis, and Alfred Adler's so-called 'individual psychology'.

There was a lot of popular nonsense talked about these theories, and especially about relativity (as still happens even today), but I was fortunate in those who introduced me to the study of this theory. We all—the small circle of students to which I belonged—were thrilled with the result of Eddington's eclipse observations which in 1919 brought the first important confirmation of Einstein's theory of gravitation. It was a great experience for us, and one which had a lasting influence on my intellectual development.

The three other theories I have mentioned were also widely discussed among students at that time. I myself happened to come into personal contact with Alfred Adler, and even to co-operate with him in his social work among the children and young people in the working-class districts of Vienna where he had established social guidance clinics.

It was during the summer of 1919 that I began to feel more and more dissatisfied with these three theories—the Marxist theory of history, psycho-analysis, and individual psychology; and I began to feel dubious about their claims to scientific status. My problem perhaps first took the simple form, 'What is wrong with Marxism, psycho-analysis, and individual psychology? Why are they so different from physical theories, from Newton's theory, and especially from the theory of relativity?'

To make this contrast clear I should explain that few of us at the time would have said that we believed in the *truth* of Einstein's theory of gravitation. This shows that it was not my doubting the *truth* of those other three theories which bothered me, but something else. Yet neither was it that I merely felt mathematical physics to be more *exact* than the sociological or psychological type of theory. Thus what worried me was neither the problem of truth, at that stage at least, nor the problem of exactness or measurability. It was rather that I felt that these other three theories, though posing as sciences, had in fact more in common with primitive myths than with science; that they resembled astrology rather than astronomy.

I found that those of my friends who were admirers of Marx, Freud, and Adler, were impressed by a number of points common to these theories, and especially by their apparent *explanatory power*. These theories appeared to be able to explain practically everything that happened within the fields to which they referred. The study of any of them seemed to have the effect of an intellectual conversion or revelation, opening your eyes to a new truth hidden from those not yet initiated. Once your eyes were thus opened you saw confirming instances everywhere: the world was full of *verifications* of the theory. Whatever happened always confirmed it. Thus its truth appeared manifest; and unbelievers were clearly people who did not want to see the manifest truth; who refused to see it, either because it was against their class interest, or because of their repressions which were still 'un-analysed' and crying out for treatment.

The most characteristic element in this situation seemed to me the incessant stream of confirmations, of observations which 'verified' the theories in question; and this point was constantly emphasized by their adherents. A Marxist could not open a newspaper without finding on every page confirming evidence for his interpretation of history; not only in the news, but also in its presentation—which revealed the class bias of the paper—and especially of course in what the paper did *not* say. The Freudian analysts emphasized that their theories were constantly verified by their 'clinical observations'. As for Adler, I was much impressed by a personal experience. Once, in 1919, I reported to him a case which to me did not seem particularly Adlerian, but which he found no difficulty in analysing in terms of his theory of inferiority feelings, although he had not even seen the child. Slightly shocked, I asked him how he could be so sure. 'Because of my thousandfold experience', he replied; whereupon I could not help saying: 'And with this new case, I suppose, your experience has become thousand-and-one-fold.'

What I had in mind was that his previous observations may not have been much sounder than this new one; that each in its turn had been interpreted in the light of 'previous experience', and at the same time counted as additional confirmation. What, I asked myself, did it confirm? No more than that a case could be interpreted in the light of the theory. But this meant very little, I reflected, since every conceivable case could be interpreted in the light of Adler's theory, or equally of Freud's. I may illustrate this by two very different examples of human behaviour: that of a man who pushes a child into the water with the intention of drowning it; and that of a man who sacrifices his life in an attempt to save the child. Each of these two cases can be explained with equal ease in Freudian and in Adlerian terms. According to Freud the first man suffered from repression (say, of some component of his Oedipus complex), while the second man had achieved sublimation. According to Adler the first man suffered from feelings of inferiority (producing perhaps the need to prove to himself that he dared to commit some crime), and so did the second man (whose need was to prove to himself that he dared to rescue the child). I could not think of any human behaviour which could not be interpreted in terms of either theory. It was precisely this fact—that they always fitted, that they were always confirmed—which in the eyes of their admirers constituted the strongest argument in favour of these theories. It began to dawn on me that this apparent strength was in fact their weakness.

With Einstein's theory the situation was strikingly different. Take one typical instance—Einstein's prediction, just then confirmed by the findings of Eddington's expedition. Einstein's gravitational theory had led to the result that light must be attracted by heavy bodies (such as the sun), precisely as material bodies were attracted. As a consequence it could be calculated that light from a distant fixed star whose apparent position was close to the sun would reach the earth from such a direction that the star would seem to be slightly shifted away from the sun; or, in other words, that stars close to the sun would look as if they had moved a little away from the sun, and from one another. This is a thing which cannot normally be observed since such stars are rendered invisible in daytime by the sun's overwhelming brightness; but during an eclipse it is possible to take photographs of them. If the same constellation is photographed at night one can measure the distances on the two photographs, and check the predicted effect.

Now the impressive thing about this case is the *risk* involved in a prediction of this kind. If observation shows that the predicted effect is definitely absent, then the theory is simply refuted. The theory is *incompatible with certain possible results of observation*—in fact with results which everybody before Einstein would have expected.[1] This is quite different from the situation I have previously described, when it turned out that the theories in question were compatible with the most divergent human behaviour, so that it was practically impossible to describe any human behaviour that might not be claimed to be a verification of these theories.

These considerations led me in the winter of 1919–20 to conclusions which I may now reformulate as follows.

(1) It is easy to obtain confirmations, or verifications, for nearly every theory—if we look for confirmations.

(2) Confirmations should count only if they are the result of *risky predictions*; that is to say, if, unenlightened by the theory in question, we

[1] This is a slight oversimplification, for about half of the Einstein effect may be derived from the classical theory, provided we assume a ballistic theory of light.

should have expected an event which was incompatible with the theory—an event which would have refuted the theory.

(3) Every 'good' scientific theory is a prohibition: it forbids certain things to happen. The more a theory forbids, the better it is.

(4) A theory which is not refutable by any conceivable event is non-scientific. Irrefutability is not a virtue of a theory (as people often think) but a vice.

(5) Every genuine *test* of a theory is an attempt to falsify it, or to refute it. Testability is falsifiability; but there are degrees of testability: some theories are more testable, more exposed to refutation, than others; they take, as it were, greater risks.

(6) Confirming evidence should not count *except when it is the result of a genuine test of the theory*; and this means that it can be presented as a serious but unsuccessful attempt to falsify the theory. (I now speak in such cases of 'corroborating evidence'.)

(7) Some genuinely testable theories, when found to be false, are still upheld by their admirers—for example by introducing *ad hoc* some auxiliary assumption, or by re-interpreting the theory *ad hoc* in such a way that it escapes refutation. Such a procedure is always possible, but it rescues the theory from refutation only at the price of destroying, or at least lowering, its scientific status. (I later described such a rescuing operation as a '*conventionalist twist*' or a '*conventionalist stratagem*'.)

One can sum up all this by saying that *the criterion of the scientific status of a theory is its falsifiability, or refutability, or testability*.

II

I may perhaps exemplify this with the help of the various theories so far mentioned. Einstein's theory of gravitation clearly satisfied the criterion of falsifiability. Even if our measuring instruments at the time did not allow us to pronounce on the results of the tests with complete assurance, there was clearly a possibility of refuting the theory.

Astrology did not pass the test. Astrologers were greatly impressed, and misled, by what they believed to be confirming evidence—so much so that they were quite unimpressed by any unfavourable evidence. Moreover, by making their interpretations and prophecies sufficiently vague they were able to explain away anything that might have been a refutation of the theory had the theory and the prophecies been more precise. In order to escape falsification they destroyed the testability of their theory. It is a typical soothsayer's trick to predict things so vaguely that the predictions can hardly fail: that they become irrefutable.

The Marxist theory of history, in spite of the serious efforts of some of its founders and followers, ultimately adopted this soothsaying practice. In some of its earlier formulations (for example in Marx's analysis of the character of the 'coming social revolution') their predictions were testable, and in fact falsified.[2] Yet instead of accepting the refutations the followers of Marx re-interpreted both the theory and the evidence in order to make them agree. In this way they rescued the theory from refutation; but they did so at the price of adopting a device which made it irrefutable. They thus gave a 'conventionalist twist' to the theory; and by this stratagem they destroyed its much advertised claim to scientific status.

The two psycho-analytic theories were in a different class. They were simply non-testable, irrefutable. There was no conceivable human behaviour which could contradict them. This does not mean that Freud and Adler were not seeing certain things correctly: I personally do not doubt that much of what they say is of considerable importance, and may well play its part one day in a psychological science which is testable. But it does mean that those 'clinical observations' which analysts naïvely believe confirm their theory cannot do this any more than the daily confirmations which astrologers find in their practice.[3] And as

[2] See, for example, my *Open Society and Its Enemies*, ch. 15, section iii, and notes 13–14.

[3] 'Clinical observations', like all other observations, are *interpretations in the light of theories* (see below, sections iv ff.); and for this reason alone they are apt to seem to support those theories in the light of which they were interpreted. But real support can be obtained only from observations undertaken as tests (by 'attempted refutations'); and for this purpose *criteria of refutation* have to be laid down beforehand: it must be agreed which observable situations, if actually observed, mean that the theory is refuted. But what kind of clinical responses would refute to the satisfaction of the analyst not merely a particular analytic diagnosis but psycho-analysis itself? And have such criteria ever been discussed or agreed upon by analysts? Is there not, on the contrary, a whole family of analytic concepts, such as 'ambivalence' (I do not suggest that there is no such thing as ambivalence), which would make it difficult, if not impossible, to agree upon such criteria? Moreover, how much headway has been made in investigating the question of the extent

for Freud's epic of the Ego, the Super-ego, and the Id, no substantially stronger claim to scientific status can be made for it than for Homer's collected stories from Olympus. These theories describe some facts, but in the manner of myths. They contain most interesting psychological suggestions, but not in a testable form.

At the same time I realized that such myths may be developed, and become testable; that historically speaking all—or very nearly all—scientific theories originate from myths, and that a myth may contain important anticipations of scientific theories. Examples are Empedocles' theory of evolution by trial and error, or Parmenides' myth of the unchanging block universe in which nothing ever happens and which, if we add another dimension, becomes Einstein's block universe (in which, too, nothing ever happens, since everything is, four-dimensionally speaking, determined and laid down from the beginning). I thus felt that if a theory is found to be non-scientific, or 'metaphysical' (as we might say), it is not thereby found to be unimportant, or insignificant, or 'meaningless', or 'nonsensical'.[4] But

to which the (conscious or unconscious) expectations and theories held by the analyst influence the 'clinical responses' of the patient? (To say nothing about the conscious attempts to influence the patient by proposing interpretations to him, etc.) Years ago I introduced the term '*Oedipus effect*' to describe the influence of a theory or expectation or prediction *upon the event which it predicts* or describes: it will be remembered that the causal chain leading to Oedipus' parricide was started by the oracle's prediction of this event. This is a characteristic and recurrent theme of such myths, but one which seems to have failed to attract the interest of the analysts, perhaps not accidentally. (The problem of confirmatory dreams suggested by the analyst is discussed by Freud, for example in *Gesammelte Schriften*, III, 1925, where he says on p. 314: 'If anybody asserts that most of the dreams which can be utilized in an analysis . . . owe their origin to [the analyst's] suggestion, then no objection can be made from the point of view of analytic theory. Yet there is nothing in this fact', he surprisingly adds, 'which would detract from the reliability of our results.')

[4] The case of astrology, nowadays a typical pseudo-science, may illustrate this point. It was attacked, by Aristotelians and other rationalists, down to Newton's day, for the wrong reason—for its now accepted assertion that the planets had an 'influence' upon terrestrial ('sublunar') events. In fact Newton's theory of gravity, and especially the lunar theory of the tides, was historically speaking an offspring of astrological lore. Newton, it seems, was most reluctant to adopt a theory which came from the same stable as for example the theory that 'influenza' epidemics are due to an astral 'influence'. And Galileo, no doubt for the same reason, actually rejected the lunar theory of the tides; and his misgivings about Kepler may easily be explained by his misgivings about astrology.

it cannot claim to be backed by empirical evidence in the scientific sense—although it may easily be, in some genetic sense, the 'result of observation'.

(There were a great many other theories of this pre-scientific or pseudo-scientific character, some of them, unfortunately, as influential as the Marxist interpretation of history; for example, the racialist interpretation of history—another of those impressive and all-explanatory theories which act upon weak minds like revelations.)

Thus the problem which I tried to solve by proposing the criterion of falsifiability was neither a problem of meaningfulness or significance, nor a problem of truth or acceptability. It was the problem of drawing a line (as well as this can be done) between the statements, or systems of statements, of the empirical sciences, and all other statements—whether they are of a religious or of a metaphysical character, or simply pseudo-scientific. Years later—it must have been in 1928 or 1929—I called this first problem of mine the '*problem of demarcation*'. The criterion of falsifiability is a solution to this problem of demarcation, for it says that statements or systems of statements, in order to be ranked as scientific, must be capable of conflicting with possible, or conceivable, observations.

III

Today I know, of course, that this *criterion of demarcation*—the criterion of testability, or falsifiability, or refutability—is far from obvious; for even now its significance is seldom realized. At that time, in 1920, it seemed to me almost trivial, although it solved for me an intellectual problem which had worried me deeply, and one which also had obvious practical consequences (for example, political ones). But I did not yet realize its full implications, or its philosophical significance. When I explained it to a fellow student of the Mathematics Department (now a distinguished mathematician in Great Britain), he suggested that I should publish it. At the time I thought this absurd; for I was convinced that my problem, since it was so important for me, must have agitated many scientists and philosophers who would surely have reached my rather obvious solution. That this was not the case I learnt from Wittgenstein's work, and from its reception; and so I published my results thirteen years later in the form of a criticism of Wittgenstein's *criterion of meaningfulness*.

Wittgenstein, as you all know, tried to show in the *Tractatus* (see for example his propositions 6.53; 6.54; and 5) that all so-called philosophical or metaphysical propositions were actually non-propositions or pseudo-propositions: that they were senseless or meaningless. All genuine (or meaningful) propositions were truth functions of the elementary or atomic propositions which described 'atomic facts'—i.e., facts which can in principle be ascertained by observation. In other words, meaningful propositions were fully reducible to elementary or atomic propositions which were simple statements describing possible states of affairs, and which could in principle be established or rejected by observation. If we call a statement an 'observation statement' not only if it states an actual observation but also if it states anything that *may* be observed, we shall have to say (according to the *Tractatus*, 5 and 4.52) that every genuine proposition must be a truth-function of, and therefore deducible from, observation statements. All other apparent propositions will be meaningless pseudo-propositions; in fact they will be nothing but nonsensical gibberish.

This idea was used by Wittgenstein for a characterization of science, as opposed to philosophy. We read (for example in 4.11, where natural science is taken to stand in opposition to philosophy): 'The totality of true propositions is the total natural science (or the totality of the natural sciences).' This means that the propositions which belong to science are those deducible from *true* observation statements; they are those propositions which can be *verified* by true observation statements. Could we know all true observation statements, we should also know all that may be asserted by natural science.

This amounts to a crude verifiability criterion of demarcation. To make it slightly less crude, it could be amended thus: 'The statements which may possibly fall within the province of science are those which may possibly be verified by observation statements; and these statements, again, coincide with the class of *all* genuine or meaningful statements.' For this approach, then, *verifiability, meaningfulness, and scientific character all coincide*.

I personally was never interested in the so-called problem of meaning; on the contrary, it appeared to me a verbal problem, a typical pseudo-problem, I was interested only in the problem of demarcation, i.e. in finding a criterion of the scientific character of theories. It was just this interest which made me see at once that Wittgenstein's verifiability criterion of meaning was intended to play the part of a criterion of demarcation as well; and which made me see that, as such, it was totally inadequate, even if all misgivings about the dubious concept of meaning were set aside. For Wittgenstein's criterion of demarcation—to use my own terminology in this context—is verifiability, or deducibility from observation statements. But this criterion is too narrow (*and* too wide): it excludes from science practically everything that is, in fact, characteristic of it (while failing in effect to exclude astrology). No scientific theory can ever be deduced from observation statements, or be described as a truth-function of observation statements.

All this I pointed out on various occasions to Wittgensteinians and members of the Vienna Circle. In 1931–2 I summarized my ideas in a largish book (read by several members of the Circle but never published; although part of it was incorporated in my *Logic of Scientific Discovery*); and in 1933 I published a letter to the Editor of *Erkenntnis* in which I tried to compress into two pages my ideas on the problems of demarcation and induction.[5] In this letter and elsewhere I described the problem of meaning as a pseudo-problem, in contrast to the problem of demarcation. But my contribution was classified by members of

[5] My *Logic of Scientific Discovery* (1959, 1960, 1961), here usually referred to as *L.Sc.D.*, is the translation of *Logik der Forschung* (1934), with a number of additional notes and appendices, including (on pp. 312–14) the letter to the Editor of *Erkenntnis* mentioned here in the text which was first published in *Erkenntnis*, **3**, 1933, pp. 426 f.

Concerning my never published book mentioned here in the text, see R. Carnap's paper 'Ueber Protokollsätze' (On Protocol-Sentences), *Erkenntnis*, **3**, 1932, pp. 215–28 where he gives an outline of my theory on pp. 223–8, and accepts it. He calls my theory 'procedure B', and says (p. 224, top): 'Starting from a point of view different from Neurath's' (who developed what Carnap calls on p. 223 'procedure A'), 'Popper developed procedure B as part of his system.' And after describing in detail my theory of tests, Carnap sums up his views as follows (p. 228): 'After weighing the various arguments here discussed, it appears to me that the second language form with procedure B—that is in the form here described—is the most adequate among the forms of scientific language at present advocated . . . in the . . . theory of knowledge.' This paper of Carnap's contained the first published report of my theory of critical testing. (See also my critical remarks in *L.Sc.D.*, note 1 to section 29, p. 104, where the date '1933' should read '1932'; and ch. 11, below, text to note 39.).

the Circle as a proposal to replace the verifiability criterion of *meaning* by a falsifiability criterion of *meaning*—which effectively made nonsense of my views.[6] My protests that I was trying to solve, not their pseudo-problem of meaning, but the problem of demarcation, were of no avail.

My attacks upon verification had some effect, however. They soon led to complete confusion in the camp of the verificationist philosophers of sense and nonsense. The original proposal of verifiability as the criterion of meaning was at least clear, simple, and forceful. The modifications and shifts which were now introduced were the very opposite.[7] This, I should say, is now seen even by the participants. But since I am usually quoted as one of them I wish to repeat that although I created this confusion I never participated in it. Neither falsifiability nor testability were proposed by me as criteria of meaning; and although I may plead guilty to having introduced both terms into the discussion, it was not I who introduced them into the theory of meaning.

Criticism of my alleged views was widespread and highly successful. I have yet to meet a criticism of my views.[8] Meanwhile, testability is being widely accepted as a criterion of demarcation.

[6] Wittgenstein's example of a nonsensical pseudo-proposition is: 'Socrates is identical'. Obviously, 'Socrates is not identical' must also be nonsense. Thus the negation of any nonsense will be nonsense, and that of a meaningful statement will be meaningful. *But the negation of a testable (or falsifiable) statement need not be testable*, as was pointed out, first in my *L.Sc.D.*, (e.g. pp. 38 f.) and later by my critics. The confusion caused by taking testability as a criterion of *meaning* rather than of *demarcation* can easily be imagined.

[7] The most recent example of the way in which the history of this problem is misunderstood is A. R. White's 'Note on Meaning and Verification', *Mind*, **63**, 1954, pp. 66 ff. J. L. Evans's article, *Mind*, **62**, 1953, pp. 1 ff., which Mr. White criticizes, is excellent in my opinion, and unusually perceptive. Understandably enough, neither of the authors can quite reconstruct the story. (Some hints may be found in my *Open Society*, notes 46, 51 and 52 to ch. 11; and a fuller analysis in ch. 11 of the present volume).

[8] In *L.Sc.D.* I discussed, and replied to, some likely objections which afterwards were indeed raised, without reference to my replies. One of them is the contention that the falsification of a natural law is just as impossible as its verification. The answer is that this objection mixes two entirely different levels of analysis (like the objection that mathematical demonstrations are impossible since checking, no matter how often repeated, can never make it quite certain that we have not overlooked a mistake). On the first level, there is a logical asymmetry: one singular statement—say about the perihelion of Mercury—can formally falsify Kepler's laws; but these cannot be formally verified by any

IV

I have discussed the problem of demarcation in some detail because I believe that its solution is the key to most of the fundamental problems of the philosophy of science. I am going to give you later a list of some of these other problems, but only one of them—the *problem of induction*—can be discussed here at any length.

I had become interested in the problem of induction in 1923. Although this problem is very closely connected with the problem of demarcation, I did not fully appreciate the connection for about five years.

I approached the problem of induction through Hume. Hume, I felt, was perfectly right in pointing out that induction cannot be logically justified. He held that there can be no valid logical[9] arguments allowing us to establish '*that those instances, of which we have had no experience, resemble those, of which we have had experience*'. Consequently '*even after the observation of the frequent or constant conjunction of objects, we have no reason to draw any inference concerning any object beyond those of which we have had experience*'. For 'shou'd it be said that we have experience'[10]—experience teaching us that objects constantly conjoined with certain other objects continue to be so conjoined—then, Hume says, 'I wou'd renew my question, *why from this experience we form any conclusion beyond those past instances, of which we have had*

number of singular statements. The attempt to minimize this asymmetry can only lead to confusion. On another level, we may hesitate to accept any statement, even the simplest observation statement; and we may point out that every statement involves *interpretation in the light of theories*, and that it is therefore uncertain. This does not affect the fundamental asymmetry, but it is important: most dissectors of the heart before Harvey observed the wrong things—those, which they expected to see. There can never be anything like a completely safe observation, free from the dangers of misinterpretation. (This is one of the reasons why the theory of induction does not work.) The 'empirical basis' consists largely of a mixture of *theories* of lower degree of universality (of 'reproducible effects'). But the fact remains that, relative to whatever basis the investigator may accept (at his peril), he can test his theory only by trying to refute it.

[9] Hume does not say 'logical' but 'demonstrative', a terminology which, I think, is a little misleading. The following two quotations are from the *Treatise of Human Nature*, Book I, Part III, sections vi and xii. (The italics are all Hume's.)

[10] This and the next quotation are from *loc. cit.*, section vi. See also Hume's *Enquiry Concerning Human Understanding*, section iv, Part II, and his *Abstract*, edited 1938 by J. M. Keynes and P. Sraffa, p. 15, and quoted in *L.Sc.D.*, new appendix *vii, text to note 6.

experience'. This 'renew'd question' indicates that an attempt to justify the practice of induction by an appeal to experience must lead to an *infinite regress*. As a result we can say that theories can never be inferred from observation statements, or rationally justified by them.

I found Hume's refutation of inductive inference clear and conclusive. But I felt completely dissatisfied with his psychological explanation of induction in terms of custom or habit.

It has often been noticed that this explanation of Hume's is philosophically not very satisfactory. Hume, however, without doubt intended it as a *psychological* rather than a philosophical theory; for it tries to give a causal explanation of a psychological fact—*the fact that we believe in laws*, in statements asserting regularities or constantly conjoined kinds of events. Hume explains this fact by asserting that it is due to (i.e. constantly conjoined with) custom or habit. But even this reformulation of Hume's theory is unacceptable; for what I have just called a 'psychological fact' may itself be described as a custom or habit—our custom or our habit of believing in laws or regularities. It is neither surprising nor enlightening to hear that such a custom or habit can be explained as due to custom or habit, or conjoined with a custom or habit (even though a different one). Only when we remember that the words 'custom' and 'habit' are used by Hume, as they are in ordinary language, not merely to *describe* regular behaviour, but rather to *theorize about its origin* (ascribed to frequent repetition), can we reformulate his psychological theory in a more satisfactory way. Hume's theory becomes then the thesis that, like other habits, *our habit of believing in laws is the product of frequent repetition*—of the repeated observation that things of a certain kind are constantly conjoined with things of another kind.

This genetic-psychological theory is, as indicated, incorporated in ordinary language, and it is therefore hardly as revolutionary as Hume thought. It is no doubt an extremely popular psychological theory—part of 'common sense', one might say. But in spite of my love of both common sense and Hume, I felt convinced that this psychological theory was mistaken; and that it was in fact refutable on purely logical grounds.

Hume's psychology, which is the popular psychology, was mistaken, I felt, about at least three different things: (*a*) the typical result of repetition; (*b*) the genesis of habits; and especially (*c*) the character of those experiences or modes of behaviour which may be described as 'believing in a law' or 'expecting a law-like succession of events'.

(*a*) The typical result of repetition—say, of repeating a difficult passage on the piano—is that movements which at first needed attention are in the end executed without attention. We might say that the process becomes radically abbreviated, and ceases to be conscious: it becomes automatized, 'physiological'. Such a development, far from creating a conscious expectation of law-like succession, or a belief in a law, may on the contrary begin with a conscious belief and destroy it by making it superfluous. In learning to ride a bicycle we may start with the belief that we can avoid falling if we steer in the direction in which we threaten to fall, and this belief may be useful for guiding our movements. After sufficient practice we may forget the rule; in any case, we do not need it any longer. On the other hand, even if it is true that repetition may create unconscious expectations, these become conscious only if something goes wrong (we may not have heard the clock tick, but we may hear that it has stopped).

(*b*) Habits or customs do not, as a rule, *originate* in repetition. Even the habit of walking, or of speaking, or of feeding at certain hours, *begins* before repetition can play any part whatever. We may say, if we like, that they deserve to be called 'habits' or 'customs' only after repetition has played its typical part described under (*a*); but we must not say that the practices in question *originated* as the result of many repetitions.

(*c*) Belief in a law is not quite the same thing as behaviour which betrays an expectation of a law-like succession of events; but these two are sufficiently closely connected to be treated together. They may, perhaps, in exceptional cases, result from a mere repetition of sense impressions (as in the case of the stopping clock). I was prepared to concede this, but I contended that normally, and in most cases of any interest, they cannot be so explained. As Hume admits, even a single striking observation may be sufficient to create a belief or an expectation—a fact which he tries to explain as due to an inductive habit, formed as the result of a vast number of long repetitive sequences which had been experienced at an earlier period of life.[11]

[11] *Treatise*, section xiii; section xv, rule 4.

But this, I contended, was merely his attempt to explain away unfavourable facts which threatened his theory; an unsuccessful attempt, since these unfavourable facts could be observed in very young animals and babies—as early, indeed, as we like. 'A lighted cigarette was held near the noses of the young puppies', reports F. Bäge. 'They sniffed at it once, turned tail, and nothing would induce them to come back to the source of the smell and to sniff again. A few days later, they reacted to the mere sight of a cigarette or even of a rolled piece of white paper, by bounding away, and sneezing.'[12] If we try to explain cases like this by postulating a vast number of long repetitive sequences at a still earlier age we are not only romancing, but forgetting that in the clever puppies' short lives there must be room not only for repetition but also for a great deal of novelty, and consequently of non-repetition.

But it is not only that certain empirical facts do not support Hume; there are decisive arguments of a *purely logical* nature against his psychological theory.

The central idea of Hume's psychological theory is that of *repetition, based upon similarity* (or 'resemblance'). This idea is used in a very uncritical way. We are led to think of the water-drop that hollows the stone: of sequences of unquestionably like events slowly forcing themselves upon us, as does the tick of the clock. But we ought to realize that in a psychological theory such as Hume's, only repetition-for-us, based upon similarity-for-us, can be allowed to have any effect upon us. We must respond to situations as if they were equivalent; *take* them as similar; *interpret* them as repetitions. In this way they become for us *functionally equal*. The clever puppies, we may assume, showed by their response, their way of acting or of reacting, that they recognized or interpreted the second situation as a repetition of the first: that they expected its main element, the objectionable smell, to be present. The situation was a repetition-for-them because they responded to it by *anticipating* its similarity to the previous one.

This apparently psychological criticism has a purely logical basis which may be summed up in the following simple argument. (It

[12] F. Bäge, 'Zur Entwicklung, etc.', *Zeitschrift f. Hundeforschung*, 1933; cp. D. Katz, *Animals and Men*, ch. VI, footnote.

happens to be the one from which I originally started my criticism.) The kind of repetition envisaged by Hume can never be perfect; the cases he has in mind cannot be cases of perfect sameness; they can only be cases of similarity. Thus *they are repetitions only from a certain point of view*. (What has the effect upon me of a repetition may not have this effect upon a spider.) But this means that, for logical reasons, there must always be a point of view—such as a system of expectations, anticipations, assumptions, or interests—*before* there can be any repetition; which point of view, consequently, cannot be merely the result of repetition. (See now also appendix *x, (1), to my *L.Sc.D.*)

We must thus replace, for the purposes of a psychological theory of the origin of our beliefs, the naïve idea of events which *are* similar by the idea of events to which we react by *interpreting* them as being similar. But if this is so (and I can see no escape from it) then Hume's psychological theory of induction leads to an infinite regress, precisely analogous to that other infinite regress which was discovered by Hume himself, and used by him to explode the logical theory of induction. For what do we wish to explain? In the example of the puppies we wish to explain behaviour which may be described as *recognizing or interpreting* a situation as a repetition of another. Clearly, we cannot hope to explain this by an appeal to earlier repetitions, once we realize that the earlier repetitions must also have been repetitions-for-them, so that precisely the same problem arises again: that of *recognizing or interpreting* a situation as a repetition of another.

To put it more concisely, similarity-for-us is the product of a response involving interpretations (which may be inadequate) and anticipations or expectations (which may never be fulfilled). It is therefore impossible to explain anticipations, or expectations, as resulting from many repetitions, as suggested by Hume. For even the first repetition-for-us must be based upon similarity-for-us, and therefore upon expectations—precisely the kind of thing we wished to explain. (Expectations must come first, *before* repetitions.)

We see that there is an infinite regress involved in Hume's psychological theory.

Hume, I felt, had never accepted the full force of his own logical analysis. Having refuted the logical idea of induction he was faced with the following problem: how do we actually obtain our knowledge, as a matter of psychological fact, if induction is a procedure which is logically invalid and rationally unjustifiable? There are two possible answers: (1) We obtain our knowledge by a non-inductive procedure. This answer would have allowed Hume to retain a form of rationalism. (2) We obtain our knowledge by repetition and induction, and therefore by a logically invalid and rationally unjustifiable procedure, so that all apparent knowledge is merely a kind of belief—belief based on habit. This answer would imply that even scientific knowledge is irrational, so that rationalism is absurd, and must be given up. (I shall not discuss here the age-old attempts, now again fashionable, to get out of the difficulty by asserting that though induction is of course logically invalid if we mean by 'logic' the same as 'deductive logic', it is not irrational by its own standards, and as inductive logic admits; as may be seen from the fact that every reasonable man applies it *as a matter of fact*. As against this, it was Hume's great achievement to break this uncritical identification of the question of fact—*quid facti?*—and the question of justification or validity—*quid juris?*. (See below, point (13) of the appendix to the present chapter.)

It seems that Hume never seriously considered the first alternative. Having cast out the logical theory of induction by repetition he struck a bargain with common sense, meekly allowing the re-entry of induction by repetition, in the guise of a psychological fact. I proposed to turn the tables upon this theory of Hume's. Instead of explaining our propensity to expect regularities as the result of repetition, I proposed to explain repetition-for-us as the result of our propensity to expect regularities and to search for them.

Thus I was led by purely logical considerations to replace the psychological theory of induction by the following view. Without waiting, passively, for repetitions to impress or impose regularities upon us, we actively try to impose regularities upon the world. We try to discover similarities in it, and to interpret it in terms of laws invented by us. Without waiting for premises we jump to conclusions. These may have to be discarded later, should observation show that they are wrong.

This was a theory of trial and error—of *conjectures and refutations*. It made it possible to understand why our attempts to force interpretations upon the world were logically prior to the observation of similarities. Since there were logical reasons behind this procedure, I thought that it would apply in the field of science also; that scientific theories were not the digest of observations, but that they were inventions—conjectures boldly put forward for trial, to be eliminated if they clashed with observations; with observations which were rarely accidental but as a rule undertaken with the definite intention of testing a theory by obtaining, if possible, a decisive refutation.

V

The belief that science proceeds from observation to theory is still so widely and so firmly held that my denial of it is often met with incredulity. I have even been suspected of being insincere—of denying what nobody in his senses can doubt.

But in fact the belief that we can start with pure observations alone, without anything in the nature of a theory, is absurd; as may be illustrated by the story of the man who dedicated his life to natural science, wrote down everything he could observe, and bequeathed his priceless collection of observations to the Royal Society to be used as inductive evidence. This story should show us that though beetles may profitably be collected, observations may not.

Twenty-five years ago I tried to bring home the same point to a group of physics students in Vienna by beginning a lecture with the following instructions: 'Take pencil and paper; carefully observe, and write down what you have observed!' They asked, of course, *what* I wanted them to observe. Clearly the instruction, 'Observe!' is absurd.[13] (It is not even idiomatic, unless the object of the transitive verb can be taken as understood.) Observation is always selective. It needs a chosen object, a definite task, an interest, a point of view, a problem. And its description presupposes a descriptive language, with property words; it presupposes similarity and classification, which in their turn presuppose interests, points of view, and problems. 'A hungry animal', writes Katz,[14] 'divides the environment into edible and inedible things. An animal in flight sees roads to escape and hiding places . . . Generally speaking, objects change . . . according to the needs of the animal.' We

[13] See section 30 of *L.Sc.D.*

[14] Katz, *loc. cit.*

may add that objects can be classified, and can become similar or dissimilar, *only* in this way—by being related to needs and interests. This rule applies not only to animals but also to scientists. For the animal a point of view is provided by its needs, the task of the moment, and its expectations; for the scientist by his theoretical interests, the special problem under investigation, his conjectures and anticipations, and the theories which he accepts as a kind of background: his frame of reference, his 'horizon of expectations'.

The problem 'Which comes first, the hypothesis (H) or the observation (O)?' is soluble; as is the problem, 'Which comes first, the hen (H) or the egg (O)?'. The reply to the latter is, 'An earlier kind of egg'; to the former, 'An earlier kind of hypothesis'. It is quite true that any particular hypothesis we choose will have been preceded by observations—the observations, for example, which it is designed to explain. But these observations, in their turn, presupposed the adoption of a frame of reference: a frame of expectations: a frame of theories. If they were significant, if they created a need for explanation and thus gave rise to the invention of a hypothesis, it was because they could not be explained within the old theoretical framework, the old horizon of expectations. There is no danger here of an infinite regress. Going back to more and more primitive theories and myths we shall in the end find unconscious, *inborn* expectations.

The theory of inborn *ideas* is absurd, I think; but every organism has inborn *reactions* or *responses*; and among them, responses adapted to impending events. These responses we may describe as 'expectations' without implying that these 'expectations' are conscious. The newborn baby 'expects', in this sense, to be fed (and, one could even argue, to be protected and loved). In view of the close relation between expectation and knowledge we may even speak in quite a reasonable sense of 'inborn knowledge'. This 'knowledge', however, is not *valid a priori*; an inborn expectation, no matter how strong and specific, may be mistaken. (The newborn child may be abandoned, and starve.)

Thus we are born with expectations; with 'knowledge' which, although not *valid a priori*, is *psychologically or genetically a priori*, i.e. prior to all observational experience. One of the most important of these expectations is the expectation of finding a regularity. It is connected with an inborn propensity to look out for regularities, or with a *need* to find regularities, as we may see from the pleasure of the child who satisfies this need.

This 'instinctive' expectation of finding regularities, which is psychologically *a priori*, corresponds very closely to the 'law of causality' which Kant believed to be part of our mental outfit and to be *a priori* valid. One might thus be inclined to say that Kant failed to distinguish between psychologically *a priori* ways of thinking or responding and *a priori* valid beliefs. But I do not think that his mistake was quite as crude as that. For the expectation of finding regularities is not only psychologically *a priori*, but also logically *a priori*: it is logically prior to all observational experience, for it is prior to any recognition of similarities, as we have seen; and all observation involves the recognition of similarities (or dissimilarities). But in spite of being logically *a priori* in this sense the expectation is not valid *a priori*. For it may fail: we can easily construct an environment (it would be a lethal one) which, compared with our ordinary environment, is so chaotic that we completely fail to find regularities. (All natural laws could remain valid: environments of this kind have been used in the animal experiments mentioned in the next section.)

Thus Kant's reply to Hume came near to being right; for the distinction between an *a priori* valid expectation and one which is both genetically *and* logically prior to observation, but not *a priori* valid, is really somewhat subtle. But Kant proved too much. In trying to show how knowledge is possible, he proposed a theory which had the unavoidable consequence that our quest for knowledge must necessarily succeed, which is clearly mistaken. When Kant said, 'Our intellect does not draw its laws from nature but imposes its laws upon nature', he was right. But in thinking that these laws are necessarily true, or that we necessarily succeed in imposing them upon nature, he was wrong.[15]

[15] Kant believed that Newton's dynamics was *a priori* valid. (See his *Metaphysical Foundations of Natural Science*, published between the first and the second editions of the *Critique of Pure Reason*.) But if, as he thought, we can explain the validity of Newton's theory by the fact that our intellect imposes its laws upon nature, it follows, I think, that our intellect *must succeed* in this; which makes it hard to understand why *a priori* knowledge such as Newton's should be so hard to come by. A somewhat fuller statement of this criticism can be found in ch. 2, especially section x, and chs. 7 and 8 of the present volume.

Nature very often resists quite successfully, forcing us to discard our laws as refuted; but if we live we may try again.

To sum up this logical criticism of Hume's psychology of induction we may consider the idea of building an induction machine. Placed in a simplified 'world' (for example, one of sequences of coloured counters) such a machine may through repetition 'learn', or even 'formulate', laws of succession which hold in its 'world'. If such a machine can be constructed (and I have no doubt that it can) then, it might be argued, my theory must be wrong; for if a machine is capable of performing inductions on the basis of repetition, there can be no logical reasons preventing us from doing the same.

The argument sounds convincing, but it is mistaken. In constructing an induction machine we, the architects of the machine, must decide *a priori* what constitutes its 'world'; what things are to be taken as similar or equal; and what *kind* of 'laws' we wish the machine to be able to 'discover' in its 'world'. In other words we must build into the machine a framework determining what is relevant or interesting in its world: the machine will have its 'inborn' selection principles. The problems of similarity will have been solved for it by its makers who thus have interpreted the 'world' for the machine.

VI

Our propensity to look out for regularities, and to impose laws upon nature, leads to the psychological phenomenon of *dogmatic thinking* or, more generally, dogmatic behaviour: we expect regularities everywhere and attempt to find them even where there are none; events which do not yield to these attempts we are inclined to treat as a kind of 'background noise'; and we stick to our expectations even when they are inadequate and we ought to accept defeat. This dogmatism is to some extent necessary. It is demanded by a situation which can only be dealt with by forcing our conjectures upon the world. Moreover, this dogmatism allows us to approach a good theory in stages, by way of approximations: if we accept defeat too easily, we may prevent ourselves from finding that we were very nearly right.

It is clear that this *dogmatic attitude*, which makes us stick to our first impressions, is indicative of a strong belief; while a *critical attitude*, which is ready to modify its tenets, which admits doubt and demands tests, is indicative of a weaker belief. Now according to Hume's theory, and to the popular theory, the strength of a belief should be a product of repetition; thus it should always grow with experience, and always be greater in less primitive persons. But dogmatic thinking, an uncontrolled wish to impose regularities, a manifest pleasure in rites and in repetition as such, are characteristic of primitives and children; and increasing experience and maturity sometimes create an attitude of caution and criticism rather than of dogmatism.

I may perhaps mention here a point of agreement with psycho-analysis. Psycho-analysts assert that neurotics and others interpret the world in accordance with a personal set pattern which is not easily given up, and which can often be traced back to early childhood. A pattern or scheme which was adopted very early in life is maintained throughout, and every new experience is interpreted in terms of it; verifying it, as it were, and contributing to its rigidity. This is a description of what I have called the dogmatic attitude, as distinct from the critical attitude, which shares with the dogmatic attitude the quick adoption of a schema of expectations—a myth, perhaps, or a conjecture or hypothesis—but which is ready to modify it, to correct it, and even to give it up. I am inclined to suggest that most neuroses may be due to a partially arrested development of the critical attitude; to an arrested rather than a natural dogmatism; to resistance to demands for the modification and adjustment of certain schematic interpretations and responses. This resistance in its turn may perhaps be explained, in some cases, as due to an injury or shock, resulting in fear and in an increased need for assurance or certainty, analogous to the way in which an injury to a limb makes us afraid to move it, so that it becomes stiff. (It might even be argued that the case of the limb is not merely analogous to the dogmatic response, but an instance of it.) The explanation of any concrete case will have to take into account the weight of the difficulties involved in making the necessary adjustments—difficulties which may be considerable, especially in a complex and changing world: we know from experiments on animals that varying degrees of neurotic behaviour may be produced at will by correspondingly varying difficulties.

I found many other links between the psychology of knowledge and psychological fields which are often considered remote from it—for example the psychology of art and music; in fact, my ideas about induction originated in a conjecture about the evolution of Western polyphony. But you will be spared this story.

VII

My logical criticism of Hume's psychological theory, and the considerations connected with it (most of which I elaborated in 1926–7, in a thesis entitled 'On Habit and Belief in Laws'[16]) may seem a little removed from the field of the philosophy of science. But the distinction between dogmatic and critical thinking, or the dogmatic and the critical attitude, brings us right back to our central problem. For the dogmatic attitude is clearly related to the tendency to *verify* our laws and schemata by seeking to apply them and to confirm them, even to the point of neglecting refutations, whereas the critical attitude is one of readiness to change them—to test them; to refute them; to *falsify* them, if possible. This suggests that we may identify the critical attitude with the scientific attitude, and the dogmatic attitude with the one which we have described as pseudo-scientific.

It further suggests that genetically speaking the pseudo-scientific attitude is more primitive than, and prior to, the scientific attitude: that it is a pre-scientific attitude. And this primitivity or priority also has its logical aspect. For the critical attitude is not so much opposed to the dogmatic attitude as super-imposed upon it: criticism must be directed against existing and influential beliefs in need of critical revision—in other words, dogmatic beliefs. A critical attitude needs for its raw material, as it were, theories or beliefs which are held more or less dogmatically.

Thus science must begin with myths, and with the criticism of myths; neither with the collection of observations, nor with the invention of experiments, but with the critical discussion of myths, and of magical techniques and practices. The scientific tradition is distinguished from the pre-scientific tradition in having two layers. Like

[16] A thesis submitted under the title '*Gewohnheit und Gesetzerlebnis*' to the Institute of Education of the City of Vienna in 1927. (Unpublished.)

the latter, it passes on its theories; but it also passes on a critical attitude towards them. The theories are passed on, not as dogmas, but rather with the challenge to discuss them and improve upon them. This tradition is Hellenic: it may be traced back to Thales, founder of the first *school* (I do not mean 'of the first *philosophical* school', but simply 'of the first school') which was not mainly concerned with the preservation of a dogma.[17]

The critical attitude, the tradition of free discussion of theories with the aim of discovering their weak spots so that they may be improved upon, is the attitude of reasonableness, of rationality. It makes far-reaching use of both verbal argument and observation—of observation in the interest of argument, however. The Greeks' discovery of the critical method gave rise at first to the mistaken hope that it would lead to the solution of all the great old problems; that it would establish certainty; that it would help to *prove* our theories, to *justify* them. But this hope was a residue of the dogmatic way of thinking; in fact nothing can be justified or proved (outside of mathematics and logic). The demand for rational proofs in science indicates a failure to keep distinct the broad realm of rationality and the narrow realm of rational certainty: it is an untenable, an unreasonable demand.

Nevertheless, the role of logical argument, of deductive logical reasoning, remains all-important for the critical approach; not because it allows us to prove our theories, or to infer them from observation statements, but because only by purely deductive reasoning is it possible for us to discover what our theories imply, and thus to criticize them effectively. Criticism, I said, is an attempt to find the weak spots in a theory, and these, as a rule, can be found only in the more remote logical consequences which can be derived from it. It is here that purely logical reasoning plays an important part in science.

Hume was right in stressing that our theories cannot be validly inferred from what we can know to be true—neither from observations nor from anything else. He concluded from this that our belief in them was irrational. If 'belief' means here our inability to doubt our natural laws, and the constancy of natural regularities, then Hume is again right: this kind of dogmatic belief has, one might say, a

[17] Further comments on these developments may be found in chs. 4 and 5, below.

physiological rather than a rational basis. If, however, the term 'belief' is taken to cover our critical acceptance of scientific theories—a *tentative* acceptance combined with an eagerness to revise the theory if we succeed in designing a test which it cannot pass—then Hume was wrong. In such an acceptance of theories there is nothing irrational. There is not even anything irrational in relying for practical purposes upon well-tested theories, for no more rational course of action is open to us.

Assume that we have deliberately made it our task to live in this unknown world of ours; to adjust ourselves to it as well as we can; to take advantage of the opportunities we can find in it; and to explain it, if possible (we need not assume that it is), and as far as possible, with the help of laws and explanatory theories. *If we have made this our task, then there is no more rational procedure than the method of trial and error—of conjecture and refutation*: of boldly proposing theories; of trying our best to show that these are erroneous; and of accepting them tentatively if our critical efforts are unsuccessful.

From the point of view here developed all laws, all theories, remain essentially tentative, or conjectural, or hypothetical, even when we feel unable to doubt them any longer. Before a theory has been refuted we can never know in what way it may have to be modified. That the sun will always rise and set within twenty-four hours is still proverbial as a law 'established by induction beyond reasonable doubt'. It is odd that this example is still in use, though it may have served well enough in the days of Aristotle and Pytheas of Massalia—the great traveller who for centuries was called a liar because of his tales of Thule, the land of the frozen sea and the *midnight sun*.

The method of trial and error is not, of course, simply identical with the scientific or critical approach—with the method of conjecture and refutation. The method of trial and error is applied not only by Einstein but, in a more dogmatic fashion, by the amoeba also. The difference lies not so much in the trials as in a critical and constructive attitude towards errors; errors which the scientist consciously and cautiously tries to uncover in order to refute his theories with searching arguments, including appeals to the most severe experimental tests which his theories and his ingenuity permit him to design.

The critical attitude might be described as the result of a conscious attempt to make our theories, our conjectures, suffer in our stead in the struggle for the survival of the fittest. It gives us a chance to survive the elimination of an inadequate hypothesis—when a more dogmatic attitude would eliminate it by eliminating us. (There is a touching story of an Indian community which disappeared because of its belief in the holiness of life, including that of tigers.) We thus obtain the fittest theory within our reach by the elimination of those which are less fit. (By 'fitness' I do not mean merely 'usefulness' but truth; see chapters 3 and 10, below.) I do not think that this procedure is irrational or in need of any further rational justification.

VIII

Let us now turn from our logical criticism of the *psychology of experience* to our real problem—the problem of *the logic of science*. Although some of the things I have said may help us here, in so far as they may have eliminated certain psychological prejudices that favour induction, my treatment of the *logical problem of induction* is completely independent of this criticism, and of all psychological considerations. Provided you do not dogmatically believe in the alleged psychological fact that we make inductions, you may now forget my whole story with the exception of two logical points: my logical remarks on testability or falsifiability as the criterion of demarcation; and Hume's logical criticism of induction.

From what I have said it is obvious that there was a close link between the two problems which interested me at that time: demarcation, and induction or scientific method. It was easy to see that the method of science is criticism, i.e. attempted falsifications. Yet it took me a few years to notice that the two problems—of demarcation and of induction—were in a sense one.

Why, I asked, do so many scientists believe in induction? I found they did so because they believed natural science to be characterized by the inductive method—by a method starting from, and relying upon, long sequences of observations and experiments. They believed that the difference between genuine science and metaphysical or pseudo-scientific speculation depended solely upon whether or not the inductive method was employed. They believed (to put it in my own terminology) that only the inductive method could provide a satisfactory *criterion of demarcation*.

I recently came across an interesting formulation of this belief in a remarkable philosophical book by a great physicist—Max Born's *Natural Philosophy of Cause and Chance*.[18] He writes: 'Induction allows us to generalize a number of observations into a general rule: that night follows day and day follows night . . . But while everyday life has no definite criterion for the validity of an induction, . . . science has worked out a code, or rule of craft, for its application.' Born nowhere reveals the contents of this inductive code (which, as his wording shows, contains a 'definite criterion for the validity of an induction'); but he stresses that 'there is no logical argument' for its acceptance: 'it is a question of faith'; and he is therefore 'willing to call induction a metaphysical principle'. But why does he believe that such a code of valid inductive rules must exist? This becomes clear when he speaks of the 'vast communities of people ignorant of, or rejecting, the rule of science, among them the members of anti-vaccination societies and believers in astrology. It is useless to argue with them; I cannot compel them to accept the same criteria of valid induction in which I believe: the code of scientific rules.' This makes it quite clear that *'valid induction' was here meant to serve as a criterion of demarcation between science and pseudo-science.*

But it is obvious that this rule or craft of 'valid induction' is not even metaphysical: it simply does not exist. No rule can ever guarantee that a generalization inferred from true observations, however often repeated, is true. (Born himself does not believe in the truth of Newtonian physics, in spite of its success, although he believes that it is based on induction.) And the success of science is not based upon rules of induction, but depends upon luck, ingenuity, and the purely deductive rules of critical argument.

I may summarize some of my conclusions as follows:

(1) Induction, i.e. inference based on many observations, is a myth. It is neither a psychological fact, nor a fact of ordinary life, nor one of scientific procedure.

(2) The actual procedure of science is to operate with conjectures:

[18] Max Born, *Natural Philosophy of Cause and Chance*, Oxford, 1949, p. 7.

to jump to conclusions—often after one single observation (as noticed for example by Hume and Born).

(3) Repeated observations and experiments function in science as *tests* of our conjectures or hypotheses, i.e. as attempted refutations.

(4) The mistaken belief in induction is fortified by the need for a criterion of demarcation which, it is traditionally but wrongly believed, only the inductive method can provide.

(5) The conception of such an inductive method, like the criterion of verifiability, implies a faulty demarcation.

(6) None of this is altered in the least if we say that induction makes theories only probable rather than certain. (See especially chapter 10, below.)

IX

If, as I have suggested, the problem of induction is only an instance or facet of the problem of demarcation, then the solution to the problem of demarcation must provide us with a solution to the problem of induction. This is indeed the case, I believe, although it is perhaps not immediately obvious.

For a brief formulation of the problem of induction we can turn again to Born, who writes: '. . . no observation or experiment, however extended, can give more than a finite number of repetitions'; therefore, 'the statement of a law—B depends on A—always transcends experience. Yet this kind of statement is made everywhere and all the time, and sometimes from scanty material.'[19]

In other words, the logical problem of induction arises from (*a*) Hume's discovery (so well expressed by Born) that it is impossible to justify a law by observation or experiment, since it 'transcends experience'; (*b*) the fact that science proposes and uses laws 'everywhere and all the time'. (Like Hume, Born is struck by the 'scanty material', i.e. the few observed instances upon which the law may be based.) To this we have to add (*c*) *the principle of empiricism* which asserts that in science, only observation and experiment may decide upon the *acceptance or rejection* of scientific statements, including laws and theories.

[19] *Natural Philosophy of Cause and Chance*, p. 6.

These three principles, (*a*), (*b*), and (*c*), appear at first sight to clash; and this apparent clash constitutes the *logical problem of induction*.

Faced with this clash, Born gives up (*c*), the principle of empiricism (as Kant and many others, including Bertrand Russell, have done before him), in favour of what he calls a 'metaphysical principle'; a metaphysical principle which he does not even attempt to formulate; which he vaguely describes as a 'code or rule of craft'; and of which I have never seen any formulation which even looked promising and was not clearly untenable.

But in fact the principles (*a*) to (*c*) do not clash. We can see this the moment we realize that the acceptance by science of a law or of a theory is *tentative only*; which is to say that all laws and theories are conjectures, or tentative *hypotheses* (a position which I have sometimes called 'hypotheticism'); and that we may reject a law or theory on the basis of new evidence, without necessarily discarding the old evidence which originally led us to accept it.[20]

The principle of empiricism (*c*) can be fully preserved, since the fate of a theory, its acceptance or rejection, is decided by observation and experiment—by the result of tests. So long as a theory stands up to the severest tests we can design, it is accepted; if it does not, it is rejected. But it is never inferred, in any sense, from the empirical evidence. There is neither a psychological nor a logical induction. *Only the falsity of the theory can be inferred from empirical evidence, and this inference is a purely deductive one.*

Hume showed that it is not possible to infer a theory from observation statements; but this does not affect the possibility of refuting a theory by observation statements. The full appreciation of this possibility makes the relation between theories and observations perfectly clear.

This solves the problem of the alleged clash between the principles (*a*), (*b*), and (*c*), and with it Hume's problem of induction.

[20] I do not doubt that Born and many others would agree that theories are accepted only tentatively. But the widespread belief in induction shows that the far-reaching implications of this view are rarely seen.

X

Thus the problem of induction is solved. But nothing seems less wanted than a simple solution to an age-old philosophical problem. Wittgenstein and his school hold that genuine philosophical problems do not exist;[21] from which it clearly follows that they cannot be solved. Others among my contemporaries do believe that there are philosophical problems, and respect them; but they seem to respect them too much; they seem to believe that they are insoluble, if not taboo; and they are shocked and horrified by the claim that there is a simple, neat, and lucid, solution to any of them. If there is a solution it must be deep, they feel, or at least complicated.

However this may be, I am still waiting for a simple, neat and lucid criticism of the solution which I published first in 1933 in my letter to the Editor of *Erkenntnis*,[22] and later in *The Logic of Scientific Discovery*.

Of course, one can invent new problems of induction, different from the one I have formulated and solved. (Its formulation was half its solution.) But I have yet to see any reformulation of the problem whose solution cannot be easily obtained from my old solution. I am now going to discuss some of these re-formulations.

One question which may be asked is this: how do we really jump from an observation statement to a theory?

Although this question appears to be psychological rather than philosophical, one can say something positive about it without invoking psychology. One can say first that the jump is not from an observation statement, but from a problem-situation, and that the theory must allow us *to explain* the observations which created the problem (that is, *to deduce* them from the theory strengthened by other accepted theories and by other observation statements, the so-called initial conditions). This leaves, of course, an immense number of possible theories, good and bad; and it thus appears that our question has not been answered.

But this makes it fairly clear that when we asked our question we had more in mind than, 'How do we jump from an observation statement to a theory?' The question we had in mind was, it now appears, 'How do we jump from an observation statement to a *good* theory?' But to this

[21] Wittgenstein still held this belief in 1946; see note 8 to ch. 2, below.

[22] See note 5 above.

the answer is: by jumping first to *any* theory and then testing it, to find whether it is good or not; i.e. by repeatedly applying the critical method, eliminating many bad theories, and inventing many new ones. Not everybody is able to do this; but there is no other way.

Other questions have sometimes been asked. The original problem of induction, it was said, is the problem of *justifying* induction, i.e. of justifying inductive inference. If you answer this problem by saying that what is called an 'inductive inference' is always invalid and therefore clearly not justifiable, the following new problem must arise: how do you justify your method of trial and error? Reply: the method of trial and error is a *method of eliminating false theories* by observation statements; and the justification for this is the purely logical relationship of deducibility which allows us to assert the falsity of universal statements if we accept the truth of singular ones.

Another question sometimes asked is this: why is it reasonable to prefer non-falsified statements to falsified ones? To this question some involved answers have been produced, for example pragmatic answers. But from a pragmatic point of view the question does not arise, since false theories often serve well enough: most formulae used in engineering or navigation are known to be false, although they may be excellent approximations and easy to handle; and they are used with confidence by people who know them to be false.

The only correct answer is the straightforward one: because we search for truth (even though we can never be sure we have found it), and because the falsified theories are known or believed to be false, while the non-falsified theories may still be true. Besides, we do not prefer *every* non-falsified theory—only one which, in the light of criticism, appears to be better than its competitors: which solves our problems, which is well tested, and of which we think, or rather conjecture or hope (considering other provisionally accepted theories), that it will stand up to further tests.

It has also been said that the problem of induction is, 'Why is it *reasonable* to believe that the future will be like the past?', and that a satisfactory answer to this question should make it plain that such a belief is, in fact, reasonable. My reply is that it is reasonable to believe that the future will be very different from the past in many vitally important respects. Admittedly it is perfectly reasonable to *act* on the assumption that it will, in many respects, be like the past, and that well-tested laws will continue to hold (since we can have no better assumption to act upon); but it is also reasonable to believe that such a course of action will lead us at times into severe trouble, since some of the laws upon which we now heavily rely may easily prove unreliable. (Remember the midnight sun!) One might even say that to judge from past experience, and from our general scientific knowledge, the future will *not* be like the past, in perhaps most of the ways which those have in mind who say that it will. Water will sometimes not quench thirst, and air will choke those who breathe it. An apparent way out is to say that the future will be like the past *in the sense that the laws of nature will not change*, but this is begging the question. We speak of a 'law of nature' only if we think that we have before us a regularity which does not change; and if we find that it changes then we shall not continue to call it a 'law of nature'. Of course our search for natural laws indicates that we hope to find them, and that we believe that there are natural laws; but our belief in any particular natural law cannot have a safer basis than our unsuccessful critical attempts to refute it.

I think that those who put the problem of induction in terms of the *reasonableness* of our beliefs are perfectly right if they are dissatisfied with a Humean, or post-Humean, sceptical despair of reason. We must indeed reject the view that a belief in science is as irrational as a belief in primitive magical practices—that both are a matter of accepting a 'total ideology', a convention or a tradition based on faith. But we must be cautious if we formulate our problem, with Hume, as one of the reasonableness of our *beliefs*. We should split this problem into three—our old problem of demarcation, or of how to *distinguish* between science and primitive magic; the problem of the rationality of the scientific or critical *procedure*, and of the role of observation within it; and lastly the problem of the rationality of our *acceptance* of theories for scientific and for practical purposes. To all these three problems solutions have been offered here.

One should also be careful not to confuse the problem of the reasonableness of the scientific procedure and the (tentative) acceptance of the results of this procedure—i.e. the scientific theories—with the problem of the rationality or otherwise *of the belief that this procedure will succeed*. In practice, in practical scientific research, this belief is no doubt unavoidable and reasonable, there being no better alternative. But the belief is certainly unjustifiable in a theoretical sense, as I have argued (in section v). Moreover, if we could show, on general logical grounds, that the scientific quest is likely to succeed, one could not understand why anything like success has been so rare in the long history of human endeavours to know more about our world.

Yet another way of putting the problem of induction is in terms of probability. Let t be the theory and *e* the evidence: we can ask for $P(t,e)$, that is to say, the probability of t, given *e*. The problem of induction, it is often believed, can then be put thus: construct a *calculus of probability* which allows us to work out for any theory t what its probability is, relative to any given empirical evidence *e*; and show that $P(t,e)$ increases with the accumulation of supporting evidence, and reaches high values—at any rate values greater than ½.

In *The Logic of Scientific Discovery* I explained why I think that this approach to the problem is fundamentally mistaken.[23] To make this clear, I introduced there the distinction between *probability* and *degree of corroboration or confirmation*. (The term 'confirmation' has lately been so much used and misused that I have decided to surrender it to the verificationists and to use for my own purposes 'corroboration' only. The term 'probability' is best used in some of the many senses which satisfy the well-known calculus of probability, axiomatized, for example, by Keynes, Jeffreys, and myself; but nothing of course depends on the choice of words, as long as we do not *assume*, uncritically, that degree of corroboration must also be a probability—that is to say, that it must satisfy the calculus of probability.)[24]

[23] L.Sc.D. (see note 5 above), ch. x, especially sections 80 to 83, also section 34 ff. See also my note 'A Set of Independent Axioms for Probability', Mind, N.S. **47**, 1938, p. 275. (This note has since been reprinted, with corrections, in the new appendix *ii of L.Sc.D. See also the next note but one to the present chapter.)

[24] A definition, in terms of probabilities (see the next note), of $C(t,e)$, i.e. of the degree of corroboration (of a theory t relative to the evidence *e*) satisfying the demands indicated in my L.Sc.D., sections 82 to 83, is the following:

$$C(t,e) = E(t,e)\ (1 + P(t)P(t,e)),$$

where $E(t,e) = (P(e,t) - P(e))/(P(e,t) + P(e))$ is a (non-additive) measure of the explanatory power of t with respect to *e*. Note that $C(t,e)$ is not a probability: it may have values between -1 (refutation of t by *e*) and $C(t,t) \leqslant +1$. Statements t which are lawlike and thus

I explained in my book why we are interested in theories with a *high degree of corroboration*. And I explained why it is a mistake to conclude from this that we are interested in *highly probable* theories. I pointed out that the probability of a statement (or set of statements) is always the greater the less the statement says: it is inverse to the content or the deductive power of the statement, and thus to its explanatory power. Accordingly every interesting and powerful statement must have a low probability; and *vice versa*: a statement with a high probability will be scientifically uninteresting, because it says little and has no explanatory power. Although we seek theories with a high degree of corroboration, *as scientists we do not seek highly probable theories but explanations; that is to say, powerful and improbable theories*. The opposite view—that science aims at high probability—is a characteristic development of verificationism: if you find that you cannot verify a theory, or make it certain by induction, you may turn to probability as a kind of '*Ersatz*' for certainty, in the hope that induction may yield at least that much.

I have discussed the two problems of demarcation and induction at some length. Yet since I set out to give you in this lecture a kind of report on the work I have done in this field I shall have to add, in the form of an *Appendix*, a few words about some other problems on which I have been working, between 1934 and 1953. I was led to most of these problems by trying to think out the consequences of the solutions to the two problems of demarcation and induction. But time does not allow me to continue my narrative, and to tell you how my new problems arose out of my old ones. Since I cannot even start a discussion of

non-verifiable cannot even reach $C(t,e) = C(t,t)$ upon empirical evidence e. $C(t,t)$ is the *degree of corroborability* of t, and is equal to the *degree of testability* of t, or to the *content* of t. Because of the demands implied in point (6) at the end of section I above, I do not think, however, that it is possible to give a complete formalization of the idea of corroboration (or, as I previously used to say, of confirmation).

(Added 1955 to the first proofs of this paper:)

See also my note 'Degree of Confirmation', *British Journal for the Philosophy of Science*, **5**, 1954, pp. 143 ff. (See also **5**, pp. 334.) I have since simplified this definition as follows (*B.J.P.S.*, 1955, **5**, p. 359):

$$C(t,e) = (P(e,t) - P(e))/(P(e,t) - P(et) + P(e))$$

For a further improvement, see *B.J.P.S.* **6**, 1955, p. 56.

these further problems now, I shall have to confine myself to giving you a bare list of them, with a few explanatory words here and there. But even a bare list may be useful, I think. It may serve to give an idea of the fertility of the approach. It may help to illustrate what our problems look like; and it may show how many there are, and so convince you that there is no need whatever to worry over the question whether philosophical problems exist, or what philosophy is really about. So this list contains, by implication, an apology for my unwillingness to break with the old tradition of trying to solve problems with the help of rational argument, and thus for my unwillingness to participate wholeheartedly in the developments, trends, and drifts, of contemporary philosophy.

APPENDIX: SOME PROBLEMS IN THE PHILOSOPHY OF SCIENCE

My first three items in this list of additional problems are connected with the calculus of probabilities.

(1) The frequency theory of probability. In *The Logic of Scientific Discovery* I was interested in developing a consistent theory of probability as it is used in science; which means, a statistical or frequency theory of probability. But I also operated there with another concept which I called 'logical probability'. I therefore felt the need for a generalization—for a formal theory of probability which allows different *interpretations*: (*a*) as a theory of the logical probability of a statement relative to any given evidence; including a theory of absolute logical probability, i.e. of the measure of the probability of a statement relative to zero evidence; (*b*) as a theory of the probability of an event relative to any given *ensemble* (or 'collective') of events. In solving this problem I obtained a simple theory which allows a number of further interpretations: it may be interpreted as a calculus of contents, or of deductive systems, or as a class calculus (Boolean algebra) or as propositional calculus; and also as a calculus of *propensities*.[25]

[25] See my note in *Mind*, *loc. cit*. The axiom system given there for elementary (i.e. non-continuous) probability can be simplified as follows ('$\bar{x}$' denotes the complement of x; 'xy' the intersection or conjunction of x and y):

(2) This problem of a *propensity interpretation of probability* arose out of my interest in Quantum Theory. It is usually believed that Quantum Theory has to be interpreted statistically, and no doubt statistics is essential for its empirical tests. But this is a point where, I believe, the

(A1) $P(xy) \geqslant P(yx)$ (Commutation)
(A2) $P(x(yz)) \geqslant P((xy)z)$ (Association)
(A3) $P(xx) \geqslant P(x)$ (Tautology)
(B1) $P(x) \geqslant P(xy)$ (Monotony)
(B2) $P(xy) + P(x\bar{y}) = P(x)$ (Addition)
(B3) $(x)(Ey)\ (P(y) \neq 0 \text{ and } P(xy) = P(x)P(y))$ (Independence)
(C1) If $P(y) \neq 0$, then $P(x,y) = P(xy)/P(y)$ (Definition of relative
(C2) If $P(y) \neq 0$, then $P(x,y) = P(x,x) = P(y,y)$ Probability)

Axiom (C2) holds, in this form, for the finitist theory only; it may be omitted if we are prepared to put up with a condition such as $P(y) \neq 0$ in most of the theorems on relative probability. For relative probability, (A1) – (B2), (C1) – (C2), and (B3) up to '*and*' suffices. For absolute probability, (A1) – (B3) is necessary and sufficient: without (B3) we cannot, for example, derive the definition of absolute in terms of relative probability,

$$P(x) = P(x,x\bar{x})$$

nor its weakened corollary

$$(x)(Ey)\ (P(y) \neq 0 \text{ and } P(x) = P(x,y))$$

from which (B3) results immediately (by substituting for '$P(x,y)$' its definiens). Thus (B3), like all other axioms with the possible exception of (C2), expresses part of the intended meaning of the concepts involved, and we must not look upon $1 \geqslant P(x)$ or $1 \geqslant P(x,y)$, which are derivable from (B1), with (B3) or with (C1) and (C2), as 'inessential conventions' (as Carnap and others have suggested).

(Added 1955 to the first proofs of this paper; see also note 31, below.)

I have since developed an axiom system for *relative probability* which holds for finite *and* infinite systems (and in which absolute probability can be defined as in the penultimate formula above). Its axioms are:

(B1) $P(x,z) \geqslant P(xy,z)$
(B2) If $P(y,y) \neq P(u,y)$ then $P(x,y) + P(\bar{x},y) = P(y,y)$
(B3) $P(xy,z) = P(x,yz)P(y,z)$
(C1) $P(x,x) = P(y,y)$
(D1) If $((u)P(x,u) = P(y,u))$ then $P(w,x) = P(w,y)$
(E1) $(Ex)\ (Ey)\ (Eu)\ (Ew)\ P(x,y) \neq P(u,w)$

This is a slight improvement on a system which I published in *B.J.P.S.*, **6**, 1955, pp. 56 f.; 'Postulate 3' is here called 'D1'. (See also *vol. cit.*, bottom of p. 176. Moreover, in line 3 of dangers of the testability theory of meaning become clear. Although the tests of the theory are statistical, and although the theory (say, Schrödinger's equation) may imply statistical consequences, it need not have a statistical meaning: and one can give examples of objective propensities (which are something like generalized forces) and of fields of propensities, which can be measured by statistical methods without being themselves statistical. (See also the last paragraph of chapter 3, below, with note 35.)

(3) The use of statistics in such cases is, in the main, to provide *empirical tests* of theories which need not be purely statistical; and this raises the question of the *refutability of statistical statements*—a problem treated, but not to my full satisfaction, in the 1934 edition of my *The Logic of Scientific Discovery*. I later found, however, that all the elements for constructing a satisfactory solution lay ready for use in that book; certain examples I had given allow a mathematical characterization of a class of infinite chance-like sequences which are, in a certain sense, the *shortest sequences* of their kind.[26] A statistical statement may now be said to be testable by comparison with these 'shortest sequences'; it is refuted if the statistical properties of the tested *ensembles* differ from the statistical properties of the initial sections of these 'shortest sequences'.

(4) There are a number of further problems connected with the interpretation of the formalism of a quantum theory. In a chapter of *The Logic of Scientific Discovery* I criticized the 'official' interpretation, and I still think that my criticism is valid in all points but one: one example which I used (in section 77) is mistaken. But since I wrote that section, Einstein, Podolsky, and Rosen have published a thought-experiment which can be substituted for my example, although their tendency

the last paragraph on p. 57, the words 'and that the limit exists' should be inserted, between brackets, before the word 'all'.)

(Added 1961 to the proofs of the present volume.)

A fairly full treatment of all these questions will now be found in the new appendices to *L.Sc.D.*

I have left this note as in the first publication because I have referred to it in various places. The problems dealt with in this and the preceding note have since been more fully treated in the new appendices to *L.Sc.D.* (To its 1961 American Edition I have added a system of only three axioms; see also *Addendum* 2 to the present volume.)

[26] Sec *L.Sc.D.*, p. 163 (section 55); see especially the new appendix *vi.

(which is deterministic) is quite different from mine. Einstein's belief in determinism (which I had occasion to discuss with him) is, I believe, unfounded, and also unfortunate: it robs his criticism of much of its force, and it must be emphasized that much of his criticism is quite independent of his determinism.

(5) As to the problem of determinism itself, I have tried to show that even classical physics, which is deterministic in a certain *prima facie* sense, is mis-interpreted if used to support a deterministic view of the physical world in Laplace's sense.

(6) In this connection, I may also mention the *problem of simplicity*—of the simplicity of a theory, which I have been able to connect with the content of a theory. It can be shown that what is usually called the simplicity of a theory is associated with its logical improbability, and not with its probability, as has often been supposed. This, indeed, allows us to deduce, from the theory of science outlined above, why it is always advantageous to try the simplest theories first. They are those which offer us the best chance to submit them to severe tests: the simpler theory has always a higher degree of testability than the more complicated one.[27] (Yet I do not think that this settles all problems about simplicity. See also chapter 10, section xviii, below.)

(7) Closely related to this problem is the problem of the *ad hoc* character of a hypothesis, and of degrees of this *ad hoc* character (of '*ad hocness*', if I may so call it). One can show that the methodology of science (and the history of science also) becomes understandable in its details if we assume that the aim of science is to get explanatory theories which are as little *ad hoc* as possible: a 'good' theory is not *ad hoc*, while a 'bad' theory is. On the other hand one can show that the probability theories of induction imply, inadvertently but necessarily, the unacceptable rule: always use the theory which is the most *ad hoc*, i.e. which transcends the available evidence as little as possible. (See also my paper 'The Aim of Science', mentioned in note 28 below.)

(8) An important problem is the problem of the *layers of explanatory hypotheses* which we find in the more developed theoretical sciences, and of the relations between these layers. It is often asserted that Newton's theory can be induced or even deduced from Kepler's and Galileo's

[27] Ibid., sections 41 to 46. But see now also ch. 10, section xviii below.

laws. But it can be shown that Newton's theory (including his theory of absolute space) strictly speaking contradicts Kepler's (even if we confine ourselves to the two-body problem[28] and neglect the mutual attraction between the planets) and also Galileo's; although approximations to these two theories can, of course, be deduced from Newton's. But it is clear that neither a deductive nor an inductive inference can lead, from consistent premises, to a conclusion which contradicts them. These considerations allow us to analyse the logical relations between 'layers' of theories, and also the idea of an *approximation*, in the two senses of (*a*) The theory *x* is an approximation to the theory *y*; and (*b*) The theory *x* is 'a good approximation to the facts'. (See also chapter 10, below.)

(9) A host of interesting problems is raised by *operationalism*, the doctrine that theoretical concepts have to be defined in terms of measuring operations. Against this view, it can be shown that *measurements presuppose theories*. There is no measurement without a theory and no operation which can be satisfactorily described in non-theoretical terms. The attempts to do so are always circular; for example, the description of the measurement of length needs a (rudimentary) theory of heat and of temperature-measurement; but these, in turn, involve measurements of length.

The analysis of operationalism shows the need for a *general theory of measurement*; a theory which does not, naïvely, take the practice of measuring as 'given', but explains it by analysing its function in the testing of scientific hypotheses. This can be done with the help of the doctrine of degrees of testability.

[28] The contradictions mentioned in this sentence of the text were pointed out, for the case of the many-body problem, by P. Duhem, *The Aim and Structure of Physical Theory* (1906; trans. by P. P. Wiener, 1954). In the case of the two-body problem, the contradictions arise in connection with Kepler's third law, which may be reformulated for the two-body problem as follows. 'Let S be any *set of pairs* of bodies such that *one* body of each pair is of the mass of our sun; then a^3/T^2 = *constant*, for any pair in S.' Clearly this contradicts Newton's theory, which yields for appropriately chosen units $a^3/T^2 = m_0 + m_1$ (where m_0 = mass of the sun = *constant*, and m_1 = mass of the second body, which varies with this body). But 'a^3/T^2 = *constant*' is, of course, an excellent approximation, *provided* the varying masses of the second bodies are all negligible compared with that of our sun. (See also my paper 'The Aim of Science', *Ratio*, 1, 1957, pp. 24 ff., and section 15 of the *Post-script* to my *Logic of Scientific Discovery*.)

Connected with, and closely parallel to, operationalism is the doctrine of *behaviourism*, i.e. the doctrine that, since all test-statements describe behaviour, our theories too must be stated in terms of possible behaviour. But the inference is as invalid as the phenomenalist doctrine which asserts that since all test-statements are observational, theories too must be stated in terms of possible observations. All these doctrines are forms of the verifiability theory of meaning; that is to say, of inductivism.

Closely related to operationalism is *instrumentalism*, i.e. the interpretation of scientific theories as practical instruments or tools for such purposes as the prediction of impending events. That theories may be used in this way cannot be doubted; but instrumentalism asserts that they can be best understood as instruments; and that this is mistaken, I have tried to show by a comparison of the *different functions* of the formulae of applied and pure science. In this context the problem of the *theoretical* (i.e. non-practical) function of predictions can also be solved. (See chapter 3, section 5, below.)

It is interesting to analyse from the same point of view the function of language—as an instrument. One immediate finding of this analysis is that we use descriptive language in order to talk *about the world*. This provides new arguments in favour of *realism*.

Operationalism and instrumentalism must, I believe, be replaced by 'theoreticism', if I may call it so: by the recognition of the fact that we are always operating within a complex framework of theories, and that we do not aim simply at correlations, but at explanations.

(10) The problem of *explanation* itself. It has often been said that scientific explanation is reduction of the unknown to the known. If pure science is meant, nothing could be further from the truth. It can be said without paradox that scientific explanation is, on the contrary, the reduction of the known to the unknown. In pure science, as opposed to an applied science which takes pure science as 'given' or 'known', explanation is always the logical reduction of hypotheses to others which are of a higher level of universality; of 'known' facts and 'known' theories to assumptions of which we know very little as yet, and which have still to be tested. The analysis of degrees of explanatory power, and of the relationship between genuine and sham explanation and between explanation and prediction, are examples of problems which are of great interest in this context.

(11) This brings me to the problem of the relationship between explanation in the natural sciences and historical explanation (a problem that, strangely enough, is logically analogous to the problem of explanation in the pure and applied sciences); and to the vast field of problems in the methodology of the social sciences, especially the problems of *historical prediction; historicism* and *historical determinism*; and *historical relativism*. These problems are linked, again, with the more general problems of determinism and relativism, including the problems of linguistic relativism.[29]

(12) A further problem of interest is the analysis of what is called 'scientific objectivity'. I have treated this problem in several places, especially in connection with a criticism of the so-called 'sociology of knowledge'.[30]

(13) One type of solution of the problem of induction should be mentioned here again (see section iv, above), in order to warn against it. (Solutions of this kind are, as a rule, put forth without a clear formulation of the problem which they are supposed to solve.) The view I have in mind may be described as follows. It is first taken for granted that nobody seriously doubts that we do, *in fact*, make inductions, and successful ones. (My suggestion that this is a myth, and that the apparent cases of induction turn out, if analysed more carefully, to be cases of the method of trial and error, is treated with the contempt which an utterly unreasonable suggestion of this kind deserves.) It is then said that the task of a theory of induction is to describe and classify our inductive policies or procedures, and perhaps to point out which of them are the most successful and reliable ones and which are less successful or reliable; and that any further question of justification is misplaced. Thus the view I have in mind is characterized by the contention that the distinction between the factual problem of describing how we argue inductively (*quid facti?*), and the problem of the justification of our inductive arguments (*quid juris?*) is a misplaced distinction. It is also said that the justification required is unreasonable,

[29] See my *Poverty of Historicism*, 1957, section 28 and note, 30 to 32; also the Addendum I to vol. ii of my *Open Society* (added to the 4th edition 1962).

[30] *Poverty of Historicism*, section 32; *L.Sc.D.*, section 8; *Open Society*, ch. 23 and Addendum to vol. ii (Fourth Edition). The passages are complementary.

since we cannot expect inductive arguments to be 'valid' in the same sense in which deductive ones may be 'valid': induction simply is not deduction, and it is unreasonable to demand from it that it should conform to the standards of logical—that is, deductive—validity. We must therefore judge it by its own standards—by inductive standards—of reasonableness.

I think that this defence of induction is mistaken. It not only takes a myth for a fact, and the alleged fact for a standard of rationality, with the result that a myth becomes a standard of rationality; but it also propagates, in this way, a principle which may be used to defend *any* dogma against *any* criticism. Moreover, it mistakes the status of formal or 'deductive' logic. (It mistakes it just as much as those who saw it as the systematization of our factual, that is, psychological, 'laws of thought'.) For deduction, I contend, is not valid because we choose or decide to adopt its rules as a standard, or decree that they shall be accepted; rather, it is valid because it adopts, and incorporates, the rules by which truth is transmitted from (logically stronger) premises to (logically weaker) conclusions, and by which falsity is re-transmitted from conclusions to premises. (This re-transmission of falsity makes formal logic the *Organon of rational criticism*—that is, of refutation.)

One point that may be conceded to those who hold the view I am criticizing here is this. In arguing from premises to the conclusion (or in what may be called the 'deductive direction'), we argue from the truth or the certainty or the probability of the premises to the corresponding property of the conclusion; while if we argue from the conclusion to the premises (and thus in what we have called the 'inductive direction'), we argue from the falsity or the uncertainty or the impossibility or the improbability of the conclusion to the corresponding property of the premises; accordingly, we must indeed concede that standards such as, more especially, *certainty*, which apply to arguments in the deductive direction, do not also apply to arguments in the inductive direction. Yet even this concession of mine turns in the end against those who hold the view which I am criticizing here; for they assume, wrongly, that we may argue in the inductive direction, though not to the certainty, yet to the *probability* of our 'generalizations'. But this assumption is mistaken, for all the intuitive ideas of probability which have ever been suggested.

This is a list of just a few of the problems of the philosophy of science to which I was led in my pursuit of the two fertile and fundamental problems whose story I have tried to tell you.[31]

[31] (13) was added in 1961. Since 1953, when this lecture was delivered, and 1955, when I read the proofs, the list given in this appendix has grown considerably, and some more recent contributions which deal with problems not listed here will be found in this volume (see especially ch. 10, below) and in my other books (see especially the new appendices to my *L.Sc.D.*, and the new *Addendum* to vol. ii of my *Open Society* which I have added to the fourth edition, 1962). See especially also my paper 'Probability Magic, or Knowledge out of Ignorance', *Dialectica*, **11**, 1957, pp. 354–374.

(Added 1989.) It is interesting that, as David Miller and I have been able to show, if probabilistic inductive support exists, it is always negative; that is, countersupport. See our paper 'Why Probabilistic Support Is Not Inductive', *Philosophical Transactions of the Royal Society of London*, series A, **321**, 1987, pp. 569–596.

2

THE NATURE OF PHILOSOPHICAL PROBLEMS AND THEIR ROOTS IN SCIENCE

I

It was after some hesitation that I decided to take as my point of departure the present position of English philosophy. For I believe that the function of a scientist or of a philosopher is to solve scientific or philosophical problems, rather than to talk about what he or other philosophers are doing or might do. Any unsuccessful attempt to solve a scientific or philosophical problem, if it is an honest and devoted attempt, appears to me more significant than a discussion of such a question as 'What is science?' or 'What is philosophy?' And even if we put this latter question, as we should, in the slightly better form, 'What is the character of philosophical problems?', I for one should not bother much about it; I should feel that it had little weight, even compared with such a minor problem of philosophy as the question

The Chairman's address, delivered at the meeting of 28th April 1952, to the Philosophy of Science Group of the British Society for the History of Science (now the British Society for the Philosophy of Science); first published in The British Journal for the Philosophy of Science, **3**, 1952.

whether every discussion or every criticism must always proceed from 'assumptions' or 'suppositions' which themselves are beyond argument.[1]

When describing 'What is the character of philosophical problems?' as a slightly better form of 'What is philosophy?' I wished to hint at one of the reasons for the futility of the current controversy concerning the nature of philosophy: the naïve belief that there is an entity such as 'philosophy', or perhaps 'philosophical activity', and that it has a certain character or essence or 'nature'. The belief that there is such a thing as physics, or biology, or archaeology, and that these 'studies' or 'disciplines' are distinguishable by the subject matter which they investigate, appears to me to be a residue from the time when one believed that a theory had to proceed from a definition of its own subject matter.[2] But subject matter, or kinds of things, do not, I hold, constitute a basis for distinguishing disciplines. Disciplines are distinguished partly for historical reasons and reasons of administrative convenience (such as the organization of teaching and of appointments), and partly because the theories which we construct to solve our problems have a tendency[3] to grow into unified systems. But all this classification and distinction is a comparatively unimportant and superficial affair. *We are not students of some subject matter but students of problems.* And problems may cut right across the borders of any subject matter or discipline.

Obvious as this fact may appear to some people, it is so important for our present discussion that it is worth while illustrating it by an example. I need hardly mention that a geologist's problem such as assessing the chances of finding deposits of oil or uranium in a certain district has to be solved with the help of theories and techniques

[1] I call this a minor problem because I believe that it can easily be solved, by refuting the ('relativistic') doctrine which gives rise to the question. (Thus the answer to the question is negative. See the Addendum to vol. ii of my *Open Society*, added to the fourth edition of 1962.)

[2] This view is part of what I have called 'essentialism'. Cf. for example my *Open Society*, chs. 2 and 11, or *The Poverty of Historicism*, section 10.

[3] This tendency can be explained by the principle that theoretical explanations are the more satisfactory the better they can be supported by *independent* evidence. For in order to be supported by mutually independent pieces of evidence, a theory must be sweeping.

usually classified as mathematical, physical and chemical. It is however less obvious that even a more 'basic' science such as atomic physics may have to make use of a geological survey, and of geological theories and techniques, to solve a problem in one of its most abstract and fundamental theories; for example the problem of testing predictions about the relative stability or instability of atoms of an even or odd atomic number.

I am quite ready to admit that many problems, even if their solution involves the most diverse disciplines, nevertheless 'belong' in some sense to one or another of the traditional disciplines; the two problems just mentioned clearly 'belong' to geology and physics respectively. This is because each of them arises out of a discussion characteristic of the tradition of the discipline in question. Each arises out of the discussion of some theory, or out of empirical tests bearing upon a theory; and theories, as opposed to subject matter, may constitute a discipline (which might be described as a somewhat loose cluster of theories undergoing challenge, change, and growth). But this does not affect my point that the classification into disciplines is comparatively unimportant, and that we are students not of disciplines but of *problems*.

But are there philosophical problems? The present position of English philosophy—my point of departure—originates, I believe, in the late Professor Ludwig Wittgenstein's doctrine that there are none; that all genuine problems are scientific problems; that the alleged problems of philosophy are pseudo-problems; that the alleged propositions or theories of philosophy are pseudo-propositions or pseudo-theories; that they are not false (if they were false, their negations would be true propositions or theories) but strictly meaningless combinations of words,[4] no more meaningful than the incoherent babbling of a child who has not yet learned to speak properly.[5]

[4] 'All animals are equal but some are more equal than others' is an excellent example of an expression which would be 'meaningless' in the technical sense of Russell and Wittgenstein, though clearly far from meaningless (in the sense of pointless) in the context of Orwell's *Animal Farm*. It is interesting that later Orwell considered the possibility of introducing a language, and enforcing its use, in which 'All men are equal' would become meaningless in Wittgenstein's technical sense.

[5] Since Wittgenstein described his own *Tractatus* as meaningless (see also the next footnote) he distinguished, at least by implication, between revealing or important and

As a consequence, philosophy cannot contain any theories. Its true nature, according to Wittgenstein, is not that of a theory, but that of an activity. The task of all genuine philosophy is to unmask philosophical nonsense, and to teach people to talk sense.

My plan is to take this doctrine[6] of Wittgenstein's as my starting point. I shall try to explain it (in section ii); to defend it, to some extent; and to criticize it (in section iii). And I shall illustrate all this (in sections iv to xi) by some examples from the history of scientific ideas.

But before proceeding to carry out this plan I wish to reaffirm my conviction that a philosopher should philosophize: he should try to solve philosophical problems, rather than talk about philosophy. If Wittgenstein's doctrine is true, then nobody can philosophize, in my sense. Were this my opinion I would give up philosophy. But it so happens that I am not only deeply interested in certain philosophical problems (I do not much care whether they are 'rightly' called 'philosophical problems'), but inspired by the hope that I may contribute—if only a little, and only by hard work—to their solution. My only excuse for talking here about philosophy—instead of philosophizing—

worthless or unimportant nonsense. But this does not affect his main doctrine which I am discussing, the non-existence of philosophical problems. (A discussion of other doctrines of Wittgenstein's can be found in the Notes to my *Open Society*, especially notes 26, 46, 51, and 52 to ch. 11.)

[6] It is easy to detect at once one flaw in this doctrine: the doctrine, it may be said, is itself a philosophic theory, claiming to be true, and not to be meaningless. This criticism, however, is perhaps a little cheap. It might be countered in at least two ways. (1) One might say that the doctrine is indeed meaningless *qua* doctrine, but not *qua* activity. (This is the view of Wittgenstein, who said at the end of his *Tractatus Logico-Philosophicus* that whoever understood the book must realize at the end that it was itself meaningless, and must discard it like a ladder, after having used it to reach the desired height.) (2) One might say that the doctrine is not a philosophical but an empirical one; that is states the historical fact that all apparent 'theories' proposed by philosophers are in fact ungrammatical; that these do not, in fact, conform to the rules inherent in those languages in which they appear to be formulated; that it turns out to be impossible to remedy this defect; and that every attempt to express them properly has led to the loss of their philosophic character (and revealed them as, for example, empirical truisms, or as false statements). These two counter arguments do, I believe, rescue the threatened consistency of the doctrine, which in this way indeed becomes 'unassailable'—to use Wittgenstein's term—by the kind of criticism referred to in this note. (See also the next note but one.)

is my hope that in carrying out my programme for this address an opportunity may turn up to do a little philosophizing after all.

II

Ever since the rise of Hegelianism there has existed a dangerous gulf between science and philosophy. Philosophers were accused—rightly, I believe—of 'philosophizing without knowledge of fact', and their philosophies were described as 'mere fancies, even imbecile fancies'.[7] Although Hegelianism was the leading influence in England and on the Continent, opposition to it, and contempt of its pretentiousness, never died out completely. Its downfall was brought about by a philosopher who like Leibniz, Berkeley, and Kant before him had a sound knowledge of science, and especially of mathematics. I am speaking of Bertrand Russell.

Russell is also the author of the classification, closely related to his famous *theory of types*, which is the basis of Wittgenstein's view of philosophy: the classification (criticized below on p. 309) of the expressions of a language into

(1) *True statements*
(2) *False statements*
(3) *Meaningless expressions*, among which there are statement-like sequences of words, the so-called 'pseudo-statements'.

Russell used this distinction to solve the problem of the logical paradoxes which he discovered. For his solution it was essential to distinguish more especially between (2) and (3). We might say, in ordinary speech, that a false statement like, '3 times 4 equals 173,' or, 'All cats are cows', is meaningless. Russell, however, reserved the term 'meaningless' for expressions such as, '3 times 4 are cows,' or 'All cats equal 173', that is for expressions of a sort which it is better not to describe as false statements. They are better not described as false

[7] The two quotations are not the words of a scientific critic but, ironically enough, Hegel's own characterization of the Natural Philosophy of his forerunner and one-time friend Schelling. Cf. my *Open Society*, note 4 (and text) to ch. 12.

because the negation of a meaningful but false statement will always be true. But the *prima facie* negation of the pseudo-statement, 'All cats equal 173', is, 'Some cats do not equal 173', and this is just as unsatisfactory a pseudo-statement as the original statement. *Negations of pseudo-statements are again pseudo-statements*, just as negations of proper statements (true or false) are proper statements (false or true, respectively).

This distinction allowed Russell to eliminate the paradoxes (which, he said, were meaningless pseudo-statements). Wittgenstein went further. Led perhaps by the feeling that what philosophers, especially Hegelian philosophers, were saying was somewhat similar to the paradoxes of logic, he used Russell's distinction in order to denounce *all philosophy* as strictly *meaningless*.

As a result there could be no genuine philosophical problems. All alleged philosophical problems could be classified under four heads:[8] (1) those which are purely logical or mathematical, to be answered by logical or mathematical propositions, and therefore not philosophical; (2) those which are factual, to be answered by some statement belonging to empirical science, and therefore again not philosophical; (3) those which are combinations of (1) and (2), and therefore again not philosophical; and (4) meaningless pseudo-problems such as, 'Do all cats equal 173?' or, 'Is Socrates identical?' or, 'Does an invisible, untouchable, and apparently altogether unknowable Socrates exist?'.

Wittgenstein's idea of eradicating philosophy (and theology) with the help of an adaptation of Russell's theory of types was ingenious and original (and more radical even than Comte's positivism, which it resembles closely).[9] This idea became the inspiration of a powerful modern school of language analysts who have inherited his belief that there are no genuine philosophical problems, and that all a philosopher

[8] Wittgenstein still upheld the doctrine of the non-existence of philosophical problems in the form here described when I saw him last (in 1946, when he presided over a stormy meeting of the Moral Sciences Club in Cambridge, on the occasion of my reading a paper on 'Are there Philosophical Problems?'). Since I had never seen any of his unpublished manuscripts which were privately circulated by some of his pupils I had been wondering whether he had modified what I here call his 'doctrine'; but on this, the most fundamental and influential part of his teaching, I found his views unchanged.

[9] Cf. note 51 (2) to ch. 11 of my *Open Society*.

can do is to unmask and dissolve the linguistic puzzles which have been proposed by traditional philosophy.

My own view of the matter is that only as long as I have genuine philosophical problems to solve shall I continue to take an interest in philosophy. I fail to understand the attraction of a philosophy without problems. I know, of course, that many people talk nonsense; and it is conceivable that it should become one's task (an unpleasant one) to unmask somebody's nonsense, for it may be dangerous nonsense. But I believe that some people have said things which were not very good sense, and certainly not very good grammar, but which were all the same highly interesting and exciting, and perhaps more worth listening to than the good sense of others. I may mention the differential and integral calculus which, especially in its early forms, was no doubt completely paradoxical and nonsensical by Wittgenstein's (and other) standards; which became, however, reasonably well founded as the result of some hundred years of great mathematical efforts; but whose foundations even at this very moment are still in need, and in the process, of clarification.[10] We might remember in this context that it was the contrast between the apparent absolute precision of mathematics and the vagueness and imprecision of philosophical language which deeply impressed the earlier followers of Wittgenstein. But had there been a Wittgenstein to use his weapons against the pioneers of the calculus, and had he succeeded in eliminating their nonsense where their contemporary critics (such as Berkeley, who was fundamentally right) failed, he would have strangled one of the most fascinating and philosophically important developments in the history of thought. Wittgenstein once wrote: 'Whereof one cannot speak, thereof one must be silent.' It was, if I remember rightly, Erwin Schrödinger who replied: 'But it is only here that speaking becomes worth while.'[10a] The

[10] I am alluding to G. Kreisel's recent construction (*Journal of Symbolic Logic*, **17**, 1952, 57) of a monotone bounded sequence of rationals every term of which can be actually computed, but which does not possess a computable limit—in contradiction to what appears to be the *prima facie* interpretation of the classical theorem of Bolzano and Weierstrass, but in agreement, it seems, with Brouwer's doubts about this theorem.

[10a] After this paper was first published Schrödinger told me that he could not remember saying this, and that he did not believe that he ever said it; but he liked the remark. (Added 1964: I have found since that its real author was my old friend Franz Urbach.)

history of the calculus—and perhaps of Schrödinger's own theory[11]—bears him out.

No doubt we should all train ourselves to speak as clearly, as precisely, as simply, and as directly, as we can. Yet I believe that there is not a classic of science, or of mathematics, or indeed a book worth reading that could not be shown, by a skilful application of the technique of language analysis, to contain many meaningless pseudo-propositions and what some people might call 'tautologies'.

Moreover, I believe that even Wittgenstein's original adaptation of Russell's theory rests upon a logical mistake. From the point of view of modern logic there no longer appears to be any justification for speaking of pseudo-statements or type mistakes or category-mistakes within ordinary, naturally grown languages (as opposed to artificial calculi) so long as the conventional rules of custom and grammar are observed. One may even say that the positivist who tells us with the air of the initiated that we are using meaningless words, or that we are talking nonsense, literally does not know what he is talking about—he simply repeats what he has heard from others who also did not know. But this raises a technical question which I cannot deal with here. (It is dealt with, however, in chapters 11 to 14, below.)

III

I have promised to say something in defence of Wittgenstein's views. What I wish to say is, first, that there is much philosophical writing (especially in the Hegelian school) which may justly be criticized as meaningless verbiage; secondly, that this kind of irresponsible writing was checked, for a time at least, by the influence of Wittgenstein and the language analysts (although it is likely that the most wholesome influence in this respect was the example of Russell who, by the incomparable charm and clarity of his writings, established the fact that subtlety of content is compatible with lucidity and unpretentiousness of style).

But I am prepared to admit more. In partial defence of Wittgenstein's views, I am prepared to adopt the following two theses.

[11] Before Max Born proposed his famous probability interpretation, Schrödinger's wave equation was, some might contend, meaningless. (This is not, however, my opinion.)

My first thesis is that every philosophy, and especially every philosophical 'school', is liable to degenerate in such a way that its problems become practically indistinguishable from pseudo-problems, and its cant, accordingly, practically indistinguishable from meaningless babble. This, I shall try to show, is a consequence of philosophical inbreeding. The degeneration of philosophical schools in its turn is the consequence of the mistaken belief that one can philosophize without having been compelled to philosophize *by problems which arise outside philosophy*—in mathematics, for example, or in cosmology, or in politics, or in religion, or in social life. In other words my first thesis is this. *Genuine philosophical problems are always rooted in urgent problems outside philosophy, and they die if these roots decay*. In their efforts to solve them, philosophers are liable to pursue what looks like a philosophical method or technique or an unfailing key to philosophical success.[12] But no such methods or techniques exist; in philosophy methods are unimportant; *any* method is legitimate if it leads to results capable of being rationally discussed. What matters is not methods or techniques but a sensitivity to problems, and a consuming passion for them; or, as the Greeks said, the gift of wonder.

There are those who feel the urge to solve a problem, those for whom a problem becomes real, like a disorder which they have to get out of their system.[13] They may make a contribution even if they bind themselves to a particular method or a technique. But there are others who do not feel this urge, who have no serious and pressing problem but who nevertheless produce exercises in fashionable methods, and for whom philosophy is *application* (of whatever insight or technique you like) rather than *search*. They are luring philosophy into the bog of

[12] It is very interesting that the imitators were always inclined to believe that the 'master' did his work with the help of a secret method or a trick. It is reported that in J. S. Bach's days some musicians believed that he possessed a secret formula for the construction of fugue themes.

It is also interesting to note that all the philosophies which have become fashionable (so far as I am aware) have offered their disciples a kind of method for producing philosophical results. This is true of Hegelian essentialism which teaches its adherents to produce essays on the essence or nature or idea of everything—the soul, the universe, or the University; it is true of Husserl's phenomenology; of existentialism; and also of language analysis.

[13] I am alluding to a remark by Professor Gilbert Ryle, who says on page 9 of his *Concept of Mind*: 'Primarily I am trying to get some disorders out of my own system.'

pseudo-problems and verbal puzzles; either by offering us pseudo-problems for real ones (the danger which Wittgenstein saw), or by persuading us to concentrate upon the endless and pointless task of unmasking what they rightly or wrongly take for pseudo-problems or 'puzzles' (the trap into which Wittgenstein fell).

My second thesis is that what appears to be the *prima facie* method of teaching philosophy is liable to produce a philosophy which answers Wittgenstein's description. What I mean by '*prima facie* method of teaching philosophy', and what would seem to be the only method, is that of giving the beginner (whom we take to be unaware of the history of mathematical, cosmological, and other ideas of science as well as of politics) the works of the great philosophers to read; the works, say, of Plato and Aristotle, Descartes and Leibniz, Locke, Berkeley, Hume, Kant and Mill. What is the effect of such a course of reading? A new world of astonishingly subtle and vast *abstractions* opens itself before the reader; abstractions on an extremely high and difficult level. Thoughts and arguments are put before his mind which sometimes are not only hard to understand, but which seem to him irrelevant because he cannot find out what they may be relevant to. Yet the student knows that these are the great philosophers, that this is the way of philosophy. Thus he will make an effort to adjust his mind to what he believes (mistakenly, as we shall see) to be their way of thinking. He will attempt to speak their queer language, to match the tortuous spirals of their argumentation, and perhaps even tie himself up in their curious knots. Some may learn these tricks in a superficial way, others may begin to become genuinely fascinated addicts. Yet I feel that we ought to respect the man who having made his effort comes ultimately to what may be described as Wittgenstein's conclusion: 'I have learned the jargon as well as anybody. It is very clever and captivating. In fact, it is dangerously captivating; for the simple truth about the matter is that it is much ado about nothing—just a lot of nonsense.'

Now I believe such a conclusion to be grossly mistaken; yet it is the almost inescapable outcome, I contend, of the *prima facie* method of teaching philosophy here described. (I do not deny, of course, that some particularly gifted students may find very much more in the works of the great philosophers than this story indicates—and without self-deception.) For the student's chance of discovering the extra-philosophical problems (mathematical, scientific, moral, and political problems) which inspired these great philosophers is very small indeed. As a rule, these problems can be discovered only by studying the history of, for example, scientific ideas, and especially the problem-situation in mathematics and the sciences during the period in question; and this in turn presupposes a considerable acquaintance with mathematics and science. Only if he understands the contemporary problem-situation in the sciences can the student of the great philosophers understand that they tried to solve urgent and concrete problems; problems which they found could not be dismissed. And only after understanding this can the student attain a different picture of the great philosophies—one which makes sense of the apparent nonsense.

I shall try to establish my two theses with the help of examples; but before turning to these examples, I wish to summarize my theses, and to balance my account with Wittgenstein.

My two theses amount to the contention that as philosophy is deeply rooted in non-philosophical problems, Wittgenstein's negative judgment is correct, by and large, so far as philosophies are concerned which have forgotten their extra-philosophical roots; and that these roots are easily forgotten by philosophers who 'study' philosophy, instead of being forced into philosophy by the pressure of non-philosophical problems.

My view of Wittgenstein's doctrine may be summed up as follows. It is perhaps true, by and large, that '*pure*' philosophical problems do not exist; for indeed the purer a philosophical problem becomes the more is lost of its original significance, and the more liable is its discussion to degenerate into empty verbalism. On the other hand there exist not only genuine scientific problems, but genuine philosophical problems. Even if, upon analysis, these problems turn out to have factual components, they need not be classified as belonging to science. And even if they should be soluble by, say, purely logical means they need not be classified as purely logical or tautological. Analogous situations arise in physics. For example, the problem of explaining series of spectral lines (with the help of a hypothesis concerning the structure of atoms) may turn out to be soluble by purely mathematical calculations. But this again does not imply that the problem belonged to pure mathematics rather than to physics. We are perfectly justified in calling a problem 'physical' if it is connected with problems and theories which have been traditionally discussed by physicists (such as the problem of the constitution of matter) even if the means used for its solution turn out to be purely mathematical. As we have seen, the solution of problems may cut through the boundary of many sciences. Similarly, a problem may rightly be called 'philosophical' if we find that although originally it arose in connection with, say, atomic theory it is more closely connected with the problems and theories which have been discussed by philosophers than with theories nowadays treated by physicists. And again, it does not matter in the least what kind of methods we use in solving such a problem. Cosmology, for example, will always be of great philosophical interest even though in some of its methods it has become closely allied with what is perhaps better called 'physics'. To say that since it deals with factual issues it must belong to science rather than to philosophy is not only pedantic but clearly the result of an epistemological, and thus of a philosophical, dogma. Similarly, there is no reason why a problem soluble by logical means should be denied the attribute 'philosophical'. It may well be typically philosophical, or physical, or biological. Logical analysis played a considerable part in Einstein's special theory of relativity; and it was partly this fact which made the theory philosophically interesting, and which gave rise to a wide range of philosophical problems connected with it.

Wittgenstein's doctrine turns out to be the result of the thesis that all genuine statements (and therefore all genuine problems) can be classified into one of two exclusive classes: factual statements (*synthetic a posteriori*), which belong to the empirical sciences, and logical statements (*analytic a priori*), which belong to pure formal logic or pure mathematics. This simple dichotomy, although extremely valuable for a rough survey, turns out to be for many purposes too simple.[14] But although it is specially designed, as it were, to exclude the existence of philosophical

[14] Already in my *L.Sc.D.* of 1934 I had pointed out that a theory such as Newton's may be *interpreted* either as factual or as consisting of implicit definitions (in the sense of Poincaré and Eddington), and that the interpretation which a physicist adopts exhibits itself in his *attitude* towards tests which go against his theory rather than in what he says. I also pointed out that there are non-analytical theories which are not testable (and therefore not *a posteriori*) but which had a great influence on science. (Examples are the early atomic theory, or the early theory of action by contact.) I called such untestable

problems, it falls considerably short of this aim; for even if we accept the dichotomy we can still claim that factual or logical or mixed problems may turn out, in certain circumstances, to be philosophical.

IV

I now turn to my first example: *Plato and the Crisis in Early Greek Atomism.*

My thesis here is that Plato's central philosophical doctrine, the so-called Theory of Forms or Ideas, cannot be properly understood except in an extra-philosophical context;[15] more especially in the context of the critical *problem situation* in Greek science[16] (mainly in the theory of

theories 'meta-physical', and asserted that they were not meaningless. The dogma of the simple dichotomy has been recently attacked, on very different lines, by F. H. Heinemann (*Proc. of the Xth Intern. Congress of Philosophy*, Fasc. **2**, 629, Amsterdam, 1949), by W. V. Quine, and by Morton G. White. It may be remarked, again from a different point of view, that the dichotomy applies in a precise sense only to a formalized language, and therefore is liable to break down for those languages in which we must speak prior to any formalization, i.e. in those languages in which all the traditional problems were conceived.

[15] In my *Open Society and its Enemies* I have tried to explain in some detail another extra-philosophical root of the same doctrine—its political root. I also discussed there (in note 9 to ch. 6 of the revised 4th edn., 1962) the problem with which I am concerned in the present section, but from a somewhat different angle. The note referred to and the present section overlap a little; but they largely supplement each other. Relevant references (especially to Plato) omitted here will be found there.

[16] There are historians who deny that the term 'science' can be properly applied to any development which is older than the sixteenth or even the seventeenth century. But quite apart from the fact that controversies about labels should be avoided, there can, I believe, no longer be any doubt nowadays about the astonishing similarity, not to say identity, of the aims, interests, activities, arguments and methods of, say, Galileo and Archimedes, or Copernicus and Plato, or Kepler and Aristarchus (the 'Copernicus of antiquity'). And any doubt concerning the extreme age of scientific observation and of careful computations based upon observation has been dispelled by the discovery of new evidence concerning the history of ancient astronomy. We can now draw a parallel not only between Tycho and Hipparchus, but even between Hansen (1857) and Cidenas the Chaldean (314 B.C.), whose computations of the 'constants for the motion of Sun and Moon' are without exception comparable in precision to those of the best nineteenth-century astronomers. 'Cidenas' value for the motion of the Sun from the Node (0″.5 too great), although inferior to Brown's, is superior to at least one of the most widely used modern values', wrote J. K. Fotheringham in 1928, in his admirable article 'The Indebtedness of Greek to Chaldean Astronomy' (*The Observatory*, 1928, **51**, No. 653), upon which my contention concerning the age of metrical astronomy is based.

matter) which developed as a result of *the discovery of the irrationality of the square root of two*. If my thesis is correct, Plato's theory has not so far been fully understood. (Whether a 'full' understanding can ever be achieved is, of course, most questionable.) But a more important consequence would be that it can never be understood by philosophers trained in accordance with the *prima facie* method described in the foregoing section—unless, of course, they are specially and *ad hoc* informed of the relevant facts. (These they may have to accept on authority—which means abandoning the *prima facie* method of teaching philosophy described above.)

It seems likely[17] that Plato's Theory of Forms is both in origin and in content closely connected with the Pythagorean theory that all things are in essence numbers. The details of this connection and the connection between Atomism and Pythagoreanism are perhaps not so well known. I will therefore briefly tell the story here, as I see it at present.

It appears that the founder of the Pythagorean order or sect was deeply impressed by two discoveries. The first was that a *prima facie* purely qualitative phenomenon such as musical harmony was, in essence, based upon the purely numerical ratios 1 : 2; 2 : 3; 3 : 4. The second was that the 'right' or 'straight' angle (obtainable for example by folding a leaf twice so that the two folds form a cross) was connected with the purely numerical ratios 3 : 4 : 5, or 5 : 12 : 13 (the sides of rectangular triangles). These two discoveries, it appears, led Pythagoras to the somewhat fantastic generalization that all things are, in essence, numbers or ratios of numbers; or that number is the ratio (*logos* = reason), the rational essence, of things, or their real nature.

Fantastic as this idea was, it proved fruitful in many ways. One of its most successful applications was to simple geometrical figures such as squares, rectangular and isosceles triangles, and also to certain simple solids such as pyramids. The treatment of some of these geometrical problems was based upon the so-called *gnōmōn*.

This can be explained as follows. If we indicate a square by four dots,

$$\begin{matrix} \cdot & \cdot \\ & \\ \cdot & \cdot \end{matrix}$$

[17] If we may trust Aristotle's famous account in his *Metaphysics*.

we may interpret this as the result of adding three dots to the one dot on the upper left corner. These three dots are the first *gnōmōn*; we may indicate it thus:

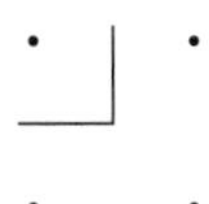

By adding a second *gnōmōn*, consisting of five more dots, we obtain

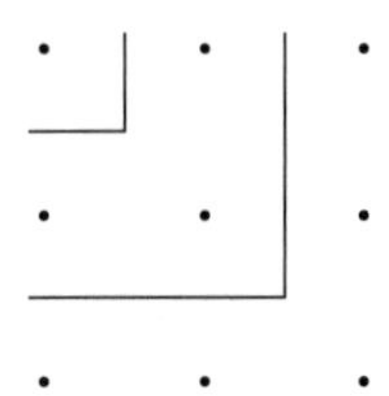

One sees at once that every number of the sequence of odd numbers, 1, 3, 5, 7 . . . , forms the *gnōmōn* of a square, and that the sums 1, 1 + 3, 1 + 3 + 5, 1 + 3 + 5 + 7, . . . are the square numbers, and that if n is the (number of dots in the) side of a square, its area (total number of dots $= n^2$) will be equal to the sum of the first n odd numbers.

As with the treatment of squares, so with the treatment of equilateral triangles. The following figure may be regarded as representing a growing triangle—growing downwards through the addition of ever new horizontal lines of dots.

Here each *gnōmōn* is a last horizontal line of dots and each element of the sequence 1, 2, 3, 4, . . . is a *gnōmōn*. The 'triangular numbers' are the sums 1 + 2; 1 + 2 + 3; 1 + 2 + 3 + 4, etc., that is, the sums of the first n natural numbers. By putting two such triangles side by side

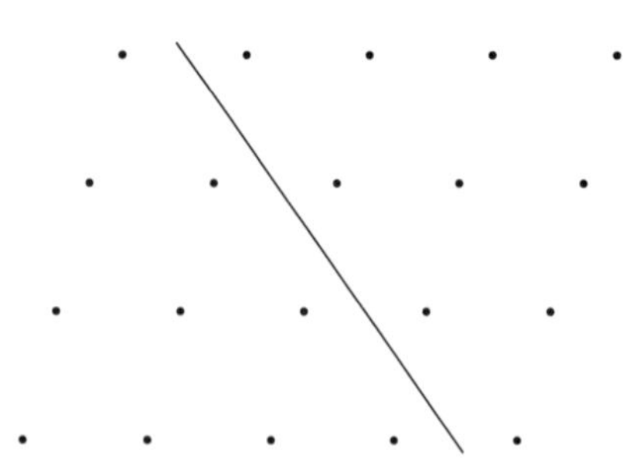

we obtain the parallelogram with the horizontal side $n + 1$ and the other side n, containing $n(n + 1)$ dots. Since it consists of two isosceles triangles its number is $2(1 + 2 + \ldots + n)$, so that we obtain the equation

$$(1) \qquad 1 + 2 + \ldots + n = \tfrac{1}{2}n(n + 1)$$

and thus

$$(2) \qquad d(1 + 2 + \ldots + n) = \tfrac{d}{2}n(n + 1).$$

From this it is easy to obtain the general formula for the sum of an arithmetical series.

We also obtain 'oblong numbers', that is the numbers of oblong rectangular figures of which the simplest is

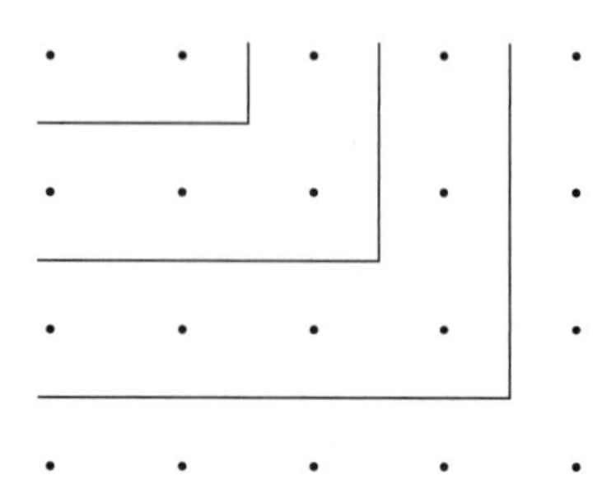

with the oblong numbers 2 + 4 + 6 . . . ; the *gnōmōn* of an oblong is an even number, and the oblong numbers are the sums of the even numbers.

These considerations were extended to solids; for example, by summing the first triangular numbers, pyramid numbers were obtained. But the main application was to plane figures, or shapes, or 'Forms'. These were believed to be characterized by the appropriate sequence of numbers, and thus by the numerical ratios of the consecutive numbers of the sequence. In other words, *'Forms' are numbers or ratios of numbers*. On the other hand, not only shapes of things, but also abstract properties, such as harmony and 'straightness' are numbers. In this way the general theory that numbers are the rational essences of all things is arrived at.

It seems probable that the development of this view was influenced by the similarity of the dot-diagrams with the diagram of a constellation such as the Lion, or the Scorpion, or the Virgo. If a Lion is an arrangement of dots it must have a number. In this way Pythagoreanism seems to be connected with the belief that the numbers, or 'Forms', are heavenly shapes of things.

V

One of the main elements of this early theory was the so-called 'Table of Opposites', based upon the fundamental distinction between odd and even numbers. It contains such things as

ONE	MANY
ODD	EVEN
MALE	FEMALE
REST (BEING)	CHANGE (BECOMING)
DETERMINATE	INDETERMINATE
SQUARE	OBLONG
STRAIGHT	CROOKED
RIGHT	LEFT
LIGHT	DARKNESS
GOOD	BAD

In reading through this strange table one gets some idea of the working of the Pythagorean mind, and why not only the 'Forms' or shapes of geometrical figures were considered to be numbers, in essence, but also abstract ideas, such as Justice and, of course, Harmony and Health, Beauty and Knowledge. The table is interesting also because it was taken over, with very little alteration, by Plato. The earliest version of Plato's famous theory of 'Forms' or 'Ideas' may indeed be described, somewhat roughly, as the doctrine that the 'Good' side of the Table of Opposites constitutes an (invisible) Universe, a Universe of Higher Reality, of the Unchanging and Determinate 'Forms' of all things; and that True and Certain Knowledge (*epistēmē* = *scientia* = *science*) can be of this Unchanging and Real Universe only, while the visible world of change and flux in which we live and die, the world of generation and destruction, the world of experience, is only a kind of reflection or copy of that Real World. It is only a world of appearance of which no True and Certain Knowledge can be obtained. All that can be obtained in the place of Knowledge (*epistēmē*) are the plausible but uncertain and prejudiced opinions (*doxa*) of fallible mortals.[18] In his interpretation of the Table of Opposites Plato was influenced by Parmenides, the man whose challenge led to the development of Democritus' atomic theory.

VI

The Pythagorean theory, with its dot-diagrams, contains no doubt the suggestion of a very primitive atomism. How far the atomic theory of Democritus was influenced by Pythagoreanism is difficult to assess. Its main influences came, so much seems certain, from the Eleatic School:

[18] Plato's distinction (*epistēmē vs. doxa*) is derived through Parmenides from Xenophanes (*truth* vs. *conjecture* or *seeming*). Plato clearly realized that *all* knowledge of the visible world, the changing world of appearance, consists of *doxa*; that it is tainted by uncertainty even if it utilizes the *epistēmē*, the knowledge of the unchanging 'Forms' and of pure mathematics, to the utmost; and even if it interprets the visible world with the help of a theory of the invisible world. Cf. *Cratylus*, 439b ff., *Republic*, 476d ff.; and especially *Timaeus*, 29b ff., where the distinction is applied to those parts of Plato's own theory which we should nowadays call 'physics' or 'cosmology', or, more generally, 'natural science'. They belong, Plato says, to the realm of *doxa* (in spite of the fact that science = *scientia* = *epistēmē*; cf. my remarks on this problem in ch. 20 below). For a different view concerning Plato's relation to Parmenides see Sir David Ross, *Plato's Theory of Ideas*, Oxford, 1951, p. 164.

from Parmenides and from Zeno. The basic problem of this school, and of Democritus, was that of the rational understanding of *change*. (I differ here from the interpretations of Cornford and others.) I think that this problem derives from Heraclitus, and thus from Ionian rather than from Pythagorean thought,[19] and that it still remains the fundamental problem of Natural Philosophy.

Although Parmenides was perhaps not a physicist (unlike his great Ionian predecessors), he may be described, I believe, as having fathered *theoretical physics*. He produced an anti-physical[20] (rather than a-physical, as Aristotle said) theory which, however, was the first hypothetico-deductive system. And it was the beginning of a long series of such systems of physical theories, each of which was an improvement on its predecessor. As a rule the improvement was found necessary because it was realized that the earlier system was falsified by certain facts of experience. Such an empirical refutation of the consequences of a deductive system leads to efforts at its reconstruction, and thus to a new and improved theory which as a rule clearly bears the marks of its ancestry, of the older theory as well as of the refuting experience.

These experiences or observations were, we shall see, very crude at first, but they became more and more subtle as the theories became more and more capable of accounting for the cruder observations. In the case of Parmenides' theory the clash with observation was so obvious that it would seem perhaps fanciful to describe the theory as the first hypothetico-deductive system of physics. We may, therefore, describe it as the last pre-physical deductive system, whose refutation

[19] Karl Reinhardt in his *Parmenides* (1916; second edition 1959, p. 220) says very forcefully: 'The history of philosophy is a history of its problems. If you want to explain Heraclitus, tell us first what his problem was.' I fully agree; and I believe, as against Reinhardt, that Heraclitus' problem was the problem of change—or more precisely, of the self-identity (*and* non-identity) of the changing thing during change. (See also my *Open Society*, ch. 2.) If we accept Reinhardt's evidence about the close link between Heraclitus and Parmenides, then this view of Heraclitus' problem makes of Parmenides' system an attempt to solve the problem of the paradoxes of change by making change unreal. As against this, Cornford and his disciples follow Burnet's doctrine that Parmenides was a (dissident) Pythagorean. This may well be true, but the evidence in its favour does not show that he did not also have an Ionian teacher. (See also ch. 5, below.)

[20] Cp. Plato, *Theaetetus*, 181a, and Sextus Empiricus, *Adv. Mathem.* (Bekker), X. 46, p. 485, 25.

or falsification gave rise to the first physical theory of matter, the atomistic theory of Democritus.

Parmenides' theory is simple. He finds it impossible to understand change or movement rationally, and concludes that there is really no change—or that change is only apparent. But before we indulge in feelings of superiority in the face of such a hopelessly unrealistic theory we should first realize that there is a serious problem here. If a thing X changes, then clearly it is no longer the same thing X. On the other hand, we cannot say that X changes without implying that X persists during the change; that it is the same thing X, at the beginning and at the end of the change. Thus it appears that we arrive at a contradiction, and that the idea of a thing that changes, and therefore the idea of change, is impossible.

All this sounds very philosophical and abstract, and so it is. But it is a fact that the difficulty here indicated has never ceased to make itself felt in the development of physics.[21] And a deterministic system such as the field theory of Einstein might even be described as a four-dimensional version of Parmenides' unchanging three-dimensional universe. For in a sense no change occurs in Einstein's four-dimensional block-universe. Everything is there just as it is, in its four-dimensional *locus*; change becomes a kind of 'apparent' change; it is 'only' the observer who as it were glides along his world-line and becomes successively conscious of the different *loci* along this world-line; that is, of his spatio-temporal surroundings . . .

To return from this new Parmenides to the older father of theoretical physics, we may paraphrase his deductive theory roughly as follows.

(1) Only what is, is.
(2) What is not does not exist.

[21] This may be seen from Emile Meyerson's *Identity and Reality*, one of the most interesting philosophical studies of the development of physical theories. Hegel (following Heraclitus, or Aristotle's account of him) took the fact of change (which he considered self-contradictory) to prove the existence of contradictions in the world, and therefore to disprove the 'law of contradiction'; i.e. the principle that our theories must avoid contradictions at all cost. Hegel and his followers (especially Engels, Lenin, and other Marxists) began to see 'contradictions' everywhere in the world, and denounced all philosophies upholding the law of contradiction as 'metaphysical', a term which they used to imply that these philosophies ignore the fact that the world changes. See ch. 15, below.

(3) Non-being, that is, the void, does not exist.
(4) The world is full.
(5) The world has no parts; it is *one* huge block (because it is full).
(6) Motion is impossible (since there is no empty space within which anything could move).

The conclusions (5) and (6) were obviously contradicted by facts. Thus Democritus argued from the falsity of the conclusion to that of the premises:

(6′) There is motion (thus motion is possible).
(5′) The world has parts; it is not one, but many.
(4′) Thus the world cannot be full.[22]
(3′) The void (or non-being) exists.

So far the theory had to be altered. With regard to being, or to the many existing things (as opposed to the void), Democritus adopted Parmenides' theory that they had no parts. They were indivisible (atoms), because they were full, because they had no void inside.

The main point about this theory is that it gives a rational account of change. The world consists of empty space (the void) with atoms in it. The atoms do not change; they are Parmenidean indivisible block universes in miniature.[23] All change is due to rearrangement of atoms in space. Accordingly *all change is movement*. Since the only kind of novelty which can arise on this view is novelty of arrangement,[24] it will be possible, in principle, *to predict all future changes in the world*, provided we

[22] The inference from the existence of motion to that of a void is not valid because Parmenides' inference from the fullness of the world to the impossibility of motion is not valid. Plato seems to have been the first to see, if only dimly, that in a full world circular or vortex-like motion is possible, provided that there is a liquid-like medium in the world. (Tea leaves can move with the vortex of tea in the cup.) This idea, first offered somewhat half-heartedly in the *Timaeus* (where *space is 'filled'*, 52c) becomes the basis of Cartesianism and of the theory of the 'luminiferous ether' as it was held down to 1905. (See also note 44, below.)
[23] Democritus' theory also admitted large block-atoms, but the vast majority of his atoms were invisibly small.
[24] Cp. *The Poverty of Historicism*, section 3.

manage to predict the motion of all atoms (or, in modern parlance, all mass-points).

Democritus' theory of change was of tremendous importance for the development of physical science. It was partly accepted by Plato, who retained much of atomism, explaining change, however, not only by unchanging yet moving atoms but also by other 'Forms' which were subject neither to change nor to motion. But it was condemned by Aristotle who taught in its stead[25] that all change was the unfolding of the inherent potentialities of essentially unchanging *substances*. Aristotle's theory of substances as the subjects of change became dominant; but it proved barren;[26] and Democritus' metaphysical theory that all change must be explained by movement became the tacitly accepted programme of work in physics down to our own day. It is still part of the philosophy of physics, in spite of the fact that physics itself has outgrown it (to say nothing of the biological and social sciences). For with Newton, in addition to moving mass-points, *forces* of changing intensity (and direction) enter the scene. True, the changes of the Newtonian forces can be explained as due to, or dependent upon, motion; that is, upon the changing position of particles. But they are nevertheless not identical with changes in the position of particles; owing to the inverse square law the dependence is not even a linear one. And with Faraday and Maxwell, changing fields of forces become as important as material atomic particles. That our modern atoms turn out to be composite is a minor matter; from Democritus' point of view not our atoms but rather our elementary particles would be the real atoms—except that these too turn out to be liable to change. Thus we have a most

[25] Inspired by Plato's *Timaeus*, 55, where the potentialities of the elements are explained by the geometrical properties (and thus the substantial forms) of the corresponding solids.

[26] The barrenness of the 'essentialist' (cf. note 2 above) theory of substance is connected with its anthropomorphism; for substances (as Locke saw) take their plausibility from the experience of a self-identical but changing and unfolding ego. But although we may welcome the fact that Aristotle's substance has disappeared from *physics*, there is nothing wrong, as Professor Hayek says, in thinking anthropomorphically about *man*. So there is no philosophical or *a priori* reason why Aristotle's substance—the psyche—ought to disappear from psychology.

interesting situation. A philosophy of change, designed to meet the difficulty of understanding change rationally, serves science for thousands of years, but is ultimately superseded by the development of science itself; and this fact passes practically unnoticed by philosophers who are busily denying the existence of philosophical problems.

Democritus' theory was a marvellous achievement. It provided a theoretical framework for the explanation of most of the empirically known properties of matter (discussed already by the Ionians), such as compressibility, degrees of hardness and resilience, rarefaction and condensation, coherence, disintegration, combustion, and many others. But the theory was important not only as an explanation of the phenomena of experience. First, it established the methodological principle that a deductive theory or explanation must 'save the phenomena';[27] that is, be in agreement with experience. Secondly, it showed that a theory may be speculative, and based upon the fundamental (Parmenidean) principle that the world as it must be understood by argumentative thought turns out to be different from the world of *prima facie* experience, from the world as seen, heard, smelled, tasted, touched;[28] and that such a speculative theory may nevertheless accept the empiricist 'criterion' that it is the visible that decides the acceptance or rejection of a theory of the invisible[29] (such as the atoms). This philosophy has remained fundamental to the whole development of physics, and has continued to conflict with all 'relativistic' and 'positivistic'[30] philosophical tendencies.

Furthermore, Democritus' theory led to the first successes of the method of exhaustion (the forerunner of the calculus of integration), since Archimedes himself acknowledged that Democritus was the first

[27] Cp. note 6 to chapter 3, below.

[28] Cf. Democritus, Diels, fragm. 11 (cf. Anaxagoras, Diels, fragm. 21; see also fragm. 7).

[29] Cf. Sextus Empiricus, *Adv. mathem.* (Bekker), vii, 140, p. 221, 23B.

[30] 'Relativistic' in the sense of philosophical relativism, e.g. of Protagoras' *homo mensura* doctrine. It is, unfortunately, still necessary to emphasize that Einstein's theory has nothing in common with this philosophical relativism.

'Positivistic' as were the tendencies of Bacon; of the theory (but fortunately not the practice) of the early *Royal Society*; and in our time of Mach (who opposed atomic theory), and of the sense-data theorists.

to formulate the theory of the volumes of cones and pyramids.[31] But perhaps the most fascinating element in Democritus' theory is his doctrine of the quantization of space and time. I have in mind the doctrine, now extensively discussed,[32] that there is a *shortest distance* and a *smallest time interval*; that is to say, that there are distances in space and time (elements of length and time, Democritus' *amerēs*[33] in contradistinction to his atoms) such that no smaller ones are possible.

VII

Democritus' atomism was developed and expounded as a point for point reply[34] to the detailed arguments of his Eleatic predecessors, of Parmenides and his pupil Zeno. Especially Democritus' theory of atomic distances and time intervals is the direct result of Zeno's arguments, or more precisely, of the rejection of Zeno's conclusions. But nowhere in what we know of Zeno is there an allusion to the *discovery of irrationals* which is of decisive importance for our story.

We do not know the date of the proof of the irrationality of the square root of two, or the date when the discovery became publicly known. Although there existed a tradition ascribing it to Pythagoras (sixth century B.C.), and although some authors[35] call it the 'theorem

[31] Cf. Diels, fragm. 155, which must be interpreted in the light of Archimedes (ed. Heiberg) II^2, p. 428 f. Cf. S. Luria's most important article 'Die Infinitesimalmethode der antiken Atomisten' (*Quellen & Studien zur Gesch. d. Math.*, B., **2**, Heft 2, 1932, p. 142).

[32] Cf. A. March, *Natur und Erkenntnis*, Vienna, 1948, p. 193 f.

[33] Cf. S. Luria, *op. cit.*, especially pp. 148 ff., 172 ff. Miss A. T. Nicols in 'Indivisible Lines' (*Class. Quarterly*, xxx, 1936, 120 f.) argues that 'two passages, one from Plutarch, the other from Simplicius' show why Democritus 'could not believe in indivisible lines'; she does not however discuss Luria's opposing views of 1932, which I find much more convincing, especially if we remember that Democritus tried to answer Zeno (see next note). But whatever Democritus' views on indivisible or atomic distances, Plato appears to have thought that Democritus' atomism needed revision in the light of the discovery of the irrationals. Heath however (*Greek Mathematics*, **1**, 1921, p. 181, referring to Simplicius and Aristotle) also believes that Democritus did not teach the existence of indivisible lines.

[34] This point for point reply is preserved in Aristotle's *On Generation and Corruption*, 316a, 14 ff., a very important passage first identified as Democritean by I. Hammer Jensen in 1910 and carefully discussed by Luria who says (*op. cit.*, 135) of Parmenides and Zeno: 'Democritus borrows their deductive arguments, but he arrives at the opposite conclusion.'

[35] Cf. G. H. Hardy and E. M. Wright, *Introduction to the Theory of Numbers*, 1938, pp. 39, 42,

of Pythagoras', there can be little doubt that the discovery was not made, and certainly not publicly known, before 450 B.C., and probably not before 420. Whether Democritus knew about it is uncertain. I now feel inclined to believe that he did not; and that the title of Democritus' two lost books, *Peri alogōn grammōn kai nastōn*, should be translated '*On Illogical Lines and Full Bodies (Atoms)*',[36] and that these two books do not contain any reference to the discovery of irrationality.[37]

My belief that Democritus was unaware of the problem of irrationals is based on the fact that there are no traces of a defence of his theory against the blow which it received from this discovery. Yet the blow was as fatal to Atomism as it was to Pythagoreanism. Both theories were based on the doctrine that all measurement is, ultimately, counting of natural units, so that every measurement must be reducible to pure numbers. The distance between any two atomic points must, therefore,

where a very interesting historical remark on Theodorus' proof, as reported in Plato's *Theaetetus*, will be found. See now also the article by A. Wasserstein, 'Theaetetus and the History of the Theory of Numbers', *Classical Quarterly*, **8**, N.S., 1958, pp. 165–79, the best discussion of the subject known to me.

[36] Rather than *On Irrational Lines and Atoms*, as I translated it in note 9 to ch. 6 of my *Open Society* (second edn.). What is probably meant by the title (considering Plato's passage mentioned in the next note) might, I think, be best rendered by '*On Crazy Lines and Atoms*'. Cf. H. Vogt, *Bibl. Math.*, 1910, **10**, 147 (against whom Heath argues, *op. cit.*, 156 f., but not I think quite successfully) and S. Luria, *op. cit.*, pp. 168 ff., where it is convincingly suggested that (Arist.) *De insec. lin.*, 968b 17 and Plutarch, *De comm. notit.*, 38, 2, p. 1078 f., contain traces of Democritus' work. According to these sources, Democritus' argument was this. *If lines are infinitely divisible* then they are composed of an infinity of ultimate units and are therefore *all* related like $\infty : \infty$, that is to say, they are all 'noncomparable' (there is no proportion). Indeed, if lines are considered as classes of points, the cardinal 'number' (potency) of the points of a line is, according to modern views, equal for all lines, whether the lines are finite or infinite. This fact has been described as 'paradoxical' (for example, by Bolzano) and might well have been described as 'crazy' by Democritus. It may be noted that according to Brouwer even the classical theory of the Lebesgue *measure* of a continuum leads to fundamentally the same results; for Brouwer asserts that all classical continua have zero measure, so that the absence of a ratio is here expressed by $0 : 0$. Democritus' result (and his theory of *amerēs*) appears to be inescapable as long as geometry is based on the Pythagorean *arithmetical method*, i.e. on the counting of dots.

[37] This would be in keeping with the fact, mentioned in the note cited from the *Open Society*, that the term '*alogos*' was, it seems, only much later used for 'irrational', and that Plato who alludes (*Republic* 534d) to Democritus' title, uses '*alogos*' there in the sense of 'crazy'; he never uses it as a synonym for '*arrhētos*' as far as I know.

consist of a certain number of atomic distances; thus all distances must be commensurable. But this turns out to be impossible even in the simple case of the distances between the corners of a square, because of the incommensurability of its diagonal d with its side a.

The English term 'incommensurable' is somewhat unfortunate. What is meant is, rather, the non-existence of a *ratio of natural numbers*; for example, what can be proved in the case of the diagonal of the unit square is that there do not exist two natural numbers, n and m, whose ratio, n/m, is equal to the diagonal of the unit square. 'Incommensurability' thus does not mean incomparability by geometrical methods, or by *measurement*, but incomparability by arithmetical methods of *counting*, or by natural numbers, including the characteristic Pythagorean method of comparing *ratios* of natural numbers and including, of course, the counting of units of length (or of 'measures').

Let us look back, for a moment, at the characteristics of this *method of natural numbers and their ratios*. Pythagoras' emphasis upon Number was fruitful from the point of view of the development of scientific ideas. This is often but somewhat loosely expressed by saying that the Pythagoreans initiated numerical scientific measurement. Now what I want to emphasize is that for the Pythagoreans all this was *counting rather than measuring*. It was the counting of numbers, of invisible essences or 'Natures' which were Numbers of little dots. They knew that we cannot count these little dots directly, since they are invisible, and that we actually do not *count* the Numbers or Natural Units, but *measure*, i.e. count arbitrary visible units. But they interpreted the significance of measurements as revealing, indirectly, the true *Ratios of the Natural Units or of the Natural Numbers*.

Thus Euclid's methods of proving the so-called 'Theorem of Pythagoras' (Euclid, 1, 47) according to which, if a is the side of a triangle opposite to its right angle between b and c,

$$a^2 = b^2 + c^2, \tag{1}$$

was foreign to the spirit of Pythagorean mathematics. It seems now accepted that the theorem was known to the Babylonians and geometrically proved by them. Yet neither Pythagoras nor Plato appear to have known Euclid's *geometrical* proof (which uses different triangles with common base and height); for the problem for which they offered solutions, the *arithmetical* one of finding the integral solutions for the sides of rectangular triangles, can, if (1) is known, be easily solved by the formula (m and n are natural numbers, and $m > n$)

$$(2) \qquad a = m^2 + n^2; \qquad b = 2mn; \qquad c = m^2 - n^2.$$

But formula (2) was apparently unknown to Pythagoras and even to Plato. This emerges from the tradition[38] according to which Pythagoras proposed the formula (obtained from (2) by putting $m = n + 1$)

$$(3) \qquad a = 2n(n + 1) + 1; \qquad b = 2n(n + 1); \qquad c = 2n + 1$$

which can be read off the *gnōmōn* of the square numbers, but which is less general than (2), since it fails, for example, for 17: 8: 15. To Plato, who is reported[39] to have improved Pythagoras' formula (3), is attributed another formula which still falls short of the general solution (2).

In order to show the difference between the Pythagorean or arithmetical method and the geometrical method, Plato's proof that the square over the diagonal of the unit square (that is, the square with the side 1 and an area of measure 1) has an area of twice the unit square (that is, an area of measure 2) may be mentioned. It consists in drawing a square with the diagonal

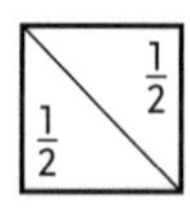

and then showing that we may extend the drawing thus

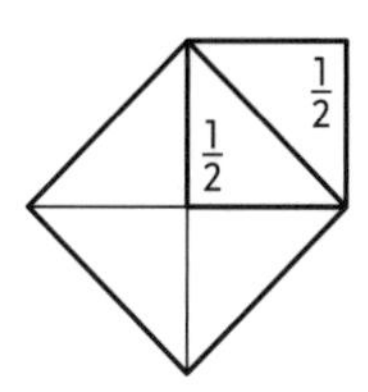

[38] *Procli Diadochi in primum Euclidis Elementorum librum commentarii*, ed. G. Friedlein, Leipzig, 1873, p. 487, 7–21.

[39] By Proclus, *op. cit.*, pp. 428, 21, to 429, 8.

from which we obtain the result by counting. But the transition from the first to the second of these figures cannot possibly be shown to be valid by the arithmetic of dots, and not even by the methods of ratios.

That this is, indeed, impossible, is established by the famous proof of the irrationality of the diagonal, that is, of the square root of 2, assumed as well-known by Plato and Aristotle. It consists in showing that the assumption

$$\sqrt{2} = n/m \tag{1}$$

that is, that $\sqrt{2}$ is equal to a ratio of any two natural numbers, n and m, leads to an absurdity.

We first note that we can assume that

(2) *not more than one* of the two numbers, n and m, is even.

For if both were even, then we could always cancel out the factor 2 so as to obtain two other natural numbers, n' and m', such that $n/m = n'/m'$, and such that at most one of the two numbers, n' and m', would be even. Now by squaring (1) we get

$$2 = n^2/m^2 \tag{3}$$

and from this

$$2m^2 = n^2 \tag{4}$$

and thus

$$n \text{ is even.} \tag{5}$$

Thus there must exist a natural number a so that

$$n = 2a \tag{6}$$

and we get from (3) and (6)

$$2m^2 = n^2 = 4a^2 \tag{7}$$

and thus

(8) $$m^2 = 2a^2$$

But this means

(9) m is even.

It is clear that (5) and (9) contradict (2). Thus the assumption that there are two natural numbers, n and m, whose ratio equals $\sqrt{2}$, leads to an absurd conclusion. Therefore $\sqrt{2}$ is not a ratio, it is 'irrational'.

This proof uses only the arithmetic of natural numbers. It therefore uses purely Pythagorean methods, and the tradition that it was discovered within the Pythagorean school need not be questioned. But it is improbable that the discovery was made by Pythagoras, or that it was made very early: Zeno does not seem to know it, nor does Democritus. Moreover, as it destroys the basis of Pythagoreanism, it is reasonable to assume that it was not made long before the order reached the height of its influence; at least not before it was well established; for it seems to have contributed to its decline. The tradition that it was made within the order but kept secret seems to me very plausible. It may be supported by considering that the old term for 'irrational'—'*arrhētos*', 'unutterable', or 'unmentionable'—may well have hinted at an unmentionable secret. Tradition has it that the member of the school who gave away the secret was killed for his treachery.[40] However this may be, there is little doubt that the realization that irrational magnitudes (they were, of course, not recognized as numbers) existed, and that their existence could be proved, undermined the faith of the Pythagorean order; it destroyed the hope of deriving cosmology, or even geometry, from the arithmetic of natural numbers.

VIII

It was Plato who realized this fact, and who in the *Laws* stressed its importance in the strongest possible terms, denouncing his

[40] The story is told of one Hippasus, a somewhat shadowy figure; he is said to have died at sea (cf. Diels[6], 4). See also A. Wasserstein's article mentioned in note 35, above.

compatriots for their failure to gauge its implications. I believe that his whole philosophy, and especially his theory of 'Forms' or 'Ideas', was influenced by it.

Plato was very close to the Pythagorean as well as to the Eleatic School; and although he appears to have felt antipathetic to Democritus he was himself a kind of atomist. (Atomist teaching remained as one of the school traditions of the Academy.[41]) This is not surprising in view of the close relation between Pythagorean and atomistic ideas. But all this was threatened by the discovery of the irrationals. I suggest that Plato's main contribution to science sprang from his realization of the problem of the irrational, and from the modification of Pythagoreanism and atomism which he undertook in order to rescue science from a catastrophic situation.

He realized that the purely arithmetical theory of nature was defeated, and that a new mathematical method for the description and explanation of the world was needed. Thus he encouraged the development of an autonomous geometrical method. It found its fulfilment in the 'Elements' of the Platonist Euclid.

What are the facts? I shall try briefly to put them all together.

(1) Pythagoreanism and atomism in Democritus' form were both fundamentally based on arithmetic; that is to say on counting.

(2) Plato emphasized the catastrophic character of the discovery of the irrationals.

(3) He inscribed over the gates of the Academy: 'Nobody Untrained in Geometry May Enter My House'. But *geometry*, according to Plato's immediate pupil Aristotle[42] as well as Euclid, typically treats of incommensurables or irrationals, in contradistinction to *arithmetic* which treats of 'the odd and the even' (i.e. of integers and their relations).

(4) Within a short time after Plato's death his school produced, in Euclid's *Elements*, a work one of whose main points was that it freed mathematics from the 'arithmetical' assumption of commensurability or rationality.

(5) Plato himself contributed to this development, and especially to the development of solid geometry.

[41] See S. Luria, especially on Plutarch, *loc. cit.*

[42] *An. Post.*, 76b9; *Metaph.*, 983a20, 1061b1. See also *Epinomis*, 990d.

(6) More especially, he gave in the *Timaeus* a specifically geometrical version of the formerly purely arithmetical atomic theory; a version which constructed the elementary particles (the famous Platonic bodies) out of triangles which incorporated the irrational square roots of two and of three. (See below.) In nearly all other respects he preserved Pythagorean ideas as well as some of the most important ideas of Democritus.[43] At the same time he tried to eliminate Democritus' void; for he realized[44] that motion remains possible even in a 'full' world, provided motion is conceived as of the character of vortices in a liquid. Thus he retained some of the most fundamental ideas of Parmenides.[45]

(7) Plato encouraged the construction of geometrical models of the world, and especially models explaining the planetary movements. And I believe that Euclid's geometry was not intended as an exercise in pure geometry (as is now usually assumed), but as an *organon* of a *theory of the world*. According to this view the 'Elements' is not a 'textbook of geometry' but an attempt to solve systematically the main problems of Plato's cosmology. This was done with such success that the problems, having been solved, disappeared and were almost forgotten; though a trace remains in Proclus who writes, 'Some have thought that the subject matter of the various books [of Euclid] pertains to the cosmos, and that they are intended to help us in our contemplation of, and theorizing about, the universe' (*op. cit.*, note 38 above, Prologus, II,

[43] Plato took over, more especially, Democritus' theory of vortices (Diels, fragm. 167, 164; cf. Anaxagoras, Diels, 9, and 12, 13; see also the next two footnotes) and his theory of what we nowadays would call gravitational phenomena (Diels, Democritus 164; Anaxagoras, 12, 13, 15, and 2)—a theory which, slightly modified by Aristotle, was ultimately discarded by Galileo.

[44] The clearest passage is *Timaeus*, 80c, where it is said that neither in the case of (rubbed) amber nor of the 'Heraclean stone' (magnet) is there any real attraction; 'there is no void and these things push themselves around, one upon another'. On the other hand Plato was not too clear on this point, since his elementary particles (other than the cube and the pyramid) cannot be packed without leaving some (empty?) space between them, as Aristotle observed in *De Caelo*, 306b5. See also note 22 above (and *Timaeus* 52e).

[45] Plato's reconciliation of atomism and the theory of the *plenum* ('nature abhors the void') became of the greatest importance for the history of physics down to our own day. For it strongly influenced Descartes, became the basis of the theory of ether and light, and thus ultimately, via Huyghens and Maxwell, of de Broglie's and of Schrödinger's wave mechanics. See my report in *Atti d. Congr. Intern. di Filosofia* (1958), **2**, 1960, pp. 367 ff.

p. 71, 2–5). Yet even Proclus does not mention in this context the main problem—that of the irrationals (of course he does mention it elsewhere); though he points out, rightly, that the 'Elements' culminate with the construction of the 'cosmic' or 'Platonic' regular polyhedra. Ever since[46] Plato and Euclid, but not before, geometry (rather than arithmetic) appears as the fundamental instrument of all physical explanations and descriptions, in the theory of matter as well as in cosmology.[47]

IX

These are the historical facts. They go a long way, I believe, towards establishing my main thesis: that what I have called the *prima facie* method of teaching philosophy cannot lead to an understanding of the problems that inspired Plato. Nor can it lead to an appreciation of what may be justly claimed to be his greatest philosophical achievement, the geometrical theory of the world. The great physicists of the Renaissance—Copernicus, Galileo, Kepler, Gilbert—who turned from Aristotle to Plato intended by this move to replace the Aristotelian qualitative substances or potentialities by a geometrical method of cosmology. Indeed, that is what the Renaissance (in science) largely meant: a renaissance of the geometrical method, which was the basis of the works of Euclid, Aristarchus, Archimedes, Copernicus, Kepler,

[46] An exception is the reappearance of arithmetical methods in Quantum Theory, e.g. in the electron shell theory of the periodic system based upon Pauli's exclusion principle; an inversion of Plato's tendency to *geometrize arithmetic* (see below).

Concerning the modern tendency towards what is sometimes called 'arithmetization of geometry' (a tendency which is by no means characteristic of all modern work on geometry), it should be noted that it shows little similarity to the Pythagorean approach since *sets*, or *infinite sequences*, of natural numbers are its main instruments, rather than the natural numbers themselves.

Only those who confine themselves to 'constructive' or 'finitist' or 'intuitionist' methods of number theory—as opposed to set-theoretic methods—might claim that their attempts to reduce geometry to number theory resemble Pythagorean or pre-Platonic ideas of arithmetization. A great step in this direction has been achieved quite recently, it seems, by the German mathematician E. de Wette.

[47] For a similar view of Plato's and Euclid's influence, see G. F. Hemens, *Proc. of the Xth Intern. Congress of Philosophy* (Amsterdam, 1949), Fasc. **2**, 847.

Galileo, Descartes, and became the basis of the works of Newton, Maxwell, and Einstein.

But is this achievement properly described as philosophical? Does it not rather belong to physics—a factual science; and to pure mathematics—a branch, as Wittgenstein's school would contend, of tautological logic?

I believe that we can at this stage see fairly clearly why Plato's achievement (although it has no doubt its physical, its logical, its mixed, and its nonsensical components) was a philosophical achievement; why at least part of his philosophy of nature and of physics has lasted and, I believe, will last.

What we find in Plato and his predecessors is the conscious construction and invention of a new approach towards the world and towards knowledge of the world. This approach transforms an originally theological idea, *the idea of explaining the visible world by a postulated invisible world*,[48] into the fundamental instrument of theoretical science. The idea was explicitly formulated by Anaxagoras and Democritus[49] as the principle of investigation into the nature of matter or body; visible matter was to be explained by hypotheses about invisibles, about an *invisible structure which is too small to be seen*. With Plato this idea is consciously accepted and generalized; the visible world of change is ultimately to be explained by an invisible world of unchanging 'Forms' (or substances, or essences, or 'natures'; that is, as I shall try to show in more detail, geometrical shapes or figures).

Is this idea about the invisible structure of matter a physical or a philosophical idea? If a physicist merely *acts* upon this theory, if he accepts it, perhaps unconsciously, by accepting the traditional problems of his subject as furnished by the problem-situation with which he is confronted; and if he, so acting, produces a new specific theory of the structure of matter, then I should not call him a philosopher. But if he reflects upon it, and, for example, rejects it (like Berkeley or Mach), preferring a phenomenological or positivistic physics to the theoretical

[48] Cf. Homer's explanation of the visible world around Troy with the help of the invisible world of the Olympus. The idea loses, with Democritus, some of its theological character (which is still strong in Parmenides, although less so in Anaxagoras) but regains it with Plato, only to lose it again soon afterwards.

[49] See note 27 above, and Anaxagoras, Fragments B4 and 17, Diels-Kranz.

and somewhat theological approach, then he may be called a philosopher. Similarly, those who consciously sought the theoretical approach, who constructed it, and who explicitly formulated it, and thus transferred the hypothetical and deductive method from theology to physics, were philosophers, even though they were physicists in so far as they acted upon their own precepts and tried to produce actual theories of the invisible structure of matter.

But I shall not pursue the question of the proper application of the label 'philosophy' any further; for this problem, which is Wittgenstein's problem, clearly turns out to be one of linguistic usage; it is indeed a pseudo-problem, and one which by now must be rapidly degenerating into a bore to my audience. Yet I wish to add a few words on Plato's theory of Forms or Ideas, or, to be more precise, on point (6) in the list of historical facts given above.

Plato's theory of the structure of matter can be found in the *Timaeus*. It has at least a superficial similarity to the modern theory of solids which interprets them as crystals. His physical bodies are composed of invisible elementary particles of various shapes, the shapes being responsible for the macroscopic properties of visible matter. The shapes of the elementary particles are determined in their turn by the shapes of the plane figures which form their sides. And these plane figures, in their turn, are ultimately all composed of *two* elementary triangles: the half-square (or *isosceles* rectangular) triangle which incorporates the *square root of two*, and the half-equilateral rectangular triangle which incorporates the *square root of three*, both of them irrationals.

These triangles, in their turn, are described as the copies[50] of unchanging 'Forms' or 'Ideas', which means that specifically *geometrical* 'Forms' are admitted into the heaven of the Pythagorean *arithmetical* Form-Numbers.

There is little doubt that the motive of this construction is the

[50] For the process by which the triangles are stamped out of space (the 'mother') by the ideas (the 'fathers'), cf. my *Open Society*, note 15 to ch. 3, and the reference there given, as well as note 9 to ch. 6. In admitting irrational triangles into his heaven of divine Forms Plato admits something 'indeterminable' in the sense of the Pythagoreans, i.e. something belonging to the 'Bad' side of the Table of Opposites. That 'Bad' things may have to be admitted seems to be first stated in Plato's *Parmenides*, 130b–e; the admission is put into the mouth of Parmenides himself.

attempt to solve the crisis of atomism by incorporating irrationals into the last elements of which the world is built. Once this has been done the difficulty arising from the existence of irrational distances is overcome.

But why did Plato choose just these two triangles? I have elsewhere[51] expressed the view, as a conjecture, that Plato believed that all other irrationals might be obtained by adding to the rationals multiples of the square roots of two and three.[52] I now feel more confident that the crucial passage in the *Timaeus* does imply this doctrine (which was mistaken, as Euclid later showed). For in the passage in question Plato says quite clearly, '*All* triangles originate from two, each having a right angle', going on to specify these two as the half-square and half-equilateral. But in the context this can only mean that *all* triangles originate somehow from these two. This seems to be a hint at the mistaken theory of the relative commensurability of all irrationals with sums of rationals and the square roots of two and three.[53]

But Plato does not pretend that he has a proof of the theory in question. On the contrary, he says that he assumes the two triangles as principles, 'in accordance with an account which combines likely conjecture with necessity'. And a little later, after explaining that he takes the half-equilateral triangle as the second of his principles, he says, 'The reason is too long a story; but if anybody should probe into this matter, and prove that it has this property' (I suppose the property that all other triangles can be composed of these two) 'then the prize is his, with all our good will'.[54] The language is somewhat obscure, and the likely reason is that Plato was conscious that he lacked a proof of his (mistaken) conjecture concerning these two triangles, and felt it should be supplied by somebody.

[51] In the last quoted note of my *Open Society*.

[52] This would mean that all geometrical distances (magnitudes) are commensurable with one of *three* 'measures' (or a sum of two or all of them) related as $1 : \sqrt{2} : \sqrt{3}$. It seems likely that Aristotle even believed that all geometrical magnitudes are commensurable with one of *two* measures, viz. 1 and $\sqrt{2}$. For he writes (*Metaphysics*, 1053a17): 'The diagonal and the side of a square and all (geometrical) magnitudes are measured by two (measures).' (Cp. Ross' note on this passage.)

[53] In note 9 to ch. 6 of my *Open Society*, mentioned above, I also conjectured that the approximation of $\sqrt{2} + \sqrt{3}$ to π encouraged Plato in his mistaken theory.

[54] The two quotations are from the *Timaeus*, 53c/d and 54a/b.

The obscurity of the passage had, it appears, the strange effect that Plato's quite clearly stated choice of triangles which introduce *irrationals* into his world of Forms escaped the notice of most of his readers and commentators in spite of Plato's emphasis upon the problem of irrationality in other places. And this in turn may perhaps explain why Plato's Theory of Forms could appear to Aristotle to be fundamentally the same as the Pythagorean theory of form-numbers,[55] and why

[55] I believe that our consideration may throw some light on the problem of Plato's famous two 'principles'—'The One' and 'The Indeterminate Dyad'. The following interpretation develops a suggestion made by van der Wielen (*De Ideegetallen van Plato*, 1941, p. 132 f.) and brilliantly defended against van der Wielen's own criticism by Ross (*Plato's Theory of Ideas*, p. 201). We assume that the 'Indeterminate Dyad' is a straight line or distance, not to be interpreted as a unit distance, or as having yet been measured at all. We assume that a point (limit, *monas*, 'One') is placed successively in such positions that it divides the Dyad according to the ratio $1 : n$, for any natural number n. Then we can describe the 'generation' of the numbers as follows. For $n = 1$, the Dyad is divided into two parts whose ratio is $1 : 1$. This may be interpreted as the 'generation' of Twoness out of Oneness ($1 : 1 = 1$) and the Dyad, since we have divided the Dyad into *two* equal parts. Having thus 'generated' the number 2, we can divide the Dyad according to the ratio $1 : 2$ (and the larger of the ensuing sections, as before, according to the ratio $1 : 1$), thus generating *three* equal parts and the number 3; generally, the 'generation' of a number n gives rise to a division of the Dyad in the ratio $1 : n$, and with this, to the 'generation' of the number $n + 1$. (And in each stage the 'One' intervenes afresh as the point which introduces a limit or form or measure into the otherwise 'indeterminate' Dyad to create the new number; this remark may strengthen Ross' case against van der Wielen's. Cp. also Toeplitz's, Stenzel's, and Becker's papers in *Quellen & Studien z. Gesch. d. Math.*, 1, 1931. None of them, however, hint at a *geometrization of arithmetic*—in spite of the figures on pp. 476f.)

Now it should be noted that this procedure, although it 'generates' (in the first instance, at least) only the series of natural numbers, nevertheless contains a *geometrical* element—the division of a line, first into two equal parts, and then into two parts according to a certain proportion $1 : n$. Both kinds of division are in need of geometrical methods, and the second, more especially, needs a method such as Eudoxus' Theory of Proportions. Now I suggest that Plato began to ask himself why he should not divide the Dyad also in the proportion of $1 : \sqrt{2}$ and of $1 : \sqrt{3}$. This, he must have felt, was a departure from the method by which the natural numbers are generated; it is less 'arithmetical' still, and it needs more specifically 'geometrical' methods. But it would 'generate', in the place of natural numbers, linear elements in the proportion $1 : \sqrt{2}$ and $1 : \sqrt{3}$, which may be identical with the 'atomic lines' (*Metaphysics*, 992a19) from which the atomic triangles are constructed. At the same time the characterization of the Dyad as 'indeterminate' would become highly appropriate, in view of the Pythagorean attitude (cf. Philolaos, Diels, fragm. 2 and 3) towards the irrational. (Perhaps the name 'The

Plato's atomism appeared to Aristotle merely as a comparatively minor variation on that of Democritus.[56] Aristotle, in spite of taking for granted both the association of arithmetic with the odd and even, and of geometry with the irrational, does not appear to have taken the problem of the irrationals seriously. Proceeding as he did from an interpretation of the *Timaeus* which identified Plato's Space with matter, he seems to have taken Plato's reform programme for geometry for granted; it had been partly carried out by Eudoxus before Aristotle entered the Academy, and Aristotle was only superficially interested in mathematics. He never alludes to the inscription over the Academy gates.

To sum up,[56a] it seems probable that Plato's theory of Forms and also his theory of matter were both restatements of the theories of his predecessors, the Pythagoreans and Democritus respectively, in the light of his realization that the irrationals demanded that geometry should come before arithmetic. By encouraging this emancipation of geometry Plato contributed to the development of Euclid's system, the most important and influential deductive theory ever constructed. By his adoption of geometry as the theory of the world he provided Aristarchus, Newton, and Einstein with their intellectual toolbox. The calamity of Greek atomism was thus transformed into a momentous achievement. But Plato's scientific interests are partly forgotten. The

Great and the Small' began to be replaced by 'The Indeterminate Dyad' when irrational proportions were generated in addition to rational ones.)

Assuming this view to be correct, we might conjecture that Plato slowly approached (beginning in the *Hippias Major*, and thus long before the *Republic*—as opposed to a remark made by Ross, *op. cit.*, top of p. 56) the view that *the irrationals are numbers* (*a*) since they are comparable with numbers (*Met.*, 1021a 4 f.) and (*b*) since both the natural numbers and the irrationals are 'generated' by similar and *essentially geometric* processes. Yet once this view is reached (and it was first reached, it appears, in the *Epinomis*, 990d-e, whether or not this work is, as I am inclined to believe, Plato's), then even the irrational triangles of the *Timaeus* become 'numbers' (i.e. characterized by numerical, if irrational, proportions). But at this point the peculiar contribution of Plato, and the difference between his and the Pythagorean theory, may become indiscernible; which may explain why it has been lost sight of even by Aristotle (who suspected both 'geometrization' *and* 'arithmetization').

[56] That this was Aristotle's view has been pointed out by Luria; see above, Note 31.

[56a] Compare with this summary the *Addendum*, I to volume I of my *Open Society*.

problem-situation in science which gave rise to his philosophical problems is little understood. And his greatest achievement, the geometrical theory of the world, has influenced our world-picture to such an extent that we unreflectingly take it for granted.

X

One example never suffices. As my second example, out of a great many interesting possibilities, I choose Kant. His *Critique of Pure Reason* is one of the most difficult books ever written. Kant wrote in great haste,[57] and about a problem which, I shall try to show, was not only insoluble but also misconceived. Nevertheless it was not a pseudo-problem, but an inescapable problem which arose out of the contemporary situation in science.

His book was written for people who knew something about Newton's stellar dynamics and who had at least some idea of his forerunners—of Copernicus, Tycho Brahe, Kepler and Galileo.

It is perhaps hard for intellectuals of our own day, spoilt and blasé as we are by the spectacle of scientific success, to realize what Newton's theory meant, not just for Kant but for any eighteenth-century thinker. After the unmatched daring with which the Ancients had tackled the riddle of the Universe there had come long periods of decay and recovery, and then a staggering success. Newton had discovered the long sought secret. His geometrical theory, based on and modelled after Euclid, had been received at first with great misgivings, even by its own originator.[58] The reason was that the gravitational force of attraction was felt to be 'occult', or at least something which needed an explanation. But although no plausible explanation was found (and Newton scorned recourse to *ad hoc* hypotheses) all misgivings had disappeared long before Kant made his own important contribution to Newtonian theory, 68 years after the *Principia*.[59] No qualified judge[60] of

[57] He was afraid that he might die before completing his work.

[58] See Newton's letters to Bentley, 1693. (Cf. note 20 to ch. 3, below.)

[59] The so-called Kant-Laplacean Hypothesis, published by Kant in 1755. (It fell dead from the press.)

[60] There had been some very pertinent criticism (especially by Leibniz and Berkeley) but in view of the success of the theory it was—I believe rightly—felt that the critics had

the situation could doubt any longer that Newton's theory was true. It had been tested by the most precise measurements, and it had always been right. It had led to the prediction of minute deviations from Kepler's laws, and to new discoveries. In a time like ours, when theories come and go like the buses in Piccadilly, and when every schoolboy has heard that Newton has long been superseded by Einstein, it is hard to recapture the sense of conviction which Newton's theory inspired, or the sense of elation, and of liberation. A *unique event* had happened in the history of thought, one which could never be repeated: the first and final discovery of the absolute truth about the universe. An age-old dream had come true. Mankind had obtained *knowledge*, real, certain, indubitable, and demonstrable knowledge—divine *scientia* or *epistēmē*, and not merely *doxa*, human opinion. This sense of conviction became—through Voltaire—the origin of the Enlightenment.[60a]

Thus for Kant Newton's theory was simply true, and the belief in its truth remained unshaken for a century after Kant's death. Kant to the end accepted what he and everybody else took for a fact, the attainment of *scientia* or *epistēmē*. At first he accepted it without question. This state he called his 'dogmatic slumber'. He was roused from it by Hume.

Hume had taught that there could be no such thing as certain knowledge of universal laws, or *epistēmē*; that all we knew was obtained with the help of observation which could be only of singular (or particular) instances, so that all theoretical knowledge was uncertain. His arguments were convincing (and he was, of course, right). Yet there was a fact, or what appeared as a fact—Newton's attainment of *epistēmē*.

Hume roused Kant to the realization of the near absurdity of what he never doubted to be a fact. Here was a problem which could not be dismissed. How could a man have got hold of such knowledge? Knowledge which was general, precise, mathematical, demonstrable, and indubitable, like Euclidean geometry, and yet capable of giving a causal explanation of observed facts?

Thus arose the central problem of the *Critique*: 'How is pure natural

somehow missed the point of the theory. We must not forget that even today the theory still stands, with only minor modifications, as an excellent first approximation (or, in view of Kepler, perhaps as a second approximation).

[60a] See below, Chapter 7, section 1.

science possible?' By 'pure natural science'—*scientia, epistēmē*—Kant simply meant Newton's theory. (This he does not say, unfortunately; and I do not see how a student reading the first *Critique*, 1781 and 1787, could possibly find out. But that Kant has Newton's theory in mind is clear from the *Metaphysical Foundations of Natural Science*, 1786, where he gives an *a priori* deduction of Newton's theory, see especially the eight theorems of the Second Main Part, with its Additions, especially Addition 2, Note 1, paragraph 2. Kant relates Newton's theory, in the fifth paragraph of the final 'General Note on Phenomenology', to the 'starry heavens'. It is also clear from the 'Conclusion' of the *Critique of Practical Reason*, 1788, where the appeal to the 'starry heavens' is explained, at the end of the second paragraph, by a reference to the *a priori* character of the new astronomy.[61])

Although the *Critique* is badly written, and although bad grammar abounds in it, Kant's central problem was not a linguistic puzzle. *Here was knowledge. How could Newton ever attain it?* The question was inescapable.[62]

But it was also insoluble. For the apparent fact of the attainment of *epistēmē* was no fact. As we now know, or believe we know, Newton's theory is no more than a marvellous *conjecture*, an astonishingly good approximation; unique indeed, but not as divine truth, only as a unique invention of a human genius: not *epistēmē*, but belonging to the realm of *doxa*. With this Kant's problem, 'How is pure natural science possible', collapses, and the most disturbing of his perplexities disappears.

Kant's proposed solution of his insoluble problem consisted of what he proudly called his 'Copernican Revolution' of the problem of knowledge. Knowledge—*epistēmē*—was possible because we are not passive receptors of sense data, but their active digestors. By digesting and assimilating them we form and organize them into a Cosmos, the Universe of Nature. In this process we impose upon the material presented to our senses the mathematical laws which are part of our digestive and organizing mechanism. Thus our intellect does not dis-

[61] Kant says there that Newton gave us so clear 'an insight into the structure of the universe that it will remain unchanged in all time; and though there is hope that our insight will ever grow through continued observation, there never need be fear of a setback'. (Added 1989.)

[62] Poincaré was still greatly troubled by it in 1909.

cover universal laws in nature, but it prescribes its own laws and imposes them upon nature.

This theory is a strange mixture of absurdity and truth. It is as absurd as the mistaken problem it attempts to solve; for it proves too much, being designed to prove too much. According to Kant's theory, 'pure natural science' is not only *possible*; although he does not always realize this, it becomes, contrary to his intention, the *necessary result* of our mental outfit. For if the fact of our attainment of *epistēmē* can be explained at all by the fact that our intellect legislates for nature, and imposes its own laws upon it, then the first of these two facts cannot be contingent any more than the second.[63] Thus the problem is no longer how Newton could make his discovery but how everybody else could have failed to make it. How is it that our digestive mechanism did not work much earlier?

This is a patently absurd consequence of Kant's idea. But to dismiss it offhand, and to dismiss his problem as a pseudo-problem, is not good enough. For we can find an element of truth in his idea (and a much needed correction of some Humean views) after reducing his problem to its proper dimensions. His question, we now know, or believe we know, should have been: 'How are successful conjectures possible?' And our answer, in the spirit of his Copernican Revolution, might, I suggest, be something like this: Because, as you said, we are not passive receptors of sense data, but active organisms. Because we react to our environment not always merely instinctively, but sometimes consciously and freely. Because we can invent myths, stories, theories; because we have a thirst for explanation, an insatiable curiosity, a wish to know. Because we not only invent stories and theories, but try them out and see whether they work and how they work. Because by a great effort, by trying hard and making many mistakes, we may sometimes, if we are lucky, succeed in hitting upon a story, an explanation, which 'saves the phenomena'; perhaps by making up a myth about 'invisibles', such as atoms or gravitational forces, which explain the

[63] A crucial requirement which any adequate theory of knowledge must satisfy is that it must not explain too much. Any non-historical theory explaining why a certain discovery had to be made must fail because it could not possibly explain why it was not made somewhat earlier.

visible. Because knowledge is an adventure of ideas. These ideas, it is true, are produced by us, and not by the world around us; they are not merely the traces of repeated sensations or stimuli or what not; here you were right. But we are more active and free than even you believed; for similar observations or similar environmental situations do not, as your theory implied, produce similar explanations in different men. Nor is the fact that we create our theories, and that we attempt to impose them upon the world, an explanation of their success,[64] as you believed. For the overwhelming majority of our theories, of our freely invented ideas, are unsuccessful; they do not stand up to searching tests, and are discarded as falsified by experience. Only a very few of them succeed, for a time, in the competitive struggle for survival.[65]

XI

Few of Kant's successors appear ever to have understood clearly the precise problem-situation which gave rise to his work. There were two such problems for him: Newton's dynamics of the heavens, and the absolute standards of human brotherhood and justice to which the French revolutionaries appealed; or, as Kant puts it, 'the starry heavens above me, and the moral law within me'. But Kant's 'starry heavens' are seldom recognized for what they were: an allusion to Newton.[66] From Fichte onward,[67] many have copied Kant's 'method' and the difficult style of parts of his *Critique*. But most of these imitators, unaware of Kant's original interests and problems, busily tried either to tighten, or else to explain away, the Gordian knot in which Kant, through no fault of his own, had tied himself up.

We must beware of mistaking the well-nigh senseless and pointless subtleties of the imitators for the pressing and genuine problems of the

[64] Applying note 63, no theory can explain why our search for explanatory theories is successful. Successful explanation must retain, on any valid theory, the probability zero, assuming that we measure this probability, approximately, by the ratio of the 'successful' explanatory hypotheses to all hypotheses which might be designed by man.

[65] The ideas of this 'answer' were elaborated in my *L.Sc.D.* (1934, 1959, and later editions).

[66] See note 61 and text, above.

[67] Cf. my *Open Society*, note 58 to ch. 12.

pioneer. We should remember that his problem, although not an empirical one in the ordinary sense, nevertheless turned out, unexpectedly, to be in some sense factual (Kant called such facts 'transcendental'), since it arose from an apparent, but non-existent, instance of *scientia* or *epistēmē*. And we should, I submit, seriously consider the suggestion that Kant's answer, in spite of its partial absurdity, contained the nucleus of a true philosophy of science.

3

THREE VIEWS CONCERNING HUMAN KNOWLEDGE

1. THE SCIENCE OF GALILEO AND ITS MOST RECENT BETRAYAL

Once upon a time there was a famous scientist whose name was Galileo Galilei. He was tried by the Inquisition, and forced to recant his teaching. This caused a great stir; and for well over two hundred and fifty years the case continued to arouse indignation and excitement—long after public opinion had won its victory, and the Church had become tolerant of science.

But this is by now a very old story, and I fear it has lost its interest. For Galilean science has no enemies left, it seems: its life hereafter is secure. The victory won long ago was final, and all is quiet on this front. So we take a detached view of the affair nowadays, having learned at last to think historically, and to understand both sides of a dispute. And nobody cares to listen to the bore who can't forget an old grievance.

What, after all, was this old case about? It was about the status of the Copernican 'System of the World' which, besides other things,

First published in *Contemporary British Philosophy*, 3rd Series, ed. H. D. Lewis, 1956.

explained the diurnal motion of the sun as only apparent, and as due to the rotation of our own earth.[1] The Church was very ready to admit that the new system was simpler than the old one: that it was a more convenient *instrument* for astronomical calculations, and for predictions. And Pope Gregory's reform of the calendar made full practical use of it. There was no objection to Galileo's teaching the mathematical theory, so long as he made it clear that its value was *instrumental* only; that it was nothing but a 'supposition', as Cardinal Bellarmino put it;[2] or a 'mathematical hypothesis'—a kind of mathematical trick, 'invented and assumed in order to abbreviate and ease the calculations'.[3] In other words there were no objections so long as Galileo was ready to fall into line with Andreas Osiander who had said in his preface to Copernicus' *De revolutionibus*: 'There is no need for these hypotheses to be true, or even to be at all like the truth; rather, one thing is sufficient for them—that they should yield calculations which agree with the observations.'

[1] I emphasize here the diurnal as opposed to the annual motion of the sun because it was the theory of the diurnal motion which clashed with Joshua 10, 12 f., and because the explanation of the diurnal motion of the sun by the motion of the earth will be one of my main examples in what follows. (This explanation is, of course, much older than Copernicus—older even than Aristarchus—and it has been repeatedly re-discovered; for example by Oresme.)

[2] '. . . Galileo will act prudently', wrote Cardinal Bellarmino (who had been one of the inquisitors in the case against Giordano Bruno) '. . . if he will speak hypothetically, *ex suppositione* . . . : to say that we give a better account of the appearances by supposing the earth to be moving, and the sun at rest, than we could if we used eccentrics and epicycles is to speak properly; there is no danger in that, and it is all that the mathematician requires.' Cf. H. Grisar, *Galileistudien*, 1882, Appendix ix. (Although this passage makes Bellarmino one of the founding fathers of the epistemology which Osiander had suggested some time before and which I am going to call 'instrumentalism', Bellarmino—unlike Berkeley—was by no means a convinced instrumentalist himself, as other passages in this letter show. He merely saw in instrumentalism one of the possible ways of dealing with inconvenient scientific hypotheses. The same might well be true of Osiander. See also note 6 below.)

[3] The quotation is from Bacon's criticism of Copernicus in the *Novum Organum*, II, 36. In the next quotation (from *De revolutionibus*) I have translated the term '*verisimilis*' by 'like the truth'. It should certainly not be translated here by 'probable'; for the whole point here is the question whether Copernicus' system is, or is not, similar in structure to the world; that is, whether it is similar to the truth, or truthlike. The question of degrees of certainty or probability does not arise. For the important *problem of truthlikeness or verisimilitude*, see also ch. 10 below, especially sections iii, x, and xiv; and *Addendum* 6.

Galileo himself, of course, was very ready to stress the superiority of the Copernican system as an *instrument of calculation*. But at the same time he conjectured, and even believed, that it was *a true description of the world*; and for him (as for the Church) this was by far the most important aspect of the matter. He had indeed some good reasons for believing in the truth of the theory. He had seen in his telescope that Jupiter and his moons formed a miniature model of the Copernican solar system (according to which the planets were moons of the sun). Moreover, if Copernicus was right, the inner planets (and they alone) should, when observed from the earth, show phases like the moon; and Galileo had seen in his telescope the phases of Venus.

The Church was unwilling to contemplate the truth of a New System of the World which seemed to contradict a passage in the Old Testament. But this was hardly its main reason. A deeper reason was clearly stated by Bishop Berkeley, about a hundred years later, in his criticism of Newton.

In Berkeley's time the Copernican System of the World had developed into Newton's Theory of gravity, and Berkeley saw in it a serious competitor to religion. He saw that a decline of religious faith and religious authority would result from the new science unless its interpretation by the 'free-thinkers' could be refuted; for they saw in its success a proof of *the power of the human intellect, unaided by divine revelation, to uncover the secrets of our world*—the reality hidden behind its appearance.

This, Berkeley felt, was to misinterpret the new science. He analysed Newton's theory with complete candour and great philosophical acumen; and a critical survey of Newton's concepts convinced him that this theory could not possibly be anything but a 'mathematical hypothesis', that is, a convenient *instrument* for the calculation and prediction of phenomena or appearances; that it could not possibly be taken as a true description of anything real.[4]

Berkeley's criticism was hardly noticed by the physicists; but it was taken up by philosophers, sceptical as well as religious. As a weapon it turned out to be a boomerang. In Hume's hands it became a threat to all belief—to all knowledge, whether human or revealed. In the hands of Kant, who firmly believed both in God and in the truth of

[4] See also ch. 6, below.

Newtonian science, it developed into the doctrine that theoretical knowledge of God is impossible, and that Newtonian science must pay for the admission of its claim to truth by the renunciation of its claim to have discovered the real world behind the world of appearance: it was a true science of nature, but *nature* was precisely the world of mere phenomena, the world as it appeared to our assimilating minds. Later certain Pragmatists based their whole philosophy upon the view that the idea of 'pure' knowledge was a mistake; that there could be no knowledge in any other sense but in the sense of *instrumental* knowledge; that knowledge was power, and that truth was usefulness.

Physicists (with a few brilliant exceptions[5]) kept aloof from all these philosophical debates, which remained completely inconclusive. Faithful to the tradition created by Galileo they devoted themselves to the search for truth, as he had understood it.

Or so they did until very recently. For all this is now past history. Today the view of physical science founded by Osiander, Cardinal Bellarmino, and Bishop Berkeley,[6] has won the battle without another

[5] The most important of them are Mach, Kirchhoff, Hertz, Duhem, Poincaré, Bridgman, and Eddington—all instrumentalists in various ways.

[6] Duhem, in his famous series of papers, '*Sōzein to phainómena*' (*Ann. de philos. chrétienne*, anneé 79, tom 6, 1908, nos. 2 to 6), claimed for instrumentalism a much older and much more illustrious ancestry than is justified by the evidence. For the postulate that, with our causal hypotheses, we ought to 'explain *the observed facts*', rather than 'do violence to them by trying to squeeze or fit them into our theories' (Aristotle, *De Caelo*, 293a25; 296b6; 297a4, b24ff; *Met.* 1073b37, 1074a1) has little to do with the instrumentalist thesis (that our theories *cannot explain* the facts). Yet this postulate is essentially the same as that we ought to '*preserve the phenomena*' or 'save' them ([*dia-*]*sōzein ta phainómena*). The phrase seems to be connected with the astronomical branch of the Platonic School tradition. (See especially the most interesting passage on Aristarchus in Plutarch's *De Facie in Orbe Lunae*, 923a; see also 933a for the 'confirmation of the cause' by the phenomena, and Cherniss' note *a* on p. 168 of his edition of this work of Plutarch's; furthermore, Simplicius' commentaries on *De Caelo* where the phrase occurs e.g. on pp. 497 1.21, 506 1.10, and 488 1.23 f, of Heiberg's edition, in commentaries on *De Caelo* 293a4 and 292b10.) We may well accept Simplicius' report that Eudoxus, under Plato's influence, in order to account for the observable phenomena of planetary motion, set himself the task of evolving an abstract geometrical system of rotating spheres *to which he did not attribute any physical reality*. (There seems to be some resemblance between this programme and that of the *Epinomis*, 990–1, where the study of abstract geometry, 990d–991b—of the theory of the irrationals—is described as a necessary preliminary to planetary theory; another such preliminary is the study of number—i.e. the odd and the even, 990c.) Yet even this

shot being fired. Without any further debate over the philosophical issue, without producing any new argument, the *instrumentalist view* (as I shall call it) has become an accepted dogma. It may well now be called the 'official view' of physical theory since it is accepted by most of our leading theorists of physics (although neither by Einstein nor by Schrödinger). And it has become part of the current teaching of physics.

2. THE ISSUE AT STAKE

All this looks like a great victory of philosophical critical thought over the 'naïve realism' of the physicists. But I doubt whether this interpretation is right.

Few if any of the physicists who have now accepted the instrumentalist view of Cardinal Bellarmino and Bishop Berkeley realize that they have accepted a philosophical theory. Nor do they realize that they have broken with the Galilean tradition. On the contrary, most of them think that they have kept clear of philosophy; and most of them no longer care anyway. What they now care about, as physicists, is (*a*) *mastery of the mathematical formalism*, i.e. of the instrument, and (*b*) its *applications*; and they care for nothing else. And they think that by thus excluding everything else they have finally got rid of all philosophical nonsense. This very attitude of being tough and not standing any nonsense prevents them from considering seriously the philosophical arguments for and against the Galilean view of science (though they will no doubt have heard of Mach[7]). Thus the victory of the instrumentalist philosophy is hardly due to the soundness of its arguments.

How then did it come about? As far as I can see, through the coincidence of two factors, (*a*) difficulties in the interpretation of the formalism of the Quantum Theory, and (*b*) the spectacular practical success of its applications.

would not mean that either Plato or Eudoxus accepted an instrumentalist epistemology: they may have consciously (and wisely) confined themselves to a preliminary problem.

[7] But they seem to have forgotten that Mach was led by his instrumentalism to fight against atomic theory—a typical example of *the obscurantism of instrumentalism* which is the topic of section 5 below.

(*a*) In 1927 Niels Bohr, one of the greatest thinkers in the field of atomic physics, introduced the so-called *principle of complementarity* into atomic physics, which amounted to a 'renunciation' of the attempt to interpret atomic theory as a description of anything. Bohr pointed out that we could avoid certain contradictions (which threatened to arise between the formalism and its various interpretations) only by reminding ourselves that the formalism as such was self-consistent, and that each single case of its application (or each kind of case) remained consistent with it. The contradictions only arose through the attempt to comprise within *one* interpretation the formalism together with more than one case, or kind of case, of its experimental application. But, as Bohr pointed out, any two of these conflicting applications were physically incapable of ever being combined in one experiment. Thus the result of *every single* experiment was consistent with the theory, and unambiguously laid down by it. This, he said, was all we could get. The claim to get more, and even the hope of ever getting more, we must renounce; physics remains consistent only if we do not try to interpret, or to understand, its theories beyond (*a*) mastering the formalism, and (*b*) relating them to each of their actually realizable cases of application separately.[8]

Thus the instrumentalist philosophy was used here *ad hoc* in order to provide an escape for the theory from certain contradictions by which it was threatened. It was used in a defensive mood—to rescue the existing theory; and the principle of complementarity has (I believe for this reason) remained completely sterile within physics. In twenty-seven years it has produced nothing except some philosophical discussions, and some arguments for the confounding of critics (especially Einstein).

I do not believe that physicists would have accepted such an *ad hoc* principle had they understood that it was *ad hoc*, or that it was a philosophical principle—part of Bellarmino's and Berkeley's instrumentalist

[8] I have explained Bohr's 'Principle of Complementarity' as I understand it after many years of effort. No doubt I shall be told that my formulation of it is unsatisfactory. But if so I am in good company; for Einstein refers to it as 'Bohr's principle of complementarity, a sharp formulation of which . . . I have been unable to attain despite much effort which I have expended on it.' Cf. *Albert Einstein: Philosopher-Scientist*, ed. by P. A. Schilpp, 1949, p. 674.

philosophy of physics. But they remembered Bohr's earlier and extremely fruitful 'principle of correspondence' and hoped (in vain) for similar results.

(*b*) Instead of results due to the principle of complementarity other and more practical results of atomic theory were obtained, some of them with a big bang. No doubt physicists were perfectly right in interpreting these successful applications as corroborating their theories. But strangely enough they took them as confirming the instrumentalist creed.

Now this was an obvious mistake. The instrumentalist view asserts that theories are *nothing but* instruments, while the Galilean view was that they are not only instruments but also—and mainly—descriptions of the world, or of certain aspects of the world. It is clear that in this disagreement even a proof showing that theories are instruments (assuming it possible to 'prove' such a thing) could not seriously be claimed to support either of the two parties to the debate, since both were agreed on this point.

If I am right, or even roughly right, in my account of the situation, then philosophers, even instrumentalist philosophers, have no reason to take pride in their victory. On the contrary, they should examine their arguments again. For at least in the eyes of those who like myself do not accept the instrumentalist view, there is much at stake in this issue.

The issue, as I see it, is this.

One of the most important ingredients of our western civilization is what I may call the 'rationalist tradition' which we have inherited from the Greeks. It is the tradition of critical discussion—not for its own sake, but in the interests of the search for truth. Greek science, like Greek philosophy, was one of the products of this tradition,[9] and of the urge to understand the world in which we live; and the tradition founded by Galileo was its renaissance.

Within this rationalist tradition science is valued, admittedly, for its practical achievements; but it is even more highly valued for its informative content, and for its ability to free our minds from old beliefs, old prejudices, and old certainties, and to offer us in their stead

[9] See ch. 4, below.

new conjectures and daring hypotheses. Science is valued for its liberalizing influence—as one of the greatest of the forces that make for human freedom.

According to the view of science which I am trying to defend here, this is due to the fact that scientists have dared (since Thales, Democritus, Plato's *Timaeus*, and Aristarchus) to create myths, or conjectures, or theories, which are in striking contrast to the everyday world of common experience, yet able to explain some aspects of this world of common experience. Galileo pays homage to Aristarchus and Copernicus precisely because they dared to go beyond this known world of our senses: 'I cannot', he writes,[10] 'express strongly enough my unbounded admiration for the greatness of mind of these men who conceived [the heliocentric system] and held it to be true . . . , in violent opposition to the evidence of their own senses. . . .' This is Galileo's testimony to the liberalizing force of science. Such theories would be important even if they were no more than exercises for our imagination. But they are more than this, as can be seen from the fact that we submit them to severe tests by trying to deduce from them some of the regularities of the known world of common experience—i.e. by trying to *explain* these regularities. And these attempts to *explain the known by the unknown* (as I have described them elsewhere[11]) have immeasurably extended the realm of the known. They have added to the facts of our everyday world the invisible air, the antipodes, the circulation of the blood, the worlds of the telescope and the microscope, of electricity, and of tracer atoms showing us in detail the movements of matter within living bodies. All these things are far from being mere instruments: they are witness to the intellectual conquest of our world by our minds.

But there is another way of looking at these matters. For some, science is still nothing but glorified plumbing, glorified gadget-making—'mechanics'; very useful, but a danger to true culture, threatening us with the domination of the near-illiterate (of Shakespeare's

[10] Salviati says so several times, with hardly a verbal variation, on the Third Day of *The Two Principal Systems*.

[11] See the Appendix, point (10) to ch. 1, above, and the penultimate paragraph of ch. 6, below.

'mechanicals'). It should never be mentioned in the same breath as literature or the arts or philosophy. Its professed discoveries are mere mechanical inventions, its theories are instruments—gadgets again, or perhaps super-gadgets. It cannot and does not reveal to us new worlds behind our everyday world of appearance; for the physical world is just surface: it has no depth. *The world is just what it appears to be. Only the scientific theories are not what they appear to be.* A scientific theory neither explains nor describes the world; it is nothing but an instrument.

I do not present this as a complete picture of modern instrumentalism, although it is a fair sketch, I think, of part of its original philosophical background. Today a much more important part of it is, I am well aware, the rise and self-assertion of the modern 'mechanic' or engineer.[12] Still, I believe that the issue should be seen to lie between a critical and adventurous rationalism—the spirit of discovery—and a narrow and defensive creed according to which we cannot and need not learn or understand more about our world than we know already. A creed, moreover, which is incompatible with the appreciation of science as one of the greatest achievements of the human spirit.

Such are the reasons why I shall try, in this paper, to uphold at least part of the Galilean view of science against the instrumentalist view. But I cannot uphold all of it. There is a part of it which I believe the instrumentalists were right to attack. I mean the view that in science we can aim at, and obtain, *an ultimate explanation by essences*. It is in its opposition to this Aristotelian view (which I have called[13] 'essentialism') that the strength and the philosophical interest of instrumentalism lies. Thus I shall have to discuss and criticize two views of human knowledge—*essentialism* and *instrumentalism*. And I shall oppose to them what I shall call *the third view*—what remains of Galileo's view after the elimination of essentialism, or more precisely, after allowance has been made for what was justified in the instrumentalist attack.

[12] The realization that natural science is not indubitable *epistēmē* (*scientia*) has led to the view that it is *technē* (technique, art, technology); but the proper view, I believe, is that it consists of *doxai* (*opinions, conjectures*), controlled by critical discussion as well as by experimental *techne*. Cf. ch. 20, below.

[13] See section 10 of my *Poverty of Historicism*, and my *Open Society and its Enemies*, vol. 1, ch. 3, section vi, and vol. II, ch. 11, sections i and ii.

3. THE FIRST VIEW: ULTIMATE EXPLANATION BY ESSENCES

Essentialism, the first of the three views of scientific theory to be discussed, is part of the Galilean philosophy of science. Within this philosophy three elements or doctrines which concern us here may be distinguished. Essentialism (our 'first view') is that part of the Galilean philosophy which I do not wish to uphold. It consists of a combination of the doctrines (2) and (3). These are the three doctrines:

(1) *The scientist aims at finding a true theory or description of the world* (and especially of its regularities or 'laws'), *which shall also be an explanation of the observable facts*. (This means that a description of these facts must be deducible from the theory in conjunction with certain statements, the so-called 'initial conditions'.)

This is a doctrine I wish to uphold. It is to form part of our 'third view'.

(2) *The scientist can succeed in finally establishing the truth of such theories beyond all reasonable doubt.*

This second doctrine, I think, needs correction. All the scientist can do, in my opinion, is to test his theories, and to eliminate all those that do not stand up to the most severe tests he can design. But he can never be quite sure whether new tests (or even a new theoretical discussion) may not lead him to modify, or to discard, his theory. In this sense all theories are, and remain hypotheses: they are conjecture (*doxa*) as opposed to indubitable knowledge (*epistēmē*).

(3) *The best, the truly scientific theories, describe the 'essences' or the 'essential natures' of things—the realities which lie behind the appearances*. Such theories are neither in need nor susceptible of further explanation: they are *ultimate explanations*, and to find them is the ultimate aim of the scientist.

This third doctrine (in connection with the second) is the one I have called 'essentialism'. I believe that like the second doctrine it is mistaken.

Now what the instrumentalist philosophers of science, from Berkeley to Mach, Duhem, and Poincaré, have in common is this. They all assert that explanation is not an aim of physical science, since physical science cannot discover 'the hidden essences of things'. The argument shows that what they have in mind is what I call *ultimate* explanation.[14] Some of them, such as Mach and Berkeley, hold this view because they do not believe that there is such a thing as an essence of anything physical: Mach, because he does not believe in essences at all; Berkeley, because he believes only in spiritual essences, and thinks that the only essential explanation of the world is God. Duhem seems to think (on lines reminiscent of Kant[15]) that there are essences but that they are undiscoverable by human science (though we may, somehow, move towards them); like Berkeley he thinks that they can be revealed by religion. But all these philosophers agree that (ultimate) scientific explanation is impossible. And from the absence of a hidden essence which scientific theories could describe they conclude that these theories (which clearly do not describe our ordinary world of common experience) describe nothing at all. Thus they are mere instruments.[16] And what may appear as the growth of theoretical knowledge is merely the improvement of instruments.

The instrumentalist philosophers therefore reject the third doctrine, i.e. the doctrine of essences. (I reject it too, but for somewhat different reasons.) At the same time they reject, and are bound to reject, the second doctrine; for if a theory is an instrument, then it cannot be true (but only convenient, simple, economical, powerful, etc.). They even frequently call the theories 'hypotheses'; but they do not, of course, mean by this what I mean: that a theory is *conjectured to be true*, that it is a descriptive though possibly a false statement; although they do mean to say that theories are uncertain: 'And as to the usefulness of hypotheses', Osiander writes (at the end of his preface), 'nobody should expect anything certain to emerge from astronomy, for nothing of the kind can ever come out of it.' Now I fully agree that there is no certainty

[14] The issue has been confused at times by the fact that the instrumentalist criticism of (ultimate) explanation was expressed by some with the help of the formula: the aim of science is *description rather than explanation*. But what was here meant by 'description' was the description *of the ordinary empirical world*; and what the formula expressed, indirectly, was that those theories which do not describe *in this sense* do not explain either, but are nothing but convenient instruments to help us in the description of ordinary phenomena.

[15] Cf. Kant's letter to Reinhold, 12.5.1789, in which the 'real essence' or 'nature' of a thing (e.g. of matter) is said to be inaccessible to human knowledge.

[16] See ch. 6, below.

about theories (which may always be refuted); and I even agree that they are instruments, although I do not agree that this is the reason why there can be no certainty about theories. (The correct reason, I believe, is simply that our tests can never be exhaustive.) There is thus a considerable amount of agreement between my instrumentalist opponents and myself over the second and third doctrines. But over the first doctrine there is complete disagreement.

To this disagreement I shall return later. In the present section I shall try to criticize (3), the essentialist doctrine of science, on lines somewhat different from the arguments of the instrumentalism which I cannot accept. For its argument that there can be no 'hidden essences' is based upon its conviction that *there can be nothing hidden* (or that if anything is hidden it can be only known by divine revelation). From what I said in the last section it will be clear that I cannot accept an argument that leads to the rejection of the claim of science to have discovered the rotation of the earth, or atomic nuclei, or cosmic radiation, or the 'radio stars'.

I therefore readily concede to essentialism that much is hidden from us, and that much of what is hidden may be discovered. (I disagree profoundly with the spirit of Wittgenstein's dictum, 'The riddle does not exist'.) And I do not even intend to criticize those who try to understand the 'essence of the world'. The essentialist doctrine I am contesting is solely *the doctrine that science aims at ultimate explanation*; that is to say, an explanation which (essentially, or by its very nature) cannot be further explained, and which is in no need of any further explanation.

Thus my criticism of essentialism does not aim at establishing the non-existence of essences; it merely aims at showing the obscurantist character of the role played by the idea of essences in the Galilean philosophy of science (down to Maxwell, who was inclined to believe in them but whose work destroyed this belief). In other words my criticism tries to show that, whether essences exist or not, the belief in them does not help us in any way and indeed is likely to hamper us; so that there is no reason why the scientist should *assume* their existence.[17]

[17] This criticism of mine is thus frankly utilitarian, and it might be described as instrumentalist; but I am concerned here with a *problem of method* which is always a problem of the fitness of means to ends.

My attacks upon *essentialism*—i.e. upon the *doctrine of ultimate explanation*—have sometimes

This, I think, can be best shown with the help of a simple example—*the Newtonian theory of gravity*.

The essentialist interpretation of Newtonian theory is due to Roger Cotes.[18] According to him Newton discovered that every particle of matter was endowed with *gravity*, i.e. with an inherent power or force to attract other matter. It was also endowed with *inertia*—an inherent power to resist a change in its state of motion (or to retain the direction and velocity of its motion). Since both gravity and inertia inhere in each particle of matter it follows that both must be strictly proportional to the amount of matter in a body, and therefore to each other; hence the law of proportionality of inert and gravitating mass. Since gravity radiates from each particle we obtain the square law of attraction. In other words, Newton's laws of motion simply describe in mathematical language the state of affairs due to the inherent properties of matter: they describe the *essential nature of matter*.

Since Newton's theory described in this way the essential nature of matter, he could explain the behaviour of matter with its help, by mathematical deduction. But Newton's theory, in its turn, is neither capable of, nor in need of, further explanation, according to Cotes—at least not within physics. (The only possible further explanation was that God has endowed matter with these essential properties.[19])

This essentialist view of Newton's theory was on the whole the accepted view until the last decades of the nineteenth century. That it

been countered by the remark that I myself operate (perhaps unconsciously) with the idea of an *essence of science* (or an *essence of human knowledge*), so that my argument, if made explicit, would run: 'It is of the essence or of the nature of human science (or human knowledge) that we cannot know, or search for, such things as essences or natures.' I have however answered, by implication, this particular objection at some length in *L.Sc.D.* (sections 9 and 10, 'The Naturalist View of Method') and I did so before it was ever raised—in fact before I ever came to describe, and to attack, essentialism. Moreover, one might adopt the view that certain *things of our own making*—such as clocks—may well be said to have 'essences', viz. their 'purposes' (and what makes them serve these 'purposes'). And science, as a human, purposeful activity (or a method), *might* therefore be claimed by some to have an 'essence', even if they deny that natural objects have essences. (This denial is not, however, implied in my criticism of essentialism.)

[18] R. Cotes' Preface to the second edition of Newton's *Principia*.

[19] There is an essentialist theory of Time and Space (similar to this theory of matter) which is due to Newton himself.

was obscurantist is clear: *it prevented fruitful questions from being raised*, such as, 'What is the cause of gravity?' or more fully, 'Can we perhaps explain gravity by deducing Newton's theory, or a good approximation of it, from a more general theory (which should be independently testable)?'

Now it is illuminating to see that Newton himself had not considered *gravity* as an essential property of matter (although he considered *inertia* to be essential, and also, with Descartes, *extension*). It appears that he had taken over from Descartes the view that the essence of a thing must be a true or absolute property of the thing (i.e. a property which does not depend on the existence of other things) such as extension, or the power to resist a change in its state of motion, and not a relational property, i.e. a property which, like gravity, determines the relations (interactions in space) between one body and other bodies. Accordingly, he strongly felt the incompleteness of this theory, and the need to explain gravity. 'That gravity', he wrote,[20] 'should be innate, inherent, and essential to matter, so that one body may act upon another at a distance . . . is to me so great an absurdity that I believe no man who has in philosophical matters a competent faculty of thinking can ever fall into it.'

It is interesting to see that Newton condemned here, in anticipation, the bulk of his followers. To them, one is tempted to remark, the properties of which they had learned in school appeared to be essential (and even self-evident), although to Newton, with his Cartesian background, the same properties had appeared to be in need of explanation (and indeed to be almost paradoxical).

Yet Newton himself was an essentialist. He had tried hard to find an acceptable ultimate explanation of gravity by trying to deduce the square law from the assumption of a mechanical push—the only kind of causal action which Descartes had permitted, since only push could be explained by the essential property of all bodies, extension.[21] But he

[20] Letter to Richard Bentley, 25th February 1692–3 (i.e. 1693); cf. also the letter of 17th January.

[21] Newton tried to explain gravity by a Cartesian *action by contact* (forerunner of an *action at vanishing distances*): his *Opticks*, Qu. 31, shows that he *did* 'consider' that 'What I call Attraction may be performed by impulse' (anticipating Lesage's explanation of gravity as an umbrella effect in a rain of particles). Qu. 21, 22, and 28 suggest that he may have been aware of the fatal excess impulse on the windscreen over the rear window.

failed. Had he succeeded we can be certain that he would have thought that his problem was finally solved—that he had found the ultimate explanation of gravity.[22] But here he would have been wrong. The question, 'Why can bodies push one another?' *can* be asked (as Leibniz first saw), and it is even an extremely fruitful question. (We now believe that they push one another because of certain repulsive electric forces.) But Cartesian and Newtonian essentialism, especially if Newton had been successful in his attempted explanation of gravity, might have prevented this question from ever being raised.

These examples, I think, make it clear that the belief in essences (whether true or false) is liable to create obstacles to thought—to the posing of new and fruitful problems. Moreover, it cannot be part of science (for even if we should, by a lucky chance, hit upon a theory describing essences, we could never be sure of it). But a creed which is likely to lead to obscurantism is certainly not one of those extra-scientific beliefs (such as a faith in the power of critical discussion) which a scientist need accept.

This concludes my criticism of essentialism.

4. THE SECOND VIEW: THEORIES AS INSTRUMENTS

The instrumentalist view has great attractions. It is modest, and it is very simple, especially if compared with essentialism.

According to essentialism we must distinguish between (i) the universe of essential reality, (ii) the universe of observable phenomena, and (iii) the universe of descriptive language or of symbolic representation. I will take each of these to be represented by a square.

[22] Newton was an essentialist for whom gravity was unacceptable as an ultimate explanation; but he was too critical to accept even his own attempts to explain it. Descartes, in this situation, would have assumed the existence of some push mechanism, proposing what he called a 'hypothesis'. But Newton, with a critical allusion to Descartes, stressed that he was 'to argue from Phaenomena without feigning [arbitrary or *ad hoc*] Hypotheses' (Qu. 28). Of course, he could not but use hypotheses all the time, and the *Opticks* overflows with bold speculations. But his explicit and repeated rejection of the method of hypotheses made a lasting impression; and Duhem used it in support of instrumentalism.

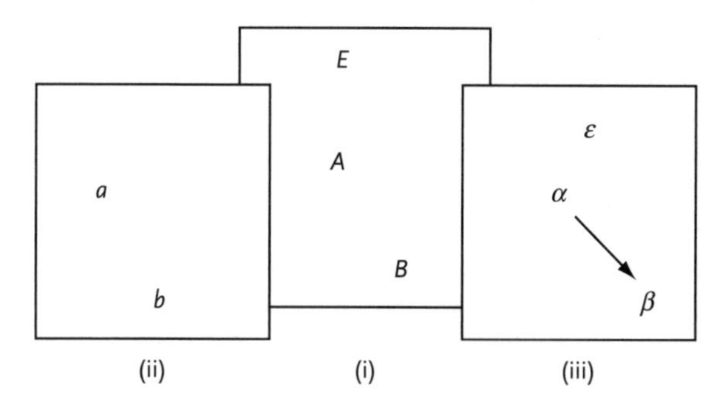

The function of a theory may here be described as follows.

a, b are phenomena; *A*, B are the corresponding realities behind these appearances; and α, β the descriptions or symbolic representations of these realities. E are the essential (relational) properties of *A*, B and ε is the theory describing E. Now from ε and α we can deduce β; this means that we can explain, with the help of our theory, why a leads to, or is the cause of, b.

A representation of instrumentalism can be obtained from this schema simply by omitting (i), i.e. the universe of the realities behind the various appearances. α then directly describes a, and β directly describes b; and ε describes nothing—it is merely an instrument which helps us to deduce β from α. (This may be expressed by saying—as Schlick did, following Wittgenstein—that a universal law or a theory is not a proper statement but rather 'a rule, or a set of instructions, for the derivation of singular statements from other singular statements'.[23])

This is the instrumentalist view. In order to understand it better we may again take Newtonian dynamics as an example. a and b may be taken to be two positions of two spots of light (or two positions of the planet Mars); α and β are the corresponding formulae of the formalism; and ε is the theory strengthened by a general description of the solar system (or by a 'model' of the solar system). Nothing

[23] For an analysis and criticism of this view see my *L.Sc.D.* especially note 7 to section 4, and my *Open Society*, note 51 to ch. 11. The idea that universal statements may function in this way can be found in Mill's *Logic*, Book II, ch. III, 3: 'All inference is from particulars to particulars.' See also G. Ryle, *The Concept of Mind* (1949), ch. v, pp. 121 ff., for a more careful and critical formulation of the same view.

corresponds to ε in the world (in the universe ii): there simply are no such things as attractive forces, for example. Newtonian forces are not entities which determine the acceleration of bodies: they are nothing but mathematical tools whose function is to allow us to deduce β from α.

No doubt we have here an attractive simplification, a radical application of Ockham's razor. But although this simplicity has converted many to instrumentalism (for example Mach) it is by no means the strongest argument in its favour.

Berkeley's strongest argument for instrumentalism was based upon his nominalistic philosophy of language. According to this philosophy the expression 'force of attraction' must be a meaningless expression, since forces of attraction can never be observed. What can be observed are movements, not their hidden alleged 'causes'. This is sufficient, on Berkeley's view of language, to show that Newton's theory cannot have any informative or descriptive content.

Now this argument of Berkeley's may perhaps be criticized because of the intolerably narrow theory of meaning which it implies. For if consistently applied it amounts to the thesis that all dispositional words are without meaning. Not only would Newtonian 'attractive forces' be without meaning, but also such ordinary dispositional words and expressions as 'breakable' (as opposed to 'broken'), or 'capable of conducting electricity' (as opposed to 'conducting electricity'). These are not names of anything observable, and they would therefore have to be treated on a par with Newtonian forces. But it would be awkward to classify all these expressions as meaningless, and *from the point of view of instrumentalism* it is quite unnecessary to do so: all that is needed is an analysis of the meaning of dispositional terms and dispositional statements. This will reveal that they have meaning. But from the point of view of instrumentalism they do not have a descriptive meaning (like non-dispositional terms and statements). Their function is not to report events, or occurrences, or 'incidents', in the world, or to describe facts. Rather, their meaning exhausts itself in the permission or licence which they give us to draw inferences or to argue from some matters of fact to other matters of fact. Non-dispositional statements which describe observable matters of fact ('this leg is broken') have cash value, as it were; dispositional statements, to which belong the laws of science, are not like cash, but rather like legal '*instruments*' creating rights to cash.

One need only proceed one step further in the same direction, it appears, in order to arrive at an instrumentalist argument which it is extremely difficult, if not impossible, to criticize; for our whole question—whether science is descriptive or instrumental—is here exposed as a pseudo-problem.[24]

The step in question consists, simply, in not only allowing meaning—an instrumental meaning—to dispositional terms, but also a kind of *descriptive meaning*. Dispositional words such as 'breakable', it may be said, certainly describe something; for to say of a thing that it is breakable is to describe it as a thing that can be broken. But to say of a thing that it is breakable, or soluble, is to describe it in a different way, and by a different method, from saying that it is broken or dissolved; otherwise we should not use the suffix 'able'. The difference is just this—that we describe, by using dispositional words, what may happen to a thing (in certain circumstances). Accordingly, dispositional descriptions *are* descriptions, but they have nevertheless a purely instrumental function. In their case, knowledge *is* power (the power to foresee). When Galileo said of the earth 'and yet, it moves', then he uttered, no doubt, a descriptive statement. But the function or meaning of this statement turns out nevertheless to be purely instrumental: it exhausts itself in the help it renders in deducing certain non-dispositional statements.

Thus the attempt to show that theories have a descriptive meaning *besides* their instrumental meaning is misconceived, according to this argument; and the whole problem—the issue between Galileo and the Church—turns out to be a pseudo-problem.

In support of the view that Galileo suffered for the sake of a pseudo-problem it has been asserted that in the light of a logically more advanced system of physics Galileo's problem has in fact dissolved into nothing. Einstein's general principle, one often hears, makes it quite

[24] I have not so far encountered in the literature this particular form of the instrumentalist argument; but if we remember the parallelism between problems concerning the *meaning* of an expression and problems concerning the *truth* of a statement (see for example the table in the Introduction above, section xii), we see that this argument closely corresponds to William James' definition of 'truth' as 'usefulness'.

clear that it is meaningless to speak of absolute motion, even in the case of rotation; for we can freely choose whatever system we wish to be (relatively) at rest. Thus Galileo's problem vanishes. Moreover, it vanishes precisely for the reasons given above. Astronomical knowledge can be nothing but knowledge of how the stars behave; thus it cannot be anything but the power to describe and predict our observations; and since these must be independent of our free choice of a co-ordinate system, we now see more clearly why Galileo's problem could not possibly be real.

I shall not criticize instrumentalism in this section, or reply to its arguments, except the very last one—the argument from general relativity. This argument is based on a mistake. From the point of view of general relativity, there is very good sense—even an absolute sense—in saying that the earth rotates: *it rotates in precisely that sense in which a bicycle wheel rotates*. It rotates, that is to say, with respect to *any* chosen local inertial system. Indeed relativity describes the solar system in such a way that from this description we can deduce that *any* observer situated on *any* sufficiently distant freely moving physical body (such as our moon, or another planet, or a star outside the system) would see the earth rotating, and could deduce, from this observation, that for its inhabitants there would be an apparent diurnal motion of the sun. But it is clear that this is precisely the sense of the words 'it moves' which was at issue; for part of the issue was whether the solar system was a system like that of Jupiter and his moons, only bigger; and whether it would look like this system, if seen from outside. On all these questions Einstein unambiguously supports Galileo.

My argument should not be interpreted as an admission that the whole question can be reduced to one of observations, or of possible observations. Admittedly both Galileo and Einstein intend, among other things, to deduce what an observer, or a possible observer, would see. But this is not their main problem. Both investigate physical systems and their movements. It is only the instrumentalist philosopher who asserts that what they discussed, or 'really meant' to discuss, were not physical systems but *only* the results of possible observations; and that their so-called 'physical systems', which *appeared* to be their objects of study, were *in reality* only instruments for predicting observations.

5. CRITICISM OF THE INSTRUMENTALIST VIEW

Berkeley's argument, we have seen, depends upon the adoption of a certain philosophy of language, convincing perhaps at first, but not necessarily true. Moreover, it hinges on the *problem of meaning*,[25] notorious for its vagueness and hardly offering hope of a solution. The position becomes even more hopeless if we consider some more recent development of Berkeley's arguments, as sketched in the preceding section. I shall try, therefore, to force a clear decision on our problem by a different approach—by way of an analysis of science rather than an analysis of language.

My proposed criticism of the instrumentalist view of scientific theories can be summarized as follows.

Instrumentalism can be formulated as the thesis that scientific theories—the theories of the so-called 'pure' sciences—are nothing but computation rules (or inference rules); of the same character, fundamentally, as the computation rules of the so-called 'applied' sciences. (One might even formulate it as the thesis that 'pure' science is a misnomer, and that all science is 'applied'.)

Now my reply to instrumentalism consists in showing that there are profound differences between 'pure' theories and technological computation rules, and that instrumentalism can give a perfect description of these rules but is quite unable to account for the difference between them and the theories. Thus instrumentalism collapses.

The analysis of the many functional differences between computation rules (for navigation, say) and scientific theories (such as Newton's) is a very interesting task, but a short list of results must suffice here. The logical relations which may hold between theories and computation rules are not symmetrical; and they are different from those which may hold between various theories, and also from those which may hold between various computation rules. The way in which computation rules are *tried out* is different from the way in which theories are *tested*; and the skill which the application of computation rules demands is quite different from that needed for their (theoretical)

[25] For this problem see my *L.Sc.D.* and my *Open Society*. See also chs. 1, 11, 13 and 14 of the present volume.

discussion, and for the (theoretical) determination of the limits of their applicability. These are only a few hints, but they may be enough to indicate the direction and the force of the argument.

I am now going to explain one of these points a little more fully, because it gives rise to an argument somewhat similar to the one I have used against essentialism. What I wish to discuss is the fact that theories are tested by *attempts to refute them* (attempts from which we learn a great deal), while there is nothing strictly corresponding to this in the case of technological rules of computation or calculation.

A theory is tested not merely by applying it, or by trying it out, but by applying it to very special cases—cases for which it yields results different from those we should have expected without that theory, or in the light of other theories. In other words we try to select for our tests those crucial cases in which we should expect the theory to fail if it is not true. Such cases are 'crucial' in Bacon's sense; they indicate the cross-roads between *two* (or more) theories. For to say that without the theory in question we should have expected a different result implies that our expectation was the result of some other (perhaps an older) theory, however dimly we may have been aware of this fact. But while Bacon believed that a crucial experiment may establish or verify a theory, we shall have to say that it can at most refute or falsify a theory.[26] It is an attempt to refute it; and if it does not succeed in refuting the theory in question—if, rather, the theory is successful with its unexpected prediction—then we say that it is corroborated by the experiment. (It is the better corroborated[27] the less expected, or the less probable, the result of the experiment has been.)

Against the view here developed one might be tempted to object (following Duhem[28]) that in every test it is not only the theory under investigation which is involved, but also the whole system of our

[26] Duhem, in his famous criticism of crucial experiments (in his *Aim and Structure of Physical Theory*), succeeds in showing that crucial experiments can never *establish* a theory. He fails to show that they cannot *refute* it.

[27] The degree of corroboration will therefore increase with the improbability (or the content) of the corroborating cases. See my 'Degree of Confirmation', *Brit. Jour. Phil. Sci.*, **5**, pp. 143 ff., now among the new appendices of my *L.Sc.D.*, and ch. 10 of the present volume (including the *Addenda*).

[28] See n. 26.

theories and assumptions—in fact, more or less the whole of our knowledge—so that we can never be certain which of all these assumptions is refuted. But this criticism overlooks the fact that if we take each of the two theories (between which the crucial experiment is to decide) *together* with all this background knowledge, as indeed we must, then we decide between two systems which differ *only* over the two theories which are at stake. It further overlooks the fact that we do not assert the refutation of the theory as such, but of the theory *together* with that background knowledge; parts of which, if other crucial experiments can be designed, may indeed one day be rejected as responsible for the failure. (Thus we may even characterize a *theory under investigation* as that part of a vast system for which we have, however vaguely, an alternative in mind, and for which we try to design crucial tests.)

Now nothing sufficiently similar to such tests exists in the case of instruments or rules of computation. An instrument may break down, to be sure, or it may become outmoded. But it hardly makes sense to say that we submit an instrument to the severest tests we can design in order to reject it if it does not stand up to them: every air frame, for example, can be 'tested to destruction', but this severe test is undertaken not in order to reject every frame when it is destroyed but to obtain information about the frame (i.e. to test a theory about it), so that it may be used *within the limits of its applicability* (or safety).

For instrumental purposes of practical application a theory may continue to be used *even after its refutation*, within the limits of its applicability: an astronomer who believes that Newton's theory has turned out to be false will not hesitate to apply its formalism within the limits of its applicability.

We may sometimes be disappointed to find that the range of applicability of an instrument is smaller than we expected at first; but this does not make us discard the instrument *qua* instrument—whether it is a theory or anything else. On the other hand a disappointment of this kind means that we have obtained new *information* through refuting a *theory*—that theory which implied that the instrument was applicable over a wider range.

Instruments, even theories *in so far as they are instruments*, cannot be refuted, as we have seen. The instrumentalist interpretation will therefore be unable to account for real tests, which are attempted refutations, and will not get beyond the assertion that *different theories have different ranges of application*. But then it cannot possibly account for scientific progress. Instead of saying (as I should) that Newton's theory was falsified by crucial experiments which failed to falsify Einstein's, and that Einstein's theory is therefore better than Newton's, the consistent instrumentalist will have to say, with reference to his 'new' point of view, like Heisenberg: 'It follows that we do not say any longer: Newton's mechanics is false. . . . Rather, we now use the following formulation: Classical mechanics . . . is everywhere exactly "right" where its concepts can be applied.'[29]

Since 'right' here means 'applicable', this assertion merely amounts to saying, 'Classical mechanics is applicable where its concepts can be applied'—which is not saying much. But be this as it may, the point is that *by neglecting falsification, and stressing application, instrumentalism proves to be as obscurantist a philosophy as essentialism*. For it is only in searching for refutations that science can hope to learn and to advance. It is only in considering how its various theories stand up to tests that it can distinguish between better and worse theories and so find a criterion of progress. (See chapter 10, below.)

Thus a mere instrument for prediction cannot be falsified. What may appear to us at first as its falsification turns out to be no more than a rider cautioning us about its limited applicability. This is why the instrumentalist view may be used *ad hoc* for rescuing a physical theory which is threatened by contradictions, as was done by Bohr (if I am right in my interpretation, given in section 2, of his principle of complementarity). If theories are mere instruments of prediction we need not discard any particular theory even though we believe that no consistent physical interpretation of its formalism exists.

Summing up we may say that instrumentalism is unable to account for the importance to pure science of testing severely even the most remote implications of its theories, since it is unable to account for the pure scientist's interest in truth and falsity. In contrast to the highly

[29] See W. Heisenberg in *Dialectica*, **2**, 1948, p. 333 f. Heisenberg's own instrumentalism is far from consistent, and he has many anti-instrumentalist remarks to his credit. But this article here quoted may be described as an out-and-out attempt to prove that his quantum theory leads of necessity to an instrumentalist philosophy, and thereby to the result that physical theory can never be unified, or even made consistent.

critical attitude requisite in the pure scientist, the attitude of instrumentalism (like that of applied science) is one of complacency at the success of applications. Thus it may well be responsible for the recent stagnation in quantum theory. (This was written before the refutation of parity.)

6. THE THIRD VIEW: CONJECTURES, TRUTH, AND REALITY

Neither Bacon nor Berkeley believed that the earth rotates, but nowadays everybody believes it, including the physicists. Instrumentalism is embraced by Bohr and Heisenberg only as a way out of the special difficulties which have arisen in quantum theory.

The motive is hardly sufficient. It is always difficult to interpret the latest theories, and they sometimes perplex even their own creators, as happened with Newton.[29a] Maxwell at first inclined towards an essentialist interpretation of his theory: a theory which ultimately contributed more than any other to the decline of essentialism. And Einstein inclined at first to an instrumentalist interpretation of relativity, giving a kind of operational analysis of the concept of simultaneity which contributed more to the present vogue for instrumentalism than anything else; but he later repented.[30]

I trust that physicists will soon come to realize that the principle of complementarity is *ad hoc*, and (what is more important) that its only function is to avoid criticism and to prevent the discussion of physical interpretations; though criticism and discussion are urgently needed for reforming any theory. They will then no longer believe that instrumentalism is forced upon them by the structure of contemporary physical theory.

Anyway, instrumentalism is, as I have tried to show, no more acceptable than essentialism. Nor is there any need to accept either of them, for there is a third view.[31]

[29a] See the quotation from Newton in the text to footnote 20, above.

[30] *Note added to the proofs.* When this paper went to press Albert Einstein was still alive, and I intended to send him a copy as soon as it was printed. My remark referred to a conversation we had on the subject in 1950. (Added 1990.) See section 28 of my *Unended Quest*.

[31] Cf. section v of ch. 6, below.

This 'third view' is not very startling or even surprising, I think. It preserves the Galilean doctrine that the scientist aims at a true description of the world, or of some of its aspects, and at a true explanation of observable facts; and it combines this doctrine with the non-Galilean view that though this remains the aim of the scientist, he can never know for certain whether his findings are true, although he may sometimes establish with reasonable certainty that a theory is false.[32]

One may formulate this 'third view' of scientific theories briefly by saying that they are *genuine conjectures*—highly informative guesses about the world which although not verifiable (i.e. capable of being shown to be true) can be submitted to severe critical tests. They are serious attempts to discover the truth. In this respect scientific hypotheses are exactly like Goldbach's famous conjecture in the theory of numbers. Goldbach thought that it might possibly be true; and it may well be true in fact, even though *we do not know, and may perhaps never know, whether it is true or not*.

I shall confine myself to mentioning only a few aspects of my 'third view', and only such aspects as distinguish it from essentialism and instrumentalism; and I shall take essentialism first.

Essentialism looks upon our ordinary world as mere appearance behind which it discovers the real world. This view has to be discarded once we become conscious of the fact that the world of each of our theories may be explained, in its turn, by further worlds which are described by further theories—theories of a higher level of abstraction, of universality, and of testability. The doctrine of an *essential or ultimate reality* collapses together with that of ultimate explanation.

Since according to our third view the new scientific theories are, like the old ones, genuine conjectures, they are genuine attempts to describe these further worlds. Thus we are led to take all these worlds, including our ordinary world, as equally real; or better, perhaps, as equally real aspects or layers of the real world. (If looking through a microscope we change its magnification, then we may see various completely different aspects or layers of the same thing, all equally real.) It is thus mistaken to say that my piano, as I know it, is real, while

[32] Cf. the discussion of this point in section 5, above, and *L.Sc.D* (*passim*); also ch. 1 above, and Xenophanes' fragments quoted towards the end of ch. 5, below.

its alleged molecules and atoms are mere 'logical constructions' (or whatever else may be indicative of their unreality); just as it is mistaken to say that atomic theory shows that the piano of my everyday world is an appearance only—a doctrine which is clearly unsatisfactory once we see that the atoms in their turn may perhaps be explained as disturbances, or structures of disturbances, in a quantized field of forces (or perhaps of probabilities). All these conjectures are equal in their claims to describe reality, although some of them are more conjectural than others.

Thus we shall not, for example, describe only the so-called 'primary qualities' of a body (such as its geometrical shape) as real, and contrast them as the essentialists once did, with its unreal and merely apparent 'secondary qualities' (such as colour). For the extension and even the shape of a body have since become *objects of explanation* in terms of theories of a higher level; of theories describing a further and deeper layer of reality—forces, and fields of forces—which are related to the primary qualities in the same way as these were believed by the essentialists to be related to the secondary ones; and the secondary qualities, such as colours, are just as real as the primary ones—though our colour experiences have to be distinguished from the colour-properties of the physical things, exactly as our geometrical-shape-experiences have to be distinguished from the geometrical-shape-properties of the physical things. From our point of view both kinds of qualities are equally real—that is, conjectured to be real; and so are forces, and fields of forces, in spite of their undoubted hypothetical or conjectural character.

Although in one sense of the word 'real', all these various levels are equally real, there is another yet closely related sense in which we might say that the higher and more conjectural levels are the *more real* ones—in spite of the fact that they are more conjectural. They are, according to our theories, more real (more stable in intention, more permanent) in the sense in which a table, or a tree, or a star, is more real than any of its aspects.

But is not just this conjectural or hypothetical character of our theories the reason why we should not ascribe reality to the worlds described by them? Should we not (even if we find Berkeley's 'to be is to be perceived' too narrow) *call only those states of affairs 'real' which are described by true statements*, rather than by conjectures which may turn out to be false? With these questions we turn to the discussion of the instrumentalist doctrine, which with its assertion that theories are mere instruments intends to deny the claim that anything like a real world is described by them.

I accept the view (implicit in the classical or correspondence theory of truth[33]) that we should call a state of affairs 'real' if, and only if, the statement describing it is true. But it would be a grave mistake to conclude from this that the uncertainty of a theory, i.e. its hypothetical or conjectural character, diminishes in any way its implicit *claim* to describe something real. For every statement *s* is equivalent to a statement claiming that *s* is true. And as to *s* being a conjecture, we must remember that, first of all, a conjecture *may* be true, and thus describe a real state of affairs. Secondly, if it is false, then it contradicts some real state of affairs (described by its true negation). Moreover, if we test our conjecture, and succeed in falsifying it, we see very clearly that there was a reality—something with which it could clash.

Our falsifications thus indicate the points where we have touched reality, as it were. And our latest and best theory is always an attempt to incorporate all the falsifications ever found in the field, by explaining them in the simplest way; and this means (as I have tried to show in *The Logic of Scientific Discovery*, sections 31 to 46) in the most testable way.

Admittedly, if we do not know how to test a theory we may be doubtful whether there is anything at all of the kind (or level)

[33] See A. Tarski's work on the *Concept of Truth* (*Der Wahrheitsbegriff, etc., Studia Philosophica*, 1935, text to note 1: 'true = in agreement with reality'). (See the English translation in A. Tarski, *Logic, Semantics, Metamathematics*, 1956, p. 153; the translation says 'corresponding' where I translated 'in agreement'.) The following remarks (and also the penultimate paragraph before the one to which this footnote is appended) have been added in an attempt to answer a friendly criticism privately communicated to me by Professor Alexander Koyré, to whom I feel greatly indebted.

I do not think that, if we accept the suggestion that 'in agreement with reality' and 'true' are equivalent, we are seriously in danger of being led up the path to idealism. I do not propose to *define* 'real' with the help of this equivalence. (And even if I did, there is no reason to believe that a definition necessarily determines the ontological status of the term defined.) What the equivalence should help us to see is that the *hypothetical character* of a statement—i.e. our *uncertainty as to its truth*—implies that we are making *guesses concerning reality*.

described by it; and if we positively know that it cannot be tested, then our doubts will grow; we may suspect that it is a mere myth, or a fairy-tale. *But if a theory is testable, then it implies that events of a certain kind cannot happen; and so it asserts something about reality*. (This is why we demand that the more conjectural a theory is, the higher should be its degree of testability.) Testable conjectures or guesses, at any rate, are thus conjectures or guesses about reality; from their uncertain or conjectural character it only follows that our knowledge concerning the reality they describe is uncertain or conjectural. And although only that is certainly real which can be known with certainty, it is a mistake to think that only that is real which is known to be certainly real. We are not omniscient and, no doubt, much is real that is unknown to us all. It is thus indeed the old Berkeleian mistake (in the form 'to be is to be known') which still underlies instrumentalism.

Theories are our own inventions, our own ideas; they are not forced upon us, but are our self-made instruments of thought: this has been clearly seen by the idealist. But some of these theories of ours can clash with reality; and when they do, we know that there is a reality; that there is something to remind us of the fact that our ideas may be mistaken. And this is why the realist is right.

Thus I agree with essentialism in its view that *science is capable of real discoveries*, and even in its view that in discovering new worlds our intellect triumphs over our sense experience. But I do not fall into the mistake of Parmenides—of denying reality to all that is colourful, varied, individual, indeterminate, and indescribable in our world.

Since I believe that science can make real discoveries I take my stand with Galileo against instrumentalism. I admit that our discoveries are conjectural. But this is even true of geographical explorations. Columbus' conjectures as to what he had discovered were in fact mistaken; and Peary could only conjecture—on the basis of theories—that he had reached the Pole. But these elements of conjecture do not make their discoveries less real, or less significant.

There is an important distinction which we can make between two kinds of scientific prediction, and which instrumentalism cannot make; a distinction which is connected with the problem of scientific discovery. I have in mind the distinction between the prediction of *events of a kind which is known*, such as eclipses or thunderstorms on the one hand and, on the other hand, the prediction of *new kinds of events* (which the physicist calls 'new effects') such as the prediction which led to the discovery of wireless waves, or of zero-point energy, or to the artificial building up of new elements not previously found in nature.

It seems to me clear that instrumentalism can account only for the first kind of prediction: if theories are instruments for prediction, then we must assume that their purpose must be determined in advance, as with other instruments. Predictions of the second kind can be fully understood only as discoveries.

It is my belief that our discoveries are guided by theory, in these as in most other cases, rather than that theories are the result of discoveries 'due to observation'; for observation itself tends to be guided by theory. Even geographical discoveries (Columbus, Franklin, the two Nordenskjölds, Nansen, Wegener, and Heyerdahl's Kon-Tiki expedition) are often undertaken with the aim of testing a theory. Not to be content with offering predictions, but to create new situations for new kinds of tests: this is a function of theories which instrumentalism can hardly explain without surrendering its main tenets.

But perhaps the most interesting contrast between the 'third view' and instrumentalism arises in connection with the latter's denial of the descriptive function of abstract words, and of disposition-words. This doctrine, by the way, exhibits an essentialist strain within instrumentalism—the belief that events or occurrences or 'incidents' (which are directly observable) must be, in a sense, more real than dispositions (which are not).

The 'third view' of this matter is different. I hold that most observations are more or less indirect, and that it is doubtful whether the distinction between directly observable incidents and whatever is only indirectly observable leads us anywhere. I cannot but think that it is a mistake to denounce Newtonian forces (the 'causes of acceleration') as occult, and to try to discard them (as has been suggested) in favour of accelerations. For accelerations cannot be observed any more directly than forces; and they are *just as dispositional*: the statement that a body's velocity is accelerated tells us that the body's velocity in the next second from now will exceed its present velocity.

In my opinion *all universals are dispositional*. If 'breakable' is dispositional, so is 'broken', considering for example how a doctor decides whether a bone is broken or not. Nor should we call a glass 'broken' if the pieces would fuse the moment they were put together: the criterion of being broken is behaviour *under certain conditions*. Similarly, 'red' is dispositional: a thing is red if it is able to reflect a certain kind of light—if it 'looks red' in certain situations. But even 'looking red' is dispositional. It describes the disposition of a thing to make onlookers agree that it looks red.

No doubt there are *degrees* of dispositional character: 'able to conduct electricity' is dispositional in a higher degree than 'conducting electricity now' which is still very highly dispositional. These degrees correspond fairly closely to those of the conjectural or hypothetical character of theories. But there is no point in denying reality to dispositions, not even if we deny reality to all universals and to all states of affairs, including incidents, and confine ourselves to using that sense of the word 'real' which, from the point of view of ordinary usage, is the narrowest and safest: to call only physical bodies 'real', and only those which are neither too small nor too big nor too distant to be easily seen and handled.

For even then we should realize (as I wrote twenty years ago[34]) that

> every description uses . . . universals; every statement has the character of a theory, a hypothesis. The statement, 'Here is a glass of water,' cannot be (completely) verified by any sense-experience, because the universals which appear in it cannot be correlated with any particular sense-experience. (An 'immediate experience' is *only once* 'immediately given'; it is unique.) By the word 'glass', for example, we denote physical bodies which exhibit a certain *law-like behaviour*; and the same holds of the word 'water'.

I do not think that a language without universals could ever work; and the use of universals commits us to asserting, and thus (at least) to conjecturing, the reality of dispositions—though not of ultimate and inexplicable ones, that is, of essences. We may express all this by saying that the customary distinction between '*observational terms*' (or

[34] See my *L.Sc.D.*, end of section 25; see also new appendix *x, (1) to (4), and ch. 1 of the present volume; also ch. 11 section v, text to notes 58–62.

'*non-theoretical terms*') and *theoretical terms* is mistaken, since all terms are theoretical to some degree, though some are more theoretical than others; just as we said that all theories are conjectural, though some are more conjectural than others.

But if we are committed, or at least prepared, to conjecture the reality of forces, and of fields of forces, then there is no reason why we should not conjecture that a die has a definite *propensity* (or disposition) to fall on one or another of its sides; that this propensity can be changed by loading it; that propensities of this kind may change continuously; and that we may operate with fields of propensities, or of entities which determine propensities. An interpretation of probability on these lines might allow us to give a new physical interpretation to quantum theory—one which differs from the purely statistical interpretation, due to Born, while agreeing with him that probability statements can be tested only statistically.[35] And this interpretation may, perhaps, be of some little help in our efforts to resolve those grave and challenging difficulties in quantum theory which today seem to imperil the Galilean tradition.[36]

[35] Concerning the propensity theory of probability, see my papers in *Observation and Interpretation*, ed. S. Körner 1957, pp. 65 ff., and in the *B.J.P.S.* **10**, 1959, pp. 25 ff.

[36] (Added 1980.) If my memory does not deceive me, then this paragraph is the first statement of the propensity interpretation that I published (in 1956; it was written in 1953, though footnote 35 was, not, of course, in the original version of this chapter).

4

TOWARDS A RATIONAL THEORY OF TRADITION

In the title of this talk the emphasis should be put on the word 'towards': I do not intend to put forward anything like a full theory. I want to explain to you and to illustrate the kind of question which a theory of tradition would have to answer, and to give in outline some ideas which may be useful for constructing it. By way of introduction I intend to say how I came to be interested in the subject, and why I think it is important; and I also intend to refer to some possible attitudes towards it.

I am a rationalist of sorts. I am not quite certain whether or not my rationalism will be acceptable to you, but that will be seen later. I am very interested in scientific method. Having studied for some time the methods of the natural sciences, I felt that it might be interesting to study also the methods of the social sciences. It was then that I first met with the problem of tradition. The anti-rationalists in the field of politics, social theory, and so on, usually suggest that this problem cannot be tackled by any kind of rational theory. Their attitude is to accept

Transcript of a lecture given at the Third Annual Conference of the Rationalist Press Association on 26th July 1948 at Magdalen College, Oxford (the Chairman was Professor A. E. Heath); first published in The Rationalist Annual, 1949.

tradition as something just given. You have to take it; you cannot rationalize it; it plays an important role in society, and you can only understand its significance and accept it. The most important name associated with this anti-rationalist view is that of Edmund Burke. He fought, as you know, against the ideas of the French Revolution, and his most effective weapon was his analysis of the importance of that irrational power which we call 'tradition'. I mention Burke because I think he has never been properly answered by rationalists. Instead rationalists tended to ignore his criticism and to persevere in their anti-traditionalist attitude without taking up the challenge. Undoubtedly there is a traditional hostility between rationalism and traditionalism. Rationalists are inclined to adopt the attitude: 'I am not interested in tradition. I want to judge everything on its own merits; I want to find out its merits and demerits, and I want to do this quite independently of any tradition. I want to judge it with my own brain, and not with the brains of other people who lived long ago.'

That the matter is not quite so simple as this attitude assumes emerges from the fact that the rationalist who says such things is himself very much bound by a rationalist tradition which traditionally says them. This shows the weakness of certain traditional attitudes towards the problem of tradition.

Our Chairman has told us today that we need not bother about the anti-rationalist reaction; that it is very weak, if not negligible. But I feel that there does exist an anti-rationalist reaction of a serious kind and among very clever men, and that it is connected with this particular problem. Quite a number of outstanding thinkers have developed the problem of tradition into a big stick with which to beat rationalism. I may instance Michael Oakeshott, a Cambridge historian, a really original thinker, who recently in the *Cambridge Journal* launched an attack on rationalism.[1] I largely disagree with his strictures; but I have to admit that the attack is a powerful one. When he launched it there was not much in the rationalist literature which could be considered an adequate answer to his arguments. Some answers may exist, but I very much doubt their adequacy. This is one of the reasons why I feel that this subject is important.

[1] Republished in M. Oakeshott, *Rationalism in Politics and other Essays*, 1962, pp. 1–36.

Another thing which induced me to take up this question was simply my own experience—my own change of social environment. I came to England from Vienna, and I found that the *atmosphere* here in England was very different from that in which I had been brought up. We heard this morning from Dr J. A. C. Brown[2] some interesting remarks about the great importance of what he calls the 'atmosphere' of a factory. I am sure that he would agree that this atmosphere has something to do with tradition. I moved from a Continental tradition or atmosphere to an English one, and later for a time to that of New Zealand. These changes have, no doubt, stimulated me to think about these matters and to try to look further into them.

Certain types of tradition of great importance are local, and cannot easily be transplanted. These traditions are precious things, and it is very difficult to restore them once they are lost. I have in mind the scientific tradition, in which I am particularly interested. I have seen that it is very difficult to transplant it from the few places where it has really taken root. Two thousand years ago this tradition was destroyed in Greece, and it did not take root again for a very long time. Similarly, recent attempts to transplant it from England overseas have not been too successful. Nothing is more striking than the lack of a research tradition in some of the countries overseas. One has a real struggle if one wants it to take root where it is missing. I may perhaps mention that at the time when I left New Zealand the Chancellor of the University undertook a thorough inquiry into the question of research. As a result of it he made an excellent critical speech in which he denounced the University for its neglect of research. But few will think that this speech means that a scientific research tradition will now be established, for this is a very hard thing to bring about. One can convince people of the need for such a tradition, but that does not mean that the tradition will take root and flourish.

I could, of course, take examples from fields other than science. To remind you that it is not only the scientific field in which tradition is important—although it is the field about which I shall mainly speak—I need only mention music. When I was in New Zealand I got hold of a

[2] The allusion is to the lecture 'Rational and Irrational Behaviour in Industrial Groups', summarized in *The Literary Guide*, October 1948.

set of American records of Mozart's 'Requiem'. When I had played these records I knew what the lack of musical tradition meant. They had been made under the directorship of a musician who was obviously untouched by the tradition which has come down from Mozart. The result was devastating. I shall not dwell on this matter; I mention it only to make it clear that when I select for my main illustration the question of the scientific or rational tradition I do not mean to convey the impression that it is either the most important or the only one.

It should be clearly understood that there are only two main attitudes possible towards tradition. One is to accept a tradition *uncritically*, often without even being aware of it. In many cases we cannot escape this; for we often just do not realize that we are faced with a tradition. If I wear my watch on my left wrist, I need not be conscious that I am accepting a tradition. Every day we do hundreds of things under the influence of traditions of which we are unaware. But if we do not know that we are acting under the influence of a tradition, then we cannot help accepting the tradition uncritically.

The other possibility is a *critical* attitude, which may result either in acceptance or in rejection, or perhaps in a compromise. Yet we have to know of and to understand a tradition before we can criticize it, before we can say: 'We reject this tradition on rational grounds.' Now I do not think that we could ever free ourselves entirely from the bonds of tradition. The so-called freeing is really only a change from one tradition to another. But we can free ourselves from the *taboos* of a tradition; and we can do that not only by rejecting it, but also by *critically* accepting it. We free ourselves from the taboo if we *think* about it, and if we ask ourselves whether we should accept it or reject it. In order to do that we have first to have the tradition clearly before us, and we have to understand in a general way what may be the function and significance of a tradition. That is why it is so important for rationalists to deal with this problem, for rationalists are those people who are ready to challenge and to criticize everything, including, I hope, their own tradition. They are ready to put question-marks to anything, at least in their minds. They will not submit blindly to any tradition.

I should say that in our invaluable rationalist tradition (which rationalists so often accept too uncritically) there are quite a few points which we ought to challenge. A part of the rationalist tradition is, for example, the metaphysical idea of determinism. People who do not agree with determinism are usually viewed with suspicion by rationalists who are afraid that if we accept indeterminism, we may be committed to accepting the doctrine of Free Will, and may thus become involved in theological arguments about the Soul and Divine Grace. I usually avoid talking about free will, because I am not clear enough about what it means, and I even suspect that our intuition of a free will may mislead us. Nevertheless, I think that determinism is a theory which is untenable on many grounds, and that we have no reason whatever to accept it. Indeed, I think that it is important for us to get rid of the determinist element in the rationalist tradition. It is not only untenable, but it creates endless trouble for us. It is, for this reason, important to realize that indeterminism—that is, the denial of determinism—does not necessarily involve us in any doctrine about our 'will' or about 'responsibility'.

Another element in the rationalist tradition which we should question is the idea of observationalism—the idea that we know about the world because we look around, open our eyes and ears, and take down what we see, hear, and so on; and that this is what constitutes the material of our knowledge. This is an extremely deep-rooted prejudice and is, I think, an idea which impedes the understanding of scientific method. I shall return to this point later. So much by way of introduction.

Now I come to a brief outline of the task of a theory of tradition. A theory of tradition must be a sociological theory, because tradition is obviously a social phenomenon. I mention this because I wish briefly to discuss with you the *task of the theoretical social sciences*. This has often been misunderstood. In order to explain what is, I think, the central task of social science, I should like to begin by describing a theory which is held by very many rationalists—a theory which I think implies exactly the opposite of the true aim of the social sciences. I shall call this theory the '*conspiracy theory of society*'. This theory, which is more primitive than most forms of theism, is akin to Homer's theory of society. Homer conceived the power of the gods in such a way that whatever happened on the plain before Troy was only a reflection of the various conspiracies on Olympus. The conspiracy theory of society is just a version of this theism, of a belief in gods whose whims and wills rule everything. It comes from abandoning God and then asking: 'Who is in his place?' His place is then filled by various powerful men and groups—sinister pressure groups, who are to be blamed for having planned the great depression and all the evils from which we suffer.

The conspiracy theory of society is very widespread, and has very little truth in it. Only when conspiracy theoreticians come into power does it become something like a theory which accounts for things which actually happen (a case of what I have called the 'Oedipus Effect'). For example, when Hitler came into power, believing in the conspiracy myth of the Learned Elders of Zion, he tried to outdo their conspiracy with his own counter-conspiracy. But the interesting thing is that *such a conspiracy never—or 'hardly ever'—turns out in the way that is intended*.

This remark can be taken as a clue to what is the true task of a social theory. Hitler, I said, made a conspiracy that failed. Why did it fail? Not just because other people conspired against Hitler. It failed, simply, because it is one of the striking things about social life that *nothing ever comes off exactly as intended*. Things always turn out a little bit differently. We hardly ever produce in social life precisely the effect that we wish to produce, and we usually get things that we do not want into the bargain. Of course, we act with certain aims in mind; but apart from these aims (which we may or may not really achieve) there are always certain unwanted consequences of our actions; and usually these unwanted consequences cannot be eliminated. To explain why they cannot be eliminated is the major task of social theory.

I will give you a very simple example. Let us say that a man in a small village must sell his house. Not long before there was a man who bought a house in that village because he needed one urgently. Now there is a seller. He will find that, under normal conditions, he will not get nearly as much for his house as the buyer had to pay when he wanted to buy a similar one. That is to say, the very fact that somebody wants to sell his house lowers the market price. And this is generally so. Whoever wants to sell something always depresses the market value of what he wants to sell; whoever wants to buy something raises the market value of what he wants to buy. This is true, of course, only for small free markets. I do not say that the economic system of free markets cannot be replaced by another one. But in a market economy this is what happens. You will agree with me that there is no need to prove that the man who wants to sell something has usually no intention of lowering the market price, and that the man who wants to buy something has no intention of raising it. We have here a typical instance of unwanted consequences.

The situation described is typical of *all social situations*. In all social situations we have individuals who do things; who want things; who have certain aims. In so far as they act in the way in which they want to act, and realize the aims which they intend to realize, no problem arises for the social sciences (except the problem whether their wants and aims can perhaps be socially explained, for example by certain traditions). The characteristic problems of the social sciences arise only out of our wish to know the *unintended consequences*, and more especially the *unwanted consequences* which may arise if we do certain things. We wish to foresee not only the direct consequences but also these unwanted indirect consequences. Why should we wish to foresee them? Either because of our scientific curiosity, or because we want to be prepared for them; we may wish, if possible, to meet them and prevent them from becoming too important. (This means, again, action, and with it the creation of further unwanted consequences.)

I think that the people who approach the social sciences with a ready-made conspiracy theory thereby deny themselves the possibility of ever understanding what the task of the social sciences is, for they assume that we can explain practically everything in society by asking who wanted it, whereas the real task of the social sciences[3] is to explain those things which nobody wants—such as, for example, a war, or a

[3] In the discussion which followed the lecture, I was criticized for rejecting the conspiracy theory, and it was asserted that Karl Marx had revealed the tremendous importance of the capitalist conspiracy for the understanding of society. In my reply I said that I should have mentioned my indebtedness to Marx, *who was one of the first critics of the conspiracy theory*, and one of the first to analyse the unintended consequences of the voluntary actions of people acting in certain social situations. Marx said quite definitely and clearly that the capitalist is as much caught in the network of the social situation (or the 'social system') as is the worker; that the capitalist cannot help acting in the way he does: he is as unfree as the worker, and the results of his actions are largely unintended. But the truly scientific (though in my opinion too deterministic) approach of Marx has been forgotten by his latter-day followers, the Vulgar Marxists, who have put forward a popular conspiracy theory of society which is no better than the myth of the Learned Elders of Zion.

depression. (Lenin's revolution, and especially Hitler's revolution and Hitler's war are, I think, exceptions. These were indeed conspiracies. But they were consequences of the fact that conspiracy theoreticians came into power—who, most significantly, failed to consummate their conspiracies.)

It is the task of social theory to explain how the unintended consequences of our intentions and actions arise, and what kind of consequences arise if people do this that or the other in a certain social situation. And it is, especially, the task of the social sciences to analyse in this way the existence and the functioning of *institutions* (such as police forces or insurance companies or schools or governments) and of social *collectives* (such as states or nations or classes or other social groups). The conspiracy theorist will believe that institutions can be understood completely as the result of conscious design; and as to collectives, he usually ascribes to them a kind of group-personality, treating them as conspiring agents, just as if they were individual men. As opposed to this view, the social theorist should recognize that the persistence of institutions and collectives creates a problem to be solved in terms of an analysis of individual social actions and their unintended (and often unwanted) social consequences, as well as their intended ones. The task of a theory of tradition must be viewed in a similar light. It is only very rarely that people consciously wish to create a tradition; and even in these cases they are not likely to succeed. On the other hand, people who never dreamt of creating a tradition may nevertheless do so, without having any such intention. Thus we arrive at one of the problems of the theory of tradition: how do traditions arise—and, more important, how do they persist—as the (possibly unintended) consequences of people's actions?

A second and more important problem is this: what is the function of tradition in social life? Has it any function which is rationally understandable, in the way in which we can give an account of the function of schools, or of the police force, or of a grocer's shop, the Stock Exchange, or other such *social institutions?* Can we analyse the functions of traditions? That is perhaps the main task of a theory of tradition. My way of approaching this task will be to analyse a particular tradition—the rational or scientific tradition—as an example, and I intend later to make use of this analysis for various purposes.

My main purpose will be to draw a parallel between, on the one side, the theories which, after being submitted to scientific tests, are held as a result of the rational or critical attitude—in the main, that is, scientific hypotheses—and the way they help us to orientate ourselves in this world; and, on the other side, beliefs, attitudes, and traditions in general, and the way they may help us to orientate ourselves, especially in the social world.

The peculiar thing which we call scientific tradition has often been discussed. People have often wondered about this queer thing that happened somehow somewhere in Greece in the sixth and fifth centuries before Christ—the invention of a rational philosophy. What did actually happen, why did it happen, and how? Some modern thinkers assert that the Greek philosophers were the first to try to *understand* what happens in nature. I shall show you why this is an unsatisfactory account.

The early Greek philosophers did indeed try to understand what happened in nature. But so did the more primitive myth-makers before them. How can we characterize that primitive type of explanation which was superseded by the standards of the early Greek philosophers—the founders of our scientific tradition? To put it crudely, the pre-scientific myth-makers said, when they saw a thunderstorm approaching: 'Oh yes, Zeus is angry.' And when they saw that the sea was rough, they said: 'Poseidon is angry.' That was the type of explanation which was found satisfactory before the rationalist tradition introduced new standards of explanation. What was really the decisive difference? One can hardly say that the new theories introduced by the Greek philosophers were more easily understood than the old ones. It is, I think, much easier to understand the statement that Zeus is angry than to understand a scientific account of a thunderstorm. And the statement that Poseidon is angry is for me a much simpler and more easily understandable explanation of the high waves of the sea than one in terms of friction between the air and the surface of the water.

I think that the innovation which the early Greek philosophers introduced was roughly this: they began to *discuss* these matters. Instead of accepting the religious tradition uncritically, and as unalterable (like children who protest if auntie alters one word of their favourite fairy-tale), instead of merely handing on a tradition, they challenged it, and sometimes even invented a new myth in place of the old one. We have, I think, to admit that the new stories which they put in the place of the old were, fundamentally, myths—just as the old stories were; but there are two things about them to be noticed.

First, they were not just repetitions or re-arrangements of the old stories, but contained new elements. Not that this in itself is a very great virtue. But the second and main thing is this: the Greek philosophers invented a *new tradition*—the tradition of adopting a critical attitude towards the myths, the tradition of discussing them; the tradition of not only telling a myth, but also of being challenged by the man to whom it is told. Telling their myth they were ready in their turn to listen to what their listener thought about it—admitting thereby the possibility that he might perhaps have a better explanation than they. This was a thing that had not happened before. A new way of asking questions arose. Together with the explanation—the myth—the question would arise: 'Can you give me a better account?'; and another philosopher might answer: 'Yes, I can.' Or he might say: 'I do not know whether I can give you a better, but I can give you a very different account which does just as well. These two accounts cannot both be true, so there must be something wrong here. We cannot simply accept these two accounts. Nor have we any reason to accept just one of them. We really want to know more about the matter. We have to discuss it further. We have to see whether our explanations really do account for the things about which we already know, and perhaps even for something we have so far overlooked.'

My thesis is that what we call 'science' is differentiated from the older myths not by being something distinct from a myth, but by being accompanied by a second-order tradition—that of critically discussing the myth. Before, there was only the first-order tradition. A definite story was handed on. Later there was still, of course, a story to be handed on, but with it went something like a silent accompanying text of a second-order character: 'I hand it on to you, but tell me what you think of it. Think it over. Perhaps you can give us a different story.' This second-order tradition was the critical or argumentative attitude. It was, I believe, a new thing, and it is still the fundamentally important thing about scientific tradition. If we understand that, then we shall have an altogether different attitude towards quite a number of problems of scientific method. We shall understand that, in a certain sense, science is myth-making just as religion is. You will say: 'But the scientific myths are so very different from the religious myths.' Certainly they are different. But why are they different? Because if one adopts this critical attitude then one's myths do become different. They change; and they change in the direction of giving a better and better account of the world—of the various things which we can observe. And they also challenge us to observe things which we would never have observed without these theories or myths.

In the critical discussions which now arose, there also arose, for the first time, something like *systematic* observation. The man to whom a myth was handed on, together with the silent but traditional request, 'What have you to say about it? Can you criticize it?'—this man would take the myth and would apply it to the various things which it was supposed to explain, such as the movement of the planets. Then he would say: 'I do not think that this myth is very good, because it does not explain the actual observable movement of the planets', or whatever it might be. Thus it is the myth or the theory which leads to, and guides, our systematic observations—observations undertaken with the intention of probing into the truth of the theory or the myth. From this point of view the growth of the theories of science should not be considered as the result of the collection, or accumulation, of observations; on the contrary, the observations and their accumulation should be considered as the result of the growth of the scientific theories. (This is what I have called the '*searchlight theory of science*'—the view that science itself throws new light on things; that it not only solves problems, but that, in doing so, it creates many more; and that it not only profits from observations, but leads to new ones.) If in this way we look out for new observations with the intention of probing into the truth of our myths, we need not be astonished if we find that myths handled in this rough manner change their character, and that in time they become what one might call more realistic or that they agree better with observable facts. In other words, under the pressure of criticism the myths are forced to adapt themselves to the task of giving us an adequate and a more detailed picture of the world in which we live. This explains why scientific myths, under the pressure of criticism, become so different from religious myths. I think, however, we should be quite clear that in their origin they remain myths or inventions, just like the others. They are not what some rationalists—the adherents of the sense-observation theory—believe: they are not digests of observations. Let me repeat this important point. Scientific theories are not just the results of observation. They are, in the main, the products of myth-making and of tests. Tests proceed partly by way of observation, and observation is thus very important; but its function is not that of producing theories. It plays its role in rejecting, eliminating, and criticizing theories; and it challenges us to produce new myths, new theories which may stand up to these observational tests. Only if we understand this can we understand the importance of tradition for science.

Those among you who hold the opposite view and who believe that scientific theories are the result of observations, I challenge to start observing here and now and to give me the scientific results of your observations. You may say that this is unfair, and that there is nothing very remarkable to observe here and now. But even if you go on to the end of your lives, notebook in hand, writing down everything you observe, and if you finally bequeath this important notebook to the Royal Society, asking them to make science out of it, then the Royal Society might preserve it as a curiosity, but decidedly not as a source of knowledge.[4] It might be lost perhaps in some cellar of the British Museum (which as you may know cannot afford to catalogue most of its treasures) but more likely it will end up on a rubbish heap.

But you may get something of scientific interest if you say: 'Here are the theories which some scientists hold today. These theories demand that such and such things should be observable under such and such conditions. Let us see whether they are observable.' In other words, if you select your observations with an eye on scientific problems and the general situation of science as it appears at the moment, then you may well be able to make a contribution to science. I do not want to be dogmatic and to deny that there are exceptions, such as the so-called chance discoveries. (Though even these very often turn out to be made under the influence of theories.) I do not say that observations are always insignificant unless they are related to theories, but I want

[4] See ch. 1, section IV.

to point out what is the main procedure in the development of science.

All this means that a young scientist who hopes to make discoveries is badly advised if his teacher tells him, 'Go round and observe', and that he is well advised if his teacher tells him: 'Try to learn what people are discussing nowadays in science. Find out where difficulties arise, and take an interest in disagreements. These are the questions which you should take up.' In other words, you should study the *problem situation* of the day. This means that you pick up, and try to continue, a line of inquiry which has the whole background of the earlier development of science behind it; you fall in with the tradition of science. It is a very simple and a decisive point, but nevertheless one that is often not sufficiently realized by rationalists—that we cannot start afresh; that we must make use of what people before us have done in science. If we start afresh, then, when we die, we shall be about as far as Adam and Eve were when they died (or, if you prefer, as far as Neanderthal man). In science we want to make progress, and this means that we must stand on the shoulders of our predecessors. We must carry on a certain tradition. From the point of view of what we want as scientists—understanding, prediction, analysis, and so on—the world in which we live is extremely complex. I should be tempted to say that it is infinitely complex, if the phrase had any meaning. We do not know where or how to start our analysis of this world. There is no wisdom to tell us. Even the scientific tradition does not tell us. It only tells us where and how other people started and where they got to. It tells us that people have already constructed in this world a kind of theoretical framework—not perhaps a very good one, but one which works more or less; it serves us as a kind of network, as a system of co-ordinates to which we can refer the various complexities of this world. We use it by checking it over, and by criticizing it. In this way we make progress.

It is necessary for us to see that of the two main ways in which we may explain the growth of science, one is rather unimportant and the other is important. The first explains science by the accumulation of knowledge: it is like a growing library (or a museum). As more and more books accumulate, so more and more knowledge accumulates. The other explains it by criticism: it grows by a method more revolutionary than accumulation—by a method which destroys, changes, and alters, the whole thing, including its most important instrument, the language in which our myths and theories are formulated.

It is interesting to see that the first method, the accumulation method, is much less important than people believe. There is much less accumulation of knowledge in science than there is revolutionary changing of scientific theories. It is a strange point, and a very interesting point, because one might at first sight believe that for the accumulative growth of knowledge tradition would be very important, and that for the revolutionary kind of development tradition would be less important. But it is exactly the other way round. If science could grow by mere accumulation, it would not matter so much if the scientific tradition were lost, because any day you could start accumulating afresh. Something would be lost, but the loss would not be serious. If, however, science advances by the tradition of changing its traditional myths, then you need something with which to start. If you have nothing to alter and to change, you can never get anywhere. Thus you need *two* beginnings for science: new myths, and a new tradition of changing them critically.[4a] But such beginnings are very rarely made. It took I do not know how many years from the invention of a descriptive language—which, we may say, was the moment when man became man—to the beginnings of science. Throughout this time language, the future instrument of science, was growing. It grew together with the growth of myth—every language incorporates and preserves countless myths and theories, even in its grammatical structure—and with the growth of the tradition which uses language for the purpose of describing facts, and for explaining and arguing about facts. (About this more later.) If these traditions were destroyed you could not even start accumulating; the instrument for it would be missing.

Having given this example of the role played by tradition in one particular field—that of science—I shall now, somewhat belatedly, proceed to the problem of a sociological theory of tradition. I again refer to Dr J. A. C. Brown, my predecessor today, who said many things which are very relevant to my topic, and especially one thing of which I have made a note. He said that if there is no discipline in a factory,

[4a] The change, though revolutionary, will have to preserve the successes and explain the failures of the preceding theory; see my *L.Sc.D.*, pp. 253 and 276 f.

then 'the workers become anxious and terrified'. Now I do not want to discuss discipline here; that is not my point. But I can put my point in this way: if they have nothing to go by, the workers become anxious and terrified. Or to put it in another and more general way: whenever we happen to be surrounded by either a natural environment or a social environment of which we know so little that we cannot predict what will happen, then we all become anxious and terrified. This is because if there is no possibility of our predicting what will happen in our environment—for example, how people will behave—then there is no possibility of reacting rationally. Whether the environment in question is a natural or a social one is more or less irrelevant.

Discipline (which was mentioned by Dr Brown) may be one of the things which help people to find their way in a certain society, but I am quite sure that Dr Brown will agree that it is only one of those things, and that there are other things, especially institutions and traditions, which may give people a clear idea of what to expect and how to proceed. This is important. What we call social life can exist only if we can know, and can have confidence, that there are things and events which must be so and cannot be otherwise.

It is here that the part played by tradition in our lives becomes understandable. We should be anxious, terrified, and frustrated, and we could not live in the social world, did it not contain a considerable amount of order, a great number of regularities to which we can adjust ourselves. The mere existence of these regularities is perhaps more important than their peculiar merits or demerits. They are needed as regularities, and therefore handed on as traditions, whether or not they are in other respects rational or necessary or good or beautiful or what you will. There is a need for tradition in social life.

Thus the creation of traditions plays a role similar to that of theories. Our scientific theories are instruments by which we try to bring some order into the chaos in which we live so as to make it rationally predictable. I do not want you to take this as a deep philosophical pronouncement. It is just a statement of one of the practical functions of our theories. Similarly, the creation of traditions, like so much of our legislation, has just that same function of bringing some order and rational predictability into the social world in which we live. It is not possible for you to act rationally in the world if you have no idea how it will respond to your actions. Every rational action assumes a certain system of reference which responds in a predictable or partly predictable way. Just as the invention of myths or theories in the field of natural science has a function—that of helping us to bring order into the events of nature—so has the creation of traditions in the field of society.

The analogy between the role of myths or theories in science and the role of traditions in society goes further. We must remember that the great significance of myths in scientific method was that they could become the objects of criticism, and that they could be changed. Similarly traditions have the important double function of not only creating a certain order or something like a social structure, but also giving us something upon which we can operate; something that we can criticize and change. This point is decisive for us, as rationalists and as social reformers. We always have new ideas for a better social world, and these new ideas have a significant function. But too many social reformers have an idea that they would like to clean the canvas, as Plato called it, of the social world, wiping off everything and starting from scratch with a brand-new rational world. This idea is nonsense and impossible to realize. If you construct a rational world afresh there is no reason to believe that it will be a happy world. There is no reason to believe that the blue-printed world will be any better than the world in which we live. Why should it be any better? An engineer does not create a motor-engine just from the blue-prints. He develops it from earlier models; he changes it; he alters it over and over again. If we wipe out the social world in which we live, wipe out its traditions and create a new world on the basis of blue-prints, then we shall very soon have to alter the new world, making little changes and adjustments. But if we are to make these little changes and adjustments, which will be needed in any case, why not start them here and now in the social world we have? It does not matter what you have and where you start. You must always make little adjustments. Since you will always have to make them, it is very much more sensible and reasonable to start with what happens to exist at the moment, because of these things which exist we at least know where the shoe pinches. We at least know of certain things that they are bad and that we want them changed. If we make our wonderful brave new world it will be quite a time before we find out what is wrong with it. Moreover the idea of canvas-cleaning (which is part of the wrong rationalist tradition) is impossible, because if the rationalist cleans the social canvas and wipes out the tradition he necessarily sweeps away with it himself and all his ideas and all his blue-prints of the future. The blue-prints have no meaning in an empty social world, in a social vacuum. They have no meaning except in a setting of traditions and institutions—such as myths, poetry, and values—which all emerge from the social world in which we live. Outside it they have no meaning at all. Therefore the very incentive and the very desire to build a new world must disappear once we have destroyed the traditions of the old world. In science it would be a tremendous loss if we were to say: 'We are not making very much progress. Let us sweep away all science and start afresh.' The rational procedure is to correct it and to revolutionize it, but not to sweep it away. You may create a new theory, but the new theory is created in order to solve those problems which the old theory did not solve. (See note 4a above.)

We have briefly examined the function of tradition in social life. What we found may now help us to answer the question how traditions arise, how they are handed on, and how they may become stereotyped—all these being unintended consequences of human actions. We can now understand why people not only try to learn the laws of their natural environment (and to teach them to others, often in the form of myth), but why they also try to learn the traditions of their social environment. We can now understand why people (especially primitive peoples and children) are inclined to cling to anything that may be or become a uniformity in their lives. They cling to myths; and they tend to cling to uniformities in their own behaviour, first, because they are afraid of irregularity and change and therefore afraid to originate irregularity and change; and secondly, because they wish to reassure others of their rationality or predictability, perhaps in the hope of making them act in a similar way. Thus they tend both to create traditions and to reaffirm those they find, by carefully conforming to them and by anxiously insisting that others conform to them also. This is how traditional taboos arise and how they are handed on.

This partly explains the strongly emotional intolerance which is characteristic of all traditionalism, an intolerance against which rationalists have always and rightly stood out. But we now see clearly that those rationalists who, because of this tendency, were led on to attack traditions as such, were mistaken. We can now say, perhaps, that what they really wanted was to replace the intolerance of the traditionalists by a new tradition—the tradition of tolerance; and, more generally, to replace the attitude of tabooism by one that considers existing traditions critically, weighing their merits against their demerits, and never forgetting the merit which lies in the fact that they are established traditions. For even if we ultimately reject them, in order to replace them by better ones (or by what we believe to be better ones), we should always remain conscious of the fact that all social criticism, and all social betterment, must refer to a framework of social traditions, of which some are criticized with the help of others, just as all progress in science must proceed within a framework of scientific theories, some of which are criticized in the light of others.

Much of what has been said here of traditions can also be said of institutions, for traditions and institutions are in most respects strikingly similar. Nevertheless, it seems desirable (although perhaps not very important) to preserve the difference which can be found in the ordinary usage of these two words, and I shall end my talk by trying to bring out the similarities and differences between these two kinds of social entity. I do not think it is a good practice to distinguish the terms 'tradition' and 'institution' by formal definitions,[5] but their use may be explained with the help of examples. In fact I have done this already, since I have mentioned schools, a police force, a grocer's shop, and the Stock Exchange as examples of social institutions, and elsewhere such things as the burning interest in scientific research, or the scientist's critical attitude, or the attitude of tolerance, or the intolerance of the traditionalist—or for that matter, of the rationalist—as examples of traditions. Institutions and traditions have much in common; among other things that they must be analysed by the social sciences in terms of individual persons, their actions, attitudes, beliefs, expectations, and interrelations. But we may say, perhaps, that we are inclined to speak of institutions wherever a (changing) body of people observe a certain set of norms or fulfil certain *prima facie* social functions (such as teaching,

[5] For a criticism of this practice, cf. ch. 11 of my book *The Open Society and its Enemies*.

policing, or selling groceries) which serve certain *prima facie* social purposes (such as the propagation of knowledge, or protection from violence or starvation), while we speak of traditions mainly when we wish to describe a uniformity of people's attitudes, or ways of behaviour, or aims or values, or tastes. Thus traditions are perhaps more closely bound up with persons and their likes and dislikes, their hopes and fears, than are institutions. They take, as it were, an intermediate place, in social theory, between persons and institutions. (We speak more naturally of a 'living tradition' than of a 'living institution'.)

The difference in question may be made clearer by reference to what I have sometimes called the 'ambivalence of social institutions', or the fact that a social institution may, in certain circumstances, function in a way which strikingly contrasts with its *prima facie* or 'proper' function. Of the perversion of boarding-schools from their 'proper' function Dickens had much to say; and it has happened that a police force, instead of protecting people from violence and blackmail, has used threats of violence or of imprisonment in order to blackmail them. Similarly the institution of a parliamentary Opposition, one of whose *prima facie* functions is to prevent the government from stealing the taxpayer's money, has worked in certain countries in a different way—by becoming an instrument for the proportional division of the spoils. The ambivalence of social institutions is connected with their character—with the fact that they perform certain *prima facie* functions and with the fact that institutions can be controlled only by persons (who are fallible) or by other institutions (which are therefore fallible also). The ambivalence can undoubtedly be much reduced by carefully constructed institutional checks, but it is impossible to eliminate it completely. The working of institutions, as of fortresses, depends ultimately upon the persons who man them; and the best that can be done by way of institutional control is to give a superior chance to those persons (if there are any) who intend to use the institutions for their 'proper' social purpose.

It is here that traditions may play an important role as intermediaries between persons and institutions. Traditions, to be sure, may also be perverted; something corresponding to the ambivalence here described affects them too. But since their character is somewhat less instrumental than that of institutions, they are less affected by this ambivalence. Institutions may also be less personal than traditions, which, in turn, are less personal and more predictable than the individuals who man the institutions. It may be said, perhaps, that the long-term 'proper' functioning of institutions depends mainly upon such traditions. It is tradition which gives the persons (who come and go) that background and that certainty of purpose which resist corruption. A tradition is, as it were, capable of extending something of the personal attitude of its founder far beyond his personal life.

From the point of view of the most typical usages of the two terms, it may be said that one of the connotations of the term 'tradition' is an allusion to *imitation*, as being either the origin of the tradition in question, or the way it is handed on. This connotation is, I think, absent from the term 'institution': an institution may or may not have its origin in imitation, and it may, or may not, continue its existence through imitation. Moreover, some of the things we call traditions may also be described as institutions—especially as institutions of that (sub-) society in which the tradition is generally followed. Thus we might say that the rationalist tradition, or the adoption of a critical attitude, is institutional within the (sub-) society of scientific workers (or that the tradition of not kicking a man when he is down is—almost—a British institution). Similarly we may say that the English language, though handed down by tradition, is an institution, while the practice of, say, avoiding split infinitives is a tradition (though it may be institutional within a certain group).

Some of these points may be further exemplified by considering certain aspects of the social institution of language. The main function of a language, communication, has been analysed by K. Bühler into three sub-functions: (1) the expressive function—i.e. the communication serves to express the emotions or thoughts of the speaker; (2) the signalling or stimulative or release function—i.e. the communication serves to stimulate or to release certain reactions in the hearer (for example, linguistic responses); and (3) the descriptive function—i.e. the communication describes a certain state of affairs. These three functions are separable in so far as each is accompanied as a rule by its preceding one but need not be accompanied by its succeeding one. The first two apply also to animal languages, while the third appears to be characteristically human. It is possible (and I believe necessary) to add a fourth to these three functions of Bühler's, and one which is particularly important from our point of view, viz. (4) the argumentative or explanatory function—i.e. the presentation and comparison of arguments or explanations in connection with certain definite questions or problems.[6] A certain language may possess the first three functions without the fourth (for example[7] that of a child at the stage when it just 'names' things). Now, in so far as language *qua* institution has these functions, it may be ambivalent. For example, it may be used by the speaker to hide his emotions or thoughts as much as to express them, or to repress rather than to stimulate argument. And there are different traditions connected with each one of these functions. For example, the different traditions of Italy and of England (where we have the tradition of under-statement) in connection with the expressive function of the respective languages are very striking. But all this becomes really important in connection with the two characteristically human functions of language—the descriptive and argumentative functions. In its descriptive function, we may speak about language as a vehicle of truth; but it may of course also become a vehicle of falsity. Without a tradition which works *against* this ambivalence and in favour of the use of language for the purpose of *correct* description (at least in all cases where there is no strong inducement to lie), the descriptive function of language would perish; for children would then never learn its descriptive use. Even more precious perhaps is the tradition that works against the ambivalence connected with the argumentative function of language, the tradition that works against that misuse of language which consists in pseudo-arguments and propaganda. This is the tradition and discipline of clear speaking and clear thinking; it is the critical tradition—the tradition of reason.

The modern enemies of reason want to destroy this tradition. They want to do this by destroying and perverting the argumentative and

[6] Compare also ch. 12, below. The reasons why I consider the argumentative and the explanatory functions as identical cannot be discussed here; they are derived from a logical analysis of *explanation and its relation to deduction* (*or argument*).

[7] An ordinary map is also an example of a description which is not argumentative; although it may of course be used to support an argument within an argumentative language.

perhaps even the descriptive functions of the human language; by a romantic reversion to its emotive functions—the expressive (there is too much talk about 'self-expression') and, perhaps, the signalling or stimulative function. We see this tendency very clearly at work in certain types of modern poetry, prose, and philosophy—in a philosophy which does not argue because it has no arguable problems. The new enemies of reason are sometimes anti-traditionalists who seek new and impressive means of self-expression or of 'communication', and sometimes traditionalists who extol the wisdom of the linguistic tradition. Both uphold implicitly a theory of language that sees no more than the first or perhaps the second of its functions, while in their practice they support the flight from reason and from the great tradition of intellectual responsibility.

5

BACK TO THE PRESOCRATICS

I

'Back to Methuselah' was a progressive programme, compared with 'Back to Thales' or 'Back to Anaximander': what Shaw offered us was an improved expectation of life—something that was in the air, at any rate when he wrote it. I have nothing to offer you, I am afraid, that is in the air today; for what I want to return to is the simple straightforward *rationality* of the Presocratics. Wherein does this much discussed 'rationality' of the Presocratics lie? The simplicity and boldness of their questions is part of it, but my thesis is that the decisive point is the critical attitude which, as I shall try to show, was first developed in the Ionian School.

The questions which the Presocratics tried to answer were primarily cosmological questions, but there were also questions of the theory of knowledge. It is my belief that philosophy must return to cosmology and to a simple theory of knowledge. There is at least one philosophical problem in which all thinking men are interested: the problem of understanding the world in which we live; and thus ourselves (who are

The Presidential Address, delivered before the meeting of the Aristotelian Society on 13th October 1958; first published in the Proceedings of the Aristotelian Society, *N.S.* **59**, *1958–9. The footnotes (and the Appendix) have been added in the present reprint of the address.*

part of that world) and the way we acquire our knowledge of it. All science is cosmology, I believe, and for me the interest of philosophy, no less than of science, lies solely in its bold attempt to add to our knowledge of the world, and to the theory of our knowledge of the world. I am interested in Wittgenstein, for example, not because of his linguistic philosophy, but because his *Tractatus* was a cosmological treatise (although a crude one), and because his theory of knowledge was closely linked with his cosmology.

For me, both philosophy and science lose all their attraction when they give up that pursuit—when they become specialisms and cease to see, and to wonder at, the riddles of our world. Specialization may be a great temptation for the scientist. For the philosopher it is the mortal sin.

II

In this paper I speak as an amateur, as a lover of the beautiful story of the Presocratics. I am not a specialist or an expert: I am completely out of my depth when an expert begins to argue which words or phrases Heraclitus might, and which he could not possibly, have used. Yet when some expert replaces a beautiful story, based on the oldest texts we possess, by one which—to me at any rate—no longer makes any sense, then I feel that even an amateur may stand up and defend an old tradition. Thus I will at least look into the expert's arguments, and examine their consistency. This seems a harmless occupation to indulge in; and if an expert or anybody else should take the trouble to refute my criticism I shall be pleased and honoured.[1]

I shall be concerned with the cosmological theories of the Presocratics, but only to the extent to which they bear upon the development of *the problem of change*, as I call it, and only to the extent to which they are needed for understanding the approach of the Presocratic philosophers to the problem of knowledge—their practical as well as their theoretical approach. For it is of considerable interest to see how their practice as well as their theory of knowledge is connected with the

[1] I am glad to be able to report that Mr G. S. Kirk has indeed replied to my address; see below, notes 4 and 5, and the *Appendix* to this paper.

cosmological and theological questions which they posed to themselves. Theirs was not a theory of knowledge that began with the question, 'How do I know that this is an orange?' or, 'How do I know that the object I am now perceiving is an orange?' Their theory of knowledge started from problems such as, 'How do we know that the world is made of water?' or, 'How do we know that the world is full of gods?' or, 'How can we know anything about the gods?'

There is a widespread belief, somewhat remotely due, I think, to the influence of Francis Bacon, that one should study the problems of the theory of knowledge in connection with our knowledge of an orange rather than our knowledge of the cosmos. I dissent from this belief, and it is one of the main purposes of my paper to convey to you some of my reasons for dissenting. At any rate it is good to remember from time to time that our Western science—and there seems to be no other—did not start with collecting observations of oranges, but with bold theories about the world.

III

Traditional empiricist epistemology and the traditional historiography of science are both deeply influenced by the Baconian myth that all science starts from observation and then slowly and cautiously proceeds to theories. That the facts are very different can be learned from studying the early Presocratics. Here we find bold and fascinating ideas, some of which are strange and even staggering anticipations of modern results, while many others are wide of the mark, from our modern point of view; but most of them, and the best of them, have nothing to do with observation. Take for example some of the theories about the shape and position of the earth. Thales said, we are told, 'that the earth is supported by water on which it rides like a ship, and when we say that there is an earthquake, then the earth is being shaken by the movement of the water'. No doubt Thales had observed earthquakes as well as the rolling of a ship before he arrived at his theory. But the point of his theory was to *explain* the support or suspension of the earth, and also earthquakes, by the conjecture that the earth floats on water; and for this conjecture (which so strangely anticipates the modern theory of continental drift) he could have had no basis in his observations.

We must not forget that the function of the Baconian myth is to explain why scientific statements are *true*, by pointing out that observation is the '*true source*' of our scientific knowledge. Once we realize that all scientific statements are hypotheses, or guesses, or conjectures, and that the vast majority of these conjectures (including Bacon's own) have turned out to be false, the Baconian myth becomes irrelevant. For it is pointless to argue that the conjectures of science—those which have proved to be false as well as those which are still accepted—all start from observation.

However this may be, Thales' beautiful theory of the support or suspension of the earth and of earthquakes, though in no sense based upon observation, is at least inspired by an empirical or observational analogy. But even this is no longer true of the theory proposed by Thales' great pupil, Anaximander. Anaximander's theory of the suspension of the earth is still highly intuitive, but it no longer uses observational analogies. In fact it may be described as counter-observational. According to Anaximander's theory, 'The earth . . . is held up by nothing, but remains stationary owing to the fact that it is equally distant from all other things. Its shape is . . . like that of a drum. . . . We walk on one of its flat surfaces, while the other is on the opposite side.' The drum, of course, is an observational analogy. But the idea of the earth's free suspension in space, and the explanation of its stability, have no analogy whatever in the whole field of observable facts.

In my opinion this idea of Anaximander's is one of the boldest, most revolutionary, and most portentous ideas in the whole history of human thought. It made possible the theories of Aristarchus and of Copernicus. But the step taken by Anaximander was even more difficult and audacious than the one taken by Aristarchus and Copernicus. To envisage the earth as freely poised in mid-space, and to say 'that it remains motionless because of its equidistance or equilibrium' (as Aristotle paraphrases Anaximander), is to anticipate to some extent even Newton's idea of immaterial and invisible gravitational forces.[2]

[2] Aristotle himself understood Anaximander in this way; for he caricatures Anaximander's 'ingenious but untrue' theory by comparing the situation of its earth to that of a man who, being equally hungry and thirsty yet equidistant from food and drink, is unable to move. (*De Caelo*, 295b32. The idea has become known by the name 'Buridan's ass'.) Clearly Aristotle conceives this man as being held in equilibrium by immaterial and

IV

How did Anaximander arrive at this remarkable theory? Certainly not by observation but by reasoning. His theory is an attempt to solve one of the problems to which his teacher and kinsman Thales, the founder of the Milesian or Ionian School, had offered a solution before him. I therefore conjecture that Anaximander arrived at his theory by criticizing Thales' theory. This conjecture can be supported, I believe, by a consideration of the structure of Anaximander's theory.

Anaximander is likely to have argued against Thales' theory (according to which the earth was floating on water) on the following lines. Thales' theory is a specimen of a type of theory which if consistently developed would lead to an infinite regress. If we explain the stable position of the earth by the assumption that it is supported by water—that it is floating on the ocean (*Okeanos*)—should we not have to explain the stable position of the ocean by an analogous hypothesis? But this would mean looking for a support for the ocean, and then for a support for this support. This method of explanation is unsatisfactory: first, because we solve our problem by creating an exactly analogous one; and also for the less formal and more intuitive reason that in any such system of supports or props failure to secure any one of the lower props must lead to the collapse of the whole edifice.

From this we see intuitively that the stability of the world cannot be secured by a system of supports or props. Instead Anaximander appeals to the internal or structural symmetry of the world, which ensures that there is no preferred direction in which a collapse can take place. He applies the principle that where there are no differences there can be no change. In this way he explains the stability of the earth by the equality of its distances from all other things.

This, it seems, was Anaximander's argument. It is important to realize that it abolishes, even though not quite consciously perhaps, and not quite consistently, the idea of an absolute direction—the absolute sense of 'upwards' and 'downwards'. This is not only contrary to all experience but notoriously difficult to grasp. Anaximenes ignored it, it

invisible attractive forces similar to Newtonian forces; and it is interesting that this 'animistic' or 'occult' character of his forces was deeply (though mistakenly) felt by Newton himself, and by his opponents, such as Berkeley, to be a blot on his theory.

seems, and even Anaximander himself did not grasp it completely. For the idea of an equal distance from all other things should have led him to the theory that the earth has the shape of a globe. Instead he believed that it had the shape of a drum, with an upper and a lower flat surface. Yet it looks as if the remark, 'We walk on one of its flat surfaces, while the other is on the opposite side', contained a hint that there was no absolute upper surface, but that on the contrary the surface on which we happened to walk was the one we might *call* the upper.

What prevented Anaximander from arriving at the theory that the earth was a globe rather than a drum? There can be little doubt: it was *observational experience* which taught him that the surface of the earth was, by and large, flat. Thus it was a speculative and critical argument, the abstract critical discussion of Thales' theory, which almost led him to the true theory of the shape of the earth; and it was observational experience which led him astray.

V

There is an obvious objection to Anaximander's theory of symmetry, according to which the earth is equally distant from all other things. The asymmetry of the universe can be easily seen from the existence of sun and moon, and especially from the fact that sun and moon are sometimes not far distant from each other, so that they are on the same side of the earth, while there is nothing on the other side to balance them. It appears that Anaximander met this objection by another bold theory—his theory of the hidden nature of the sun, the moon, and the other heavenly bodies.

He envisages the rims of two huge chariot wheels rotating round the earth, one 27 times the size of the earth, the other 18 times its size. Each of these rims or circular pipes is filled with fire, and each has a breathing-hole through which the fire is visible. These holes we call the sun and the moon respectively. The rest of the wheel is invisible, presumably because it is dark (or misty) and far away. The fixed stars (and presumably the planets) are also holes on wheels which are nearer to the earth than the wheels of the sun and the moon. The wheels of the fixed stars rotate on a common axis (which we now call the axis of the earth) and together they form a sphere round the earth, so the postulate of equal distance from the earth is (roughly) satisfied. This makes Anaximander also a founder of the *theory of the spheres*. (For its relation to the wheels or circles see Aristotle, *De Caelo*, 289b10 to 290b10.)

VI

There can be no doubt whatever that Anaximander's theories are critical and speculative rather than empirical: and considered as approaches to truth his critical and abstract speculations served him better than observational experience or analogy.

But, a follower of Bacon may reply, this is precisely why Anaximander was not a scientist. This is precisely why we speak of early Greek *philosophy* rather than of early Greek *science*. Philosophy is speculative: everybody knows this. And as everybody knows, science begins only when the speculative method is replaced by the observational method, and when deduction is replaced by induction.

This reply, of course, amounts to the thesis that, by definition, theories are (or are not) *scientific* according to their origin in observations, or in so-called 'inductive procedures'. Yet I believe that few, if any, physical theories would fall under this definition. And I do not see why the question of origin should be important in this connection. What is important about a theory is its explanatory power, and whether it stands up to criticism and to tests. The question of its origin, of how it is arrived at—whether by an 'inductive procedure', as some say, or by an act of intuition—may be extremely interesting, especially for the biographer of the man who invented the theory, but it has little to do with its scientific status or character.

VII

As to the Presocratics, I assert that there is the most perfect possible continuity of thought between their theories and the later developments in physics. Whether they are called philosophers, or pre-scientists, or scientists, matters very little, I think. But I do assert that Anaximander's theory cleared the way for the theories of Aristarchus, Copernicus, Kepler, and Galileo. It is not that he merely 'influenced' these later thinkers; 'influence' is a very superficial category. I would rather put it like this: Anaximander's achievement is valuable in itself, like a work of art. Besides, his achievement made other achievements possible, among them those of the great scientists mentioned.

But are not Anaximander's theories false, and therefore non-scientific? They are false, I admit; but so are many theories, 'based' upon countless experiments, which modern science accepted until recently, and whose scientific character nobody would dream of denying, even though they are now believed to be false. (An example is the theory that the typical chemical properties of hydrogen belong only to one kind of atom—the lightest of all atoms.) There were historians of science who tended to regard as unscientific (or even as superstitious) any view no longer accepted at the time they were writing; but this is an untenable attitude. A false theory may be as great an achievement as a true one. And many false theories have been more helpful in our search for truth than some less interesting theories which are still accepted. For false theories can be helpful in many ways; they may for example suggest some more or less radical modifications, and they may stimulate criticism. Thus Thales' theory that the earth floats on water reappeared in a modified form in Anaximenes, and in more recent times in the form of Wegener's theory of continental drift. How Thales' theory stimulated Anaximander's criticism has been shown already.

Anaximander's theory, similarly, suggested a modified theory—the theory of an earth globe, freely poised in the centre of the universe, and surrounded by spheres on which heavenly bodies were mounted. And by stimulating criticism it also led to the theory that the moon shines by reflecting light; to the Pythagorean theory of a central fire; and ultimately to the heliocentric world-system of Aristarchus and Copernicus.

VIII

I believe that the Milesians, like their oriental predecessors who took the world for a tent, envisaged the world as a kind of house, the home of all creatures—our home. Thus there was no need to ask what it was for. But there was a real need to inquire into its architecture. The questions of its structure, its ground-plan, and its building material, constitute the three main problems of Milesian cosmology. There is also a speculative interest in its origin, the question of cosmogony. It seems to me that the cosmological interest of the Milesians far exceeded their cosmogonical interest, especially if we consider the strong cosmogonical tradition, and the almost irresistible tendency to describe a thing by describing how it has been made, and thus to present a cosmological account in a cosmogonical form. The cosmological interest must be very strong, as compared with the cosmogonical one, if the presentation of a cosmological theory is even partially free from these cosmogonical trappings.

I believe that it was Thales who first discussed the architecture of the cosmos—its structure, ground-plan, and building material. In Anaximander we find answers to all three questions. I have briefly mentioned his answer to the question of structure. As to the question of the ground-plan of the world, he studied and expounded this too, as indicated by the tradition that he drew the first map of the world. And of course he had a theory about its building material—the 'endless' or 'boundless' or 'unbounded' or 'unformed'—the '*apeiron*'.

In Anaximander's world all kinds of *changes* were going on. There was a fire which needed air and breathing-holes, and these were at times blocked up ('obstructed'), so that the fire was smothered:[3] this was his theory of eclipses, and of the phases of the moon. There were winds, which were responsible for the changing weather.[4] And there were the

[3] I do not suggest that the smothering is due to blocking breathing-in holes: according to the phlogiston theory, for example, fire is smothered by obstructing breathing-out holes. But I do not wish to ascribe to Anaximander either a phlogiston theory of combustion, or an anticipation of Lavoisier.

[4] In my address, as it was originally published, I continued here 'and indeed for all other changes within the cosmic edifice', relying on Zeller who wrote (appealing to the testimony of Aristotle's *Meteor.* 353b6): 'Anaximander, it seems, explained the motion of the heavenly bodies by the currents of the air which are responsible for the turning of the stellar spheres.' (*Phil. d. Griechen*, 5th edn., vol. 1, 1892, p. 223; see also p. 220, n. 2; Heath, *Aristarchus*, 1913, p. 33; and Lee's edition of the *Meteorologica*, 1952, p. 125.) But I should perhaps not have interpreted Zeller's 'currents of air' as 'winds', especially as Zeller should have said 'vapours' (they are evaporations resulting from a process of drying up). I have twice inserted 'vapours and' before 'winds', and 'almost' before 'all' in the second paragraph of section ix; and I have replaced, in the third paragraph of section ix, 'winds' by 'vapours'. I have made these changes in the hope of meeting Mr G. S. Kirk's criticism on p. 332 of his article (discussed in the appendix to the present chapter).

vapours, resulting from the drying up of water and air, which were the cause of the winds and of the 'turnings' of the sun (the solstices) and of the moon.

We have here the first hint of what was soon to come: of the *general problem of change*, which became the central problem of Greek cosmology, and which ultimately led, with Leucippus and Democritus, to *a general theory of change* that was accepted by modern science almost up to the beginning of the twentieth century. (It was given up only with the breakdown of Maxwell's models of the ether, an historic event that was little noticed before 1905.)

This *general problem of change* is a philosophical problem; indeed in the hands of Parmenides and Zeno it almost turns into a logical one. *How is change possible*—logically possible, that is? How can a thing change, without losing its identity? If it remains the same, it does not change; yet if it loses its identity, then it is no longer that thing which has changed.

IX

The exciting story of the development of the problem of change appears to me in danger of being completely buried under the mounting heap of the minutiae of textual criticism. The story cannot, of course, be fully told in one short paper, and still less in one of its many sections. But in briefest outline, it is this.

For Anaximander, our own world, our own cosmic edifice, was only one of an infinity of worlds—an infinity without bounds in space and time. This system of worlds was eternal, and so was motion. There was thus no need to explain motion, no need to offer a *general* theory of change (in the sense in which we shall find a general problem and a general theory of change in Heraclitus; see below). But there was a need to explain the well-known changes occurring in our world. The most obvious changes—the change of day and night, of winds and of weather, of the seasons, from sowing to harvesting, and of the growth of plants and animals and men—all were connected with the contrast of temperatures, with the opposition between the hot and the cold, and with that between the dry and the wet. 'Living creatures came into being from moisture evaporated by the sun', we are told; and the hot and the cold also administer to the genesis of our own world edifice. The hot and the cold were also responsible for the vapours and winds which in their turn were conceived as the agents of almost all other changes.

Anaximenes, a pupil of Anaximander and his successor, developed these ideas in much detail. Like Anaximander he was interested in the oppositions of the hot and the cold and of the moist and the dry, and he explained the transitions between these opposites by a theory of condensation and rarefaction. Like Anaximander he believed in eternal motion and in the action of the winds; and it seems not unlikely that one of the two main points in which he deviated from Anaximander was reached by a criticism of the idea that what was completely boundless and formless (the *apeiron*) could yet be in motion. At any rate, he replaced the *apeiron* by air—something that was almost boundless and formless, and yet, according to Anaximander's old theory of vapours, not only capable of motion, but the main agent of motion and change. A similar unification of ideas was achieved by Anaximenes' theory that 'the sun consists of earth, and that it gets very hot owing to the rapidity of its motion'.

Anaximenes' replacement of the more abstract theory of the unbounded *apeiron* by the less abstract and more commonsense theory of air is matched by the replacement of Anaximander's bold theory of the stability of the earth by the more commonsense idea that the earth's 'flatness is responsible for its stability; for it . . . covers like a lid the air beneath it'. Thus the earth rides on air as the lid of a pot may ride on steam, or as a ship may ride on water; Thales' question and Thales' answer are both re-instituted, and Anaximander's epoch-making argument is not understood. Anaximenes is an eclectic, a systematizer, an empiricist, a man of common sense. Of the three great Milesians he is least productive of revolutionary new ideas; he is the least philosophically minded.

The three Milesians all looked on our world as our home. There was movement, there was change in this home, there were hot and cold, fire and moisture. There was a fire in the hearth, and on it a kettle with water. The house was exposed to the winds, and a bit draughty, to be sure; but it was home, and it meant security and stability of a sort. But for Heraclitus the house was on fire.

There was no stability left in the world of Heraclitus. 'Everything is in flux, and nothing is at rest.' *Everything* is in flux, even the beams, the timber, the building material of which the world is made: earth and rocks, or the bronze of a cauldron—they are all in flux. The beams are rotting, the earth is washed away and blown away, the very rocks split and wither, the bronze cauldron turns into green patina, or into verdigris: 'All things are in motion all the time, even though . . . this escapes our senses', as Aristotle expressed it. Those who do not know and do not think believe that only the fuel is burned, while the bowl in which it burns (cp. DK, A 4) remains unchanged; for we do not see the bowl burning. And yet it burns; it is eaten up by the fire it holds. We do not *see* our children grow up, and change, and grow old, but they do.

Thus there are no solid bodies. Things are not really things, they are processes, they are in flux. They are like fire, like a flame which, though it may have a definite shape, is a process, a stream of matter, a river. All things are flames: fire is the very building material of our world; and the apparent stability of things is merely due to the laws, the measures, which the processes in our world are subject to.

This, I believe, is Heraclitus' story; it is his 'message', the 'true word' (the *logos*), to which we ought to listen: 'Listening not to me but to the true account, it is wise to admit that all things are one': they are 'an everlasting fire, flaring up in measures, and dying down in measures'.

I know very well that the traditional interpretation of Heraclitus' philosophy here restated is not generally accepted at present. But the critics have put nothing in its place—nothing, that is, of philosophical interest. I shall briefly discuss their new interpretation in the next section. Here I wish only to stress that Heraclitus' philosophy, by appealing to thought, to the word, to argument, to reason, and by pointing out that we are living in a world of things whose changes escape our senses, though we *know* that they do change, created two new problems—*the problem of change* and *the problem of knowledge*. These problems were the more urgent as his own account of change was difficult to understand. But this, I believe, is due to the fact that he saw more clearly than his predecessors the difficulties that were involved in the very idea of change.

For all change is the change of something: change presupposes something that changes. And it presupposes that, while changing, this something must remain the same. We may say that a green leaf changes when it turns brown; but we do not say that the green leaf changes when we substitute for it a brown leaf. It is essential to the idea of change that the thing which changes retains its identity while changing. And yet it must become something else: it was green, and it becomes brown; it was moist, and it becomes dry; it was hot, and it becomes cold.

Thus every change is the transition of a thing into something with, in a way, opposite qualities (as Anaximander and Anaximenes had seen). And yet, while changing, the changing thing must remain identical with itself.

This is the problem of change. It led Heraclitus to a theory which (partly anticipating Parmenides) distinguishes between reality and appearance. 'The real nature of things loves to hide itself. An unapparent harmony is stronger than the apparent one.' Things are *in appearance* (and for us) opposites, but in truth (and for God) they are the same.

> Life and death, being awake and being asleep, youth and old age, all these are the same . . . for the one turned round is the other and the other turned round is the first. . . . The path that leads up and the path that leads down are the same path. . . . Good and bad are identical. . . . For God all things are beautiful and good and just, but men assume some things to be unjust, and others to be just. . . . It is not in the nature or character of man to possess true knowledge, though it is in the divine nature.

Thus in truth (and for God) the opposites are identical; it is only to man that they appear as non-identical. And all things are one—they are all part of the process of the world, the everlasting fire.

This theory of change appeals to the 'true word', to the *logos*, to reason; nothing is more real for Heraclitus than change. Yet his doctrine of the oneness of the world, of the identity of opposites, and of appearance and reality threatens his doctrine of the reality of change.

For change is the transition from one opposite to the other. Thus if in truth the opposites are identical, though they appear different, then change itself might be only apparent. If in truth, and for God, all things are one, there might, in truth, be no change.

This consequence was drawn by Parmenides, the pupil (*pace* Burnet and others) of the monotheist Xenophanes who said of the one God: 'He always remains in the same place, never moving. It is not fitting that He should go to different places at different times . . . He is in no way similar to mortal men, neither in body nor in thought.'

Xenophanes' pupil Parmenides taught that the real world was one, and that it always remained in the same place, never moving. It was not *fitting* that it should go to different places at different times. It was in no way similar to what it appeared to be to mortal men. The world was one, an undivided whole, without parts, homogeneous and motionless: motion was impossible in such a world. In truth there was no change. The world of change was an illusion.

Parmenides based this theory of an unchanging reality on something like a logical proof; a proof which can be presented as proceeding from the single premiss, 'What is not is not'. From this we can derive that the nothing—that which is not—does not exist; a result which Parmenides interprets to mean that the void does not exist. Thus the world is full: it consists of one undivided block, since any division into parts could only be due to separation of the parts by the void. (This is 'the well-rounded truth' which the goddess revealed to Parmenides.) In this full world there is no room for motion.

Only the delusive belief in the reality of opposites—the belief that not only *what is* exists but also *what is not*—leads to the illusion of a world of change.

Parmenides' theory may be described as the first hypothetico-deductive theory of the world. The atomists—Leucippus and Democritus—took it as such; and they asserted that it was refuted by experience, since motion does exist. Accepting the formal validity of Parmenides' argument, they inferred from the falsity of his conclusion the falsity of his premises. But this meant that the nothing—the void, or empty space—existed. Consequently there was now no need to assume that 'what is'—the full, that which fills some space—had no parts; for its parts could now be separated by the void. Thus there are many parts, each of which is 'full': there are full particles in the world, separated by empty space, and able to move in empty space, each of them being 'full', undivided, indivisible, and unchanging. Thus what exists is *atoms and the void*. In this way the atomists arrived at a *theory of change*—a theory that dominated scientific thought until 1900. It is the theory that *all change, and especially all qualitative change, has to be explained by the spatial movement of unchanging bits of matter—by atoms moving in the void.*

The next great step in our cosmology and the theory of change was made when Maxwell, developing certain ideas of Faraday's, replaced this theory by a theory of changing intensities of fields.

X

I have sketched the story, as I see it, of the Presocratic theory of change. I am of course well aware of the fact that my story (which is based on Plato, Aristotle, and the doxographic tradition) clashes at many points with the views of some experts, English as well as German, and especially with the views expressed by G. S. Kirk and J. E. Raven in their book, *The Presocratic Philosophers*, 1957. I cannot of course examine their arguments in detail here, and especially not their minute exegeses of various passages some of which are relevant to the differences between their interpretation and mine. (See, for example, Kirk and Raven's discussion of the question whether there is a reference to Heraclitus in Parmenides; cf. their note 1 on pp. 193 f., and note 1 on p. 272.) But I wish to say that I have examined their arguments and that I have found them unconvincing and often quite unacceptable.

I will mention here only some points regarding Heraclitus (although there are other points of equal importance, such as their comments on Parmenides).

The traditional view, according to which Heraclitus' central doctrine was that all things are in flux, was attacked forty years ago by Burnet. His main argument (discussed by me at length in note 2 to ch. 2 of my *Open Society*) was that the theory of change was not new, and that only a new message could explain the urgency with which Heraclitus speaks. This argument is repeated by Kirk and Raven when they write (pp. 186 f.): 'But all Presocratic thinkers were struck by the predominance of change in the world of our experience.' About this attitude I said in my *Open Society*: 'Those who suggest . . . that the doctrine of universal flux was not new . . . are, I feel, unconscious witnesses to Heraclitus' originality, for they fail now, after 2,400 years, to grasp his main point.' In brief, they do not see the difference between the Milesian message,

'There is a fire in the house', and Heraclitus' somewhat more urgent message, 'The house is on fire'. An implicit reply to this criticism can be found on p. 197 of the book by Kirk and Raven, where they write: 'Can Heraclitus really have thought that a rock or a bronze cauldron, for example, was invariably undergoing invisible changes of material? Perhaps so; but nothing in the extant fragments suggests that he did.' But is this so? Heraclitus' extant fragments about the fire (Kirk and Raven, fragm. 220–2) are interpreted by Kirk and Raven themselves as follows (p. 200): 'Fire is the archetypal form of matter.' Now I am not at all sure what 'archetypal' means here (especially in view of the fact that we read a few lines later, 'Cosmogony . . . is not to be found in Heraclitus'). But whatever 'archetypal' may mean, it is clear that once it is admitted that Heraclitus says in the extant fragments that all matter is somehow (whether archetypally or otherwise) fire, he also says that all matter, like fire, is a process; which is precisely the theory denied to Heraclitus by Kirk and Raven.

Immediately after saying that 'nothing in the extant fragments suggests' that Heraclitus believed in continuous invisible changes, Kirk and Raven make the following methodological remark: 'It cannot be too strongly emphasized that [in texts] before Parmenides and his apparent proof that the senses were completely fallacious . . . gross departures from common sense must only be accepted when the evidence for them is extremely strong.' This is intended to mean that the doctrine that bodies (of any substance) constantly undergo invisible changes represents a gross departure from common sense, a departure which one ought not to expect in Heraclitus.

But to quote Heraclitus: 'He who does not expect the unexpected will not detect it: for him it will remain undetectable, and unapproachable' (DK, B 18). In fact Kirk and Raven's last argument is invalid on many grounds. Long before Parmenides we find ideas far removed from common sense in Anaximander, Pythagoras, Xenophanes, and especially in Heraclitus. Indeed the suggestion that we should test the historicity of the ideas ascribed to Heraclitus—as we might indeed test the historicity of those ascribed to Anaximenes—by standards of 'common sense' is a little surprising (whatever 'common sense' may mean here). For this suggestion runs counter not only to Heraclitus' notorious obscurity and oracular style, confirmed by Kirk and Raven, but also to his burning interest in antinomy and paradox. And it runs counter, last but not least, to the (in my view quite absurd) doctrine which Kirk and Raven finally attribute to Heraclitus (the italics are mine): '. . . that natural changes of all kinds [and thus presumably also earthquakes and great fires] are regular *and balanced*, and *that the cause of this balance is fire, the common constituent of things that was also termed their Logos*.' But why, I ask, should fire be 'the cause' of any balance—either 'this balance' or any other? And where does Heraclitus say such things? Indeed, had this been Heraclitus' philosophy, then I could see no reason to take any interest in it; at any rate, it would be much further removed from common sense (as I see it) than the inspired philosophy which tradition ascribes to Heraclitus and which, in the name of common sense, is rejected by Kirk and Raven.

But the decisive point is, of course, that this inspired philosophy is *true*, for all we know.[5] With his uncanny intuition Heraclitus saw that things are processes, that our bodies are flames, that 'a rock or a bronze cauldron . . . was invariably undergoing invisible changes'. Kirk and Raven say (p. 197, note 1; the argument reads like an answer to Melissus): 'Every time the finger rubs, it rubs off an invisible portion of iron; yet when it does not rub, what reason is there to think that the iron is still changing?' The reason is that the wind rubs, and that there is always wind; or that iron turns invisibly into rust—by oxidation, and this means by slow burning; or that old iron looks different from new iron, just as an old man looks different from a child (cp. DK, B 88). This was Heraclitus' teaching, as the extant fragments show.

I suggest that Kirk and Raven's methodological principle 'that gross departures from common sense must only be accepted when the evidence for them is extremely strong' might well be replaced by the clearer and more important principle that *gross departures from the historical tradition must only be accepted when the evidence for them is extremely strong*. This, in fact, is a universal principle of historiography. Without it history

[5] This should establish that it makes sense, at any rate. I hope it is clear from the text that I appeal to truth here in order (*a*) to make clear that my interpretation at least makes sense, and (*b*) to refute the arguments of Kirk and Raven (discussed later in this paragraph) that the theory is absurd. An answer to G. S. Kirk which was too long to be appended here (although it refers to the present passage and to the present paragraph) will be found in the *Appendix* at the end of this paper.

would be impossible. Yet it is constantly violated by Kirk and Raven: when, for example, they try to make Plato's and Aristotle's evidence suspect, with arguments which are partly circular and partly (like the one from common sense) in contradiction to their own story. And when they say that 'little serious attempt seems to have been made by Plato and Aristotle to penetrate his [i.e. Heraclitus'] real meaning' then I can only say that the philosophy outlined by Plato and Aristotle seems to me a philosophy that has real meaning and real depth. It is a philosophy worthy of a great philosopher. Who, if not Heraclitus, was the great thinker who first realized that men are flames and that things are processes? Are we really to believe that this great philosophy was a 'post-Heraclitean exaggeration' (p. 197), and that it may have been suggested to Plato, 'in particular, perhaps, by Cratylus'? Who, I ask, was this unknown philosopher—perhaps the greatest and the boldest thinker among the Presocratics? Who was he, if not Heraclitus?

XI

The early history of Greek philosophy, especially the history from Thales to Plato, is a splendid story. It is almost too good to be true. In every generation we find at least one new philosophy, one new cosmology of staggering originality and depth. How was this possible? Of course one cannot explain originality and genius. But one can try to throw some light on them. What was the secret of the ancients? I suggest that it was a *tradition—the tradition of critical discussion*.

I will try to put the problem more sharply. In all or almost all civilizations we find something like religious and cosmological teaching, and in many societies we find schools. Now schools, especially primitive schools, all have, it appears, a characteristic structure and function. Far from being places of critical discussion they make it their task to impart a definite doctrine, and to preserve it, pure and unchanged. It is the task of a school to hand on the tradition, the doctrine of its founder, its first master, to the next generation, and to this end the most important thing is to keep the doctrine inviolate. A school of this kind never admits a new idea. New ideas are heresies, and lead to schisms; should a member of the school try to change the doctrine, then he is expelled as a heretic. But the heretic claims, as a rule, that his is the true doctrine of the founder. Thus not even the inventor admits that he has introduced an invention; he believes, rather, that he is returning to the true orthodoxy which has somehow been perverted.

In this way all changes of doctrine—if any—are surreptitious changes. They are all presented as re-statements of the true sayings of the master, of his own words, his own meaning, his own intentions.

It is clear that in a school of this kind we cannot expect to find a history of ideas, or even the material for such a history. For new ideas are not admitted to be new. Everything is ascribed to the master. All we might reconstruct is a history of schisms, and perhaps a history of the defence of certain doctrines against the heretics.

There cannot, of course, be any rational discussion in a school of this kind. There may be arguments against dissenters and heretics, or against some competing schools. But in the main it is with assertion and dogma and condemnation rather than argument that the doctrine is defended.

The great example of a school of this kind among the Greek philosophical schools is the Italian School founded by Pythagoras. Compared with the Ionian school, or with that of Elea, it had the character of a religious order, with a characteristic way of life and a secret doctrine. The story that a member, Hippasus of Metapontum, was drowned at sea because he revealed the secret of the irrationality of certain square roots, is characteristic of the atmosphere surrounding the Pythagorean school, whether or not there is any truth in this story.

But among Greek philosophic schools the early Pythagoreans were an exception. Leaving them aside, we could say that the character of Greek Philosophy, and of the philosophical schools, is strikingly different from the dogmatic type of school here described. I have shown this by an example: *the story of the problem of change which I have told is the story of a critical debate, of a rational discussion*. New ideas are propounded as such, and arise as the result of open criticism. There are few, if any, surreptitious changes. Instead of anonymity we find a history of ideas and of their originators.

Here is a unique phenomenon, and it is closely connected with the astonishing freedom and creativeness of Greek philosophy. How can we explain this phenomenon? *What we have to explain is the rise of a tradition.*

It is a tradition that allows or encourages critical discussions between various schools and, more surprisingly still, within one and the same school. For nowhere outside the Pythagorean school do we find a school devoted to the preservation of a doctrine. Instead we find changes, new ideas, modifications, and outright criticism of the master.

(In Parmenides we even find, at an early date, a most remarkable phenomenon—that of a philosopher who propounds *two* doctrines, one which he says is true, and one which he himself describes as false. Yet he makes the false doctrine not simply an object of condemnation or of criticism; rather he presents it as the best possible account of the delusive opinion of mortal men, and of the world of mere appearance—the best account which a mortal man can give.)

How and where was this critical tradition founded? This is a problem deserving serious thought. This much is certain: Xenophanes who brought the Ionian tradition to Elea was fully conscious of the fact that his own teaching was purely conjectural, and that others might come who would know better. I shall come back to this point again in my next and last section.

If we look for the first signs of this new critical attitude, this new freedom of thought, we are led back to Anaximander's criticism of Thales. Here is a most striking fact: Anaximander criticizes his master and kinsman, one of the Seven Sages, the founder of the Ionian school. He was, according to tradition, only about fourteen years younger than Thales, and he must have developed his criticism and his new ideas while his master was alive. (They seem to have died within a few years of each other.) But there is no trace in the sources of a story of dissent, of any quarrel, or of any schism.

This suggests, I think, that it was Thales who founded the new tradition of freedom—based upon a new relation between master and pupil—and who thus created a new type of school, utterly different from the Pythagorean school. He seems to have been able to tolerate criticism. And what is more, he seems to have created the tradition that one ought to tolerate criticism.

Yet I like to think that he did even more than this. I can hardly imagine a relationship between master and pupil in which the master merely tolerates criticism without actively encouraging it. It does not seem to me possible that a pupil who is being trained in the dogmatic attitude would ever dare to criticize the dogma (least of all that of a famous sage) and to voice his criticism. And it seems to me an easier and simpler explanation to assume that the master encouraged a critical attitude—possibly not from the outset, but only after he was struck by the pertinence of some questions asked, by the pupils perhaps, without any critical intention.

However this may be, the conjecture that Thales actively encouraged criticism in his pupils would explain the fact that the critical attitude towards the master's doctrine became part of the Ionian school tradition. I like to think that Thales was the first teacher who said to his pupils: 'This is how I see things—how I believe that things are. Try to improve upon my teaching.' (Those who believe that it is 'unhistorical' to attribute this undogmatic attitude to Thales may again be reminded of the fact that only two generations later we find a similar attitude consciously and clearly formulated in the fragments of Xenophanes.) At any rate, there is the historical fact that the Ionian school was the first in which pupils criticized their masters, in one generation after the other. There can be little doubt that the Greek tradition of philosophical criticism had its main source in Ionia.

It was a momentous innovation. It meant a break with the dogmatic tradition which permits only *one* school doctrine, and the introduction in its place of a tradition that admits a *plurality* of doctrines which all try to approach the truth by means of critical discussion.

It thus leads, almost by necessity, to the realization that our attempts to see and to find the truth are not final, but open to improvement; that our knowledge, our doctrine, is conjectural; that it consists of guesses, of hypotheses, rather than of final and certain truths; and that criticism and critical discussion are our only means of getting nearer to the truth. It thus leads to the tradition of bold conjectures and of free criticism, the tradition which created the rational or scientific attitude, and with it our Western civilization, the only civilization which is based upon science (though of course not upon science alone).

In this rationalist tradition bold changes of doctrine are not forbidden. On the contrary, innovation is encouraged, and is regarded as success, as improvement, if it is based on the result of a critical discussion of its predecessors. The very boldness of an innovation is admired; for it can be controlled by the severity of its critical examination. This is why changes of doctrine, far from being made surreptitiously, are traditionally handed down together with the older doctrines and the names of the innovators. And the material for a history of ideas becomes part of the school tradition.

To my knowledge the critical or rationalist tradition was invented only once. It was lost after two or three centuries, perhaps owing to the rise of the Aristotelian doctrine of *epistēmē*, of certain and demonstrable knowledge (a development of the Eleatic and Heraclitean distinction between certain truth and mere guesswork). It was rediscovered and consciously revived in the Renaissance, especially by Galileo Galilei.

XII

I now come to my last and most central contention. It is this. The rationalist tradition, the tradition of critical discussion, represents the only practicable way of expanding our knowledge—conjectural or hypothetical knowledge, of course. There is no other way. More especially, there is no way that starts from observation or experiment. In the development of science observations and experiments play only the role of critical arguments. And they play this role alongside other, non-observational arguments. It is an important role; but the significance of observations and experiments depends *entirely* upon the question whether or not they may be used to *criticize theories*.

According to the theory of knowledge here outlined there are in the main only two ways in which theories may be superior to others: they may explain more; and they may be better tested—that is, they may be more fully and more critically discussed, in the light of all we know, of all the objections we can think of, and especially also in the light of observational or experimental tests which were designed with the aim of criticizing the theory.

There is only one element of rationality in our attempts to know the world: it is the critical examination of our theories. These theories themselves are guesswork. We do not know, we only guess. If you ask me, 'How do you know?' my reply would be, 'I don't; I only propose a guess. If you are interested in my problem, I shall be most happy if you criticize my guess, and if you offer counter-proposals, I in turn will try to criticize them.'

This, I believe, is the true theory of knowledge (which I wish to submit for your criticism): the true description of a practice which arose in Ionia and which is incorporated in modern science (though there are many scientists who still believe in the Baconian myth of induction): the theory that knowledge proceeds by way of *conjectures and refutations*.

Two of the greatest men who clearly saw that there was no such thing as an inductive procedure, and who clearly understood what I regard as the true theory of knowledge, were Galileo and Einstein. Yet the ancients also knew it. Incredible as it sounds, we find a clear recognition and formulation of this theory of rational knowledge almost immediately after the practice of critical discussion had begun. Perhaps our oldest extant fragments in this field are those of Xenophanes. I will present here five of them in an order that suggests that it was the boldness of his attack and the gravity of his problems which made him conscious of the fact that all our knowledge was guesswork, yet that we may nevertheless, by searching for that knowledge 'which is the better', find it in the course of time. Here are the five fragments (DK, B 16 and 15; 18; 35; and 34) from Xenophanes' writings.

The Ethiops say that their gods are pug-nosed and black
While the Thracians say that theirs have blue eyes and red hair.

Yet if cattle or horses or lions had hands and could draw
And could sculpture like men, then the horses would draw their gods
Like horses, and cattle like cattle, and each would then shape
Bodies of gods in the likeness, each kind, of its own.

The gods did not reveal, from the beginning,
All things to us; but in the course of time,
Through seeking we may learn, and know things better . . .

These things, we conjecture, are somehow like the truth.

But as for certain truth, no man has known it,
Nor will he know it; neither of the gods,
Nor yet of all the things of which I speak.
And even if perchance he were to utter
The perfect truth, he would himself not know it:
For all is but a woven web of guesses.

To show that Xenophanes was not alone I may also repeat here two of Heraclitus' sayings (DK, B 78 and 18) which I have quoted before in a different context. Both express the conjectural character of human knowledge, and the second refers to its daring, to the need to anticipate boldly what we do not know.

> It is not in the nature or character of man to possess true knowledge, though it is in the divine nature . . . He who does not expect the unexpected will not detect it: for him it will remain undetectable, and unapproachable.

My last quotation is a very famous one from Democritus (DK, B 117):

> But in fact, nothing do we know from having seen it; for the truth is hidden in the deep.

This is how the critical attitude of the Presocratics foreshadowed, and prepared for, the ethical rationalism of Socrates: his belief that the search for truth through critical discussion was a way of life—the best he knew.

APPENDIX: HISTORICAL CONJECTURES AND HERACLITUS ON CHANGE

In an article entitled 'Popper on Science and the Presocratics' (*Mind*, NS. **69**, July 1960, pp. 318 to 339), Mr G. S. Kirk has responded to a challenge, and to a criticism, which formed part of my presidential address to the Aristotelian Society 'Back to the Presocratics'. Mr Kirk's article is not, however, mainly devoted to the task of replying to my criticism. It is, largely, devoted to another task: it tries to explain how and why I am the victim of a fundamentally mistaken 'attitude to

This Appendix, a reply to Mr Kirk's article in Mind, *was published in part under the title 'Kirk on Heraclitus, and on Fire as the Cause of Balance', in* Mind, *N.S.* **72**, *July 1963, pp. 386–92. I wish to thank the editor of* Mind *for his permission to publish here the whole article as originally submitted to him. (In the second and some later editions of this book I have made some additions to this Appendix.)*

scientific methodology' which has made me come forward with mistaken assertions about the Presocratics and with mistaken principles of historiography.

A counter-attack of this kind might, to be sure, have its intrinsic merits and interest. And the fact that Mr Kirk has adopted this procedure shows at any rate that he and I agree at least on two points: that the fundamental issue between us is a philosophical one; and that the philosophical attitude we adopt can have a decisive influence on our interpretation of the historical evidence—such as, for example, the evidence concerning the Presocratics.

Now Mr Kirk does not accept my general philosophical attitude any more than I do his. Thus he rightly feels that he should give reasons for rejecting mine.

I do not think that he has offered any reason for rejecting my views; simply because Mr Kirk's views on what he believes to be my views, and the devastating conclusions he draws from these views, are unrelated to my actual views, as I shall show.

There is another difficulty. The method of counter-attack which he has adopted has its own peculiar drawback: it does not seem to lend itself easily to furthering the discussion of the definite points of criticism made in my address. Kirk does not, for instance, state very clearly which of my points he accepts (for he does accept some) and which he rejects; instead, acceptance and rejection are submerged in a general rejection of what he believes to be my 'attitude to scientific methodology', and of some of the consequences of this imaginary attitude.

I

My first task will be to give some evidence for my allegation that Kirk's treatment of my 'attitude to scientific methodology' is largely based upon misunderstandings and misreadings of what I have written, and upon popular inductivist misconceptions about natural science, which were fully discussed and dismissed in my book, *The Logic of Scientific Discovery* (*L.Sc.D.*).

Kirk rightly presents me as an opponent of the widely accepted *dogma of inductivisim*—of the view that science starts from observation and proceeds, by induction, to generalizations, and ultimately to theories. But he is mistaken in believing that, since I am an opponent of *induction*, I must be an adherent of *intuition*, and that my approach must be due to an attempt to defend an intuitionist philosophy, which he calls 'traditional philosophy', against modern empiricism. For although I do not believe in induction, I do not believe in intuition either. Inductivists are inclined to think that intuition is the only alternative to induction. But they are simply mistaken: there are other possible approaches besides these two. And my own view may be fairly described as critical *empiricism*.

But Kirk ascribes to me an almost Cartesian intuitionism when he presents the situation as follows (p. 319): 'Philosophy of the traditional type had assumed that philosophical truths were metaphysical in content and could be apprehended by intuition. The positivists of the Vienna Circle denied this. In disagreeing with them Popper was asserting his belief in something not far distant from the classical conception of the role of philosophy.' Whatever one may say about this, there certainly exists a 'traditional philosophy'—that of Descartes or Spinoza, for example—which treats '*intuition*' as a source of knowledge; but I have always opposed this philosophy.[1] From this passage on, Kirk writes 'intuition', in the sense in which he uses it here, several times in quotes (pp. 320, 321, 322, 327) and several times out of quotes (pp. 318, 319, 320, 324, 327, 332, 337), yet always apparently under the impression—and certainly creating the impression—that he is citing me when ascribing to me intuitionist views; views which I have never held in my life. In fact, the only time the word '*intuition*' occurs in my address,[2] it is used in a context which is anti-inductivist *and* anti-intuitionist at the same time. For I write there (p. 7; this volume p. 189)

[1] Kirk quotes on p. 322 my *L.Sc.D.*, p. 32, but a reading of what precedes my reference there to Bergson will show that my admission that every discovery contains (among other elements) 'an irrational element' or a 'creative intuition' is neither irrationalist nor intuitionist in the sense of any 'traditional philosophy'. See also my Introduction to the present volume, 'On the Sources of Knowledge and of Ignorance', especially pp. 36 ff.

[2] There are also casual occurrences such as 'uncanny intuition', 'less formal and more intuitive reasons' and 'From this we see intuitively', on pp. 17 and 5 (this volume, pp. 199 and 187). In all cases the word is used in a non-technical and *almost* deprecatory sense.

about the problem of the scientific character of a theory (italics not in the original): 'What is important about a theory is its explanatory power, and whether it stands up to criticism and to tests. The question of its origin, of how it is arrived at—*whether by an "inductive procedure", as some say, or by an act of intuition*— . . . has little to do with its scientific [status or] character.'[3]

Now Kirk quotes this passage and discusses it. But the undeniable fact that this passage indicates that I am a believer in *neither* induction *nor* intuition does not prevent him from constantly ascribing to me intuitionist views. He does so, for example, in the passage on p. 319 quoted above; or on p. 324, when he discusses the question whether to accept my alleged 'premise that science starts from intuitions' (while I say it starts from problems; see below); or on pp. 326 f. when he writes: 'Are we therefore to infer with Popper that Thales's theory must have been based on a non-empirical intuition?'

Now my own view is very different from all this. As to the starting point of science, I do not say that science starts from intuitions but that *it starts from problems*; that we arrive at a new theory, in the main, by trying to solve problems; that these problems arise in our attempts to understand the world as we know it—the world of our 'experience' (where 'experience' consists largely of expectations or theories and partly also of observational knowledge—although I happen to believe that there does not exist anything like *pure* observational knowledge, untainted by expectations or theories). A few of these problems—and some of the most interesting ones—arise from the conscious criticism of theories uncritically accepted hitherto, or from the conscious criticism of the theory of a predecessor. One of the main things I set out to do in my paper on the Presocratics was to suggest that Anaximander's theory may well have originated in an attempt to criticize Thales; and that this may well have been the origin of the rationalist tradition, which I identify with the tradition of critical discussion.

I do not think that a view of this kind has much similarity to traditional intuitionistic philosophy. And I was surprised to find that Kirk suggests that my mistaken approach might be explained as that

[3] The two words in square brackets have now been added by me in order to make my meaning still more obvious.

of a speculative philosopher not sufficiently intimately acquainted with scientific practice; he suggests for example on p. 320: 'It seems possible that his [Popper's] view of science was not the result of an initial objective observation of how scientists proceed, but was itself, in an early application of Popper's developed theory, an "intuition" closely related to current philosophical difficulties and subsequently compared with actual scientific procedure.'[4] (I should have thought that even a reader who knows very little about science might have noticed that some at least of my problems originated within the physical sciences themselves, and that my own acquaintance with scientific practice and research was not wholly second-hand.)

The kind of critical discussion I have in mind is, of course, a discussion in which experience plays a major role: observation and experiment are constantly appealed to as *tests* of our theories. Yet Kirk surprisingly goes so far (p. 332; italics mine) as to speak of '*Popper's thesis that all scientific theories are entirely based on intuitions*'.

Like most philosophers I am quite used to seeing my views distorted and caricatured. But this is hardly a caricature (which must always rely on a recognizable similarity to the original). I may remark that none of my empiricist and positivist friends, opponents, and critics, has ever criticized me for holding or for reviving an intuitionist epistemology, and that, on the contrary, they usually say that my epistemology does not significantly deviate from theirs.

It will be seen from the foregoing that Kirk offers several conjectures, not only about the content of my philosophy, but also about its origin. But he does not seem to be aware of the conjectural character of these constructions. On the contrary, he believes that he has some textual evidence for them. For he says of me that my 'own attitude of scientific methodology . . . was formed as he [Popper] writes in the 1958 preface to *The Logic of Scientific Discovery*, in reaction against the attempts of the Vienna Circle to base all philosophical [*sic*] and scientific truth upon verification by experience' (Kirk, p. 319). I need not comment here on this mistaken description of the Wittgensteinian philosophy of the Vienna Circle. But since it is a historian of

[4] It is Kirk who puts the word 'intuition' in quotes, thereby suggesting that it is I who use 'intuition' in this sense.

philosophy who writes here about what I have written, I feel I must nip in the bud a historical myth about what I have written. For in the preface to which Kirk refers, I do not say a word about how I formed my views or my attitude; nor do I say a word about the Vienna Circle. Indeed, I could not have written anything resembling Kirk's account, because the facts are otherwise. (Part of the story, first published in 1957, Mr Kirk might have found in a Cambridge lecture of mine, now in this volume under the title 'Science: Conjectures and Refutations', in which I tell how I developed my 'attitude . . . in reaction against the attempts of' Marx, Freud, and Adler, none of whom was either a positivist or a member of the Vienna Circle.) It seems unlikely that it was the Heraclitean obscurity of my style which caused this quite inexplicable misreading by Mr Kirk, for in comparing it with 'Back to the Presocratics' he describes (p. 318) the same 1958 preface to which his above quoted passage refers as 'a more lucid statement'.

Another example of misreading *The Logic of Scientific Discovery* is equally inexplicable—at least for anybody who has read the book as far as p. 61 (not to mention pp. 274 or 276) where I refer to the problem of truth, and to Alfred Tarski's theory of truth. Kirk says that 'Popper abandons the concept of absolute scientific truth' (p. 320). He does not seem to see that, when I say that we cannot know, even of a well-corroborated scientific theory, whether or not it is true, I am actually assuming a 'concept of absolute scientific truth'; just as somebody who says 'I did not succeed in reaching the winning post' operates with an 'absolute concept of a winning post'—that is, one whose existence is assumed *independently of its being reached*.

It is surprising to find these obvious misunderstandings, and these occasional misquotations, in a paper by an outstanding scholar and historian of philosophy. They make a philosophical defence of my real views about science unnecessary.

II

Thus I may now turn to something more to the point—to the history of the Presocratics. In this section I shall confine myself to straightening out two of Kirk's mistakes concerning my historical method, and my views on historical method. In section III I will deal with our real disagreements.

(1) Kirk discusses on p. 325 a remark of mine which I made in order to disclaim any competence regarding such matters as text emendation. The passage he quotes reads: 'I am completely out of my depth when an expert begins to argue what words or phrases Heraclitus might have used, and what words or phrases he could not possibly have used.'

Commenting on this disclaimer of competence, Kirk exclaims: 'As though "what words or phrases Heraclitus might have used", for example, is irrelevant to the assessment of what he thought!'

But I never said or suggested that these matters are 'irrelevant'. I merely confessed that I had not studied the linguistic usages of Heraclitus (and others) sufficiently deeply to feel myself equipped to discuss the work done in this field by such scholars as, say, Burnet or Diels or Reinhardt, and, more recently, Vlastos or Kirk himself.

Yet Kirk goes on to say that

> It is these 'words and phrases', and the other *verbatim* fragments of the Presocratics themselves, and not the reports of Plato, Aristotle, and the doxographers, as Popper appears to think, that are 'the oldest texts we possess' . . . It should in fact be obvious even to an 'amateur' that the reconstruction of Presocratic thought must be based both upon the later tradition and upon the surviving fragments.

I cannot imagine how my disclaimer of competence in the field of linguistic criticism can have induced Kirk to suggest that such things are not 'obvious' even to the particular amateur in question. Moreover, he might have noticed that fairly frequently I quote, translate, and discuss, the fragments themselves (much more than the reports of Plato and Aristotle, though we now seem to agree that these are quite relevant also), both in 'Back to the Presocratics' and in my *Open Society*, where I discussed, for example, a considerable number of the surviving fragments of Heraclitus. Kirk refers to this book on p. 324. Why then does he, on p. 325, interpret my disclaimer in the sense that I disclaim interest in the surviving fragments, or in the problem of their historical status?

(2) As an example of the way, unsatisfactory in my opinion, in

which Kirk answers the criticisms I made in 'Back to the Presocratics', I now quote the end of his reply (p. 339). He says:

> More startling still, he [Popper] applies the criterion of possible *truth* as the test of the historicity of a theory. On page 16 he [Popper] finds that 'the suggestion that we should test the historicity of Heraclitus' ideas . . . by standards of "common sense" is a little surprising.' Shall we [Kirk] not find his [Popper's] own 'test' much more surprising— 'But the decisive point is, of course, that this inspired philosophy [i.e. that man is a flame, etc.] is *true*, for all we know' (p. 17 [in this volume, p. 199])?

The simple answer to this is that I neither said nor implied that the truth, or the possible truth, of a theory is a 'test' of its historicity. (This may be seen from pages 16 and 17 of my address—in this volume, pp. 198 f.—and the second paragraph of section vii; incidentally, did Kirk forget his thesis that I have abandoned the idea of truth?) And when Kirk here puts 'test' in quotes—thereby indicating that I have used the term 'test' in this context, or in this sense—then he clearly misquotes me. For all I have said or implied is that the *truth* of that theory of change which has been traditionally, and I think correctly, attributed to Heraclitus, shows that *this attribution at least makes sense* of Heraclitus' philosophy—while I at any rate could not make sense of the philosophy attributed to Heraclitus by Kirk. Incidentally, I do think that it is an important and even an obvious principle of the historiography and interpretation of ideas that we should always try to attribute to a thinker an interesting and a true theory rather than an uninteresting or a false one, *provided of course the transmitted historical evidence allows us to do so*. This is neither a criterion nor a 'test', to be sure; but he who does not try to apply this principle of historiography is unlikely to understand a great thinker such as Heraclitus.

III

The most important disagreement between Kirk and me as far as the Presocratics are concerned was over the interpretation of Heraclitus. And here I claim that Kirk, perhaps unconsciously, has almost ceded my two main points which I am going to discuss below under (1) and (2).

My general approach to Heraclitus may be put in the words of Karl Reinhardt: 'The history of philosophy is the history of its problems. If you want to explain Heraclitus, tell us first what his problem was.'[5]

My answer to this challenge was that Heraclitus' problem is the *problem of change*—the *general* problem, *How is change possible?* How can *a thing* change without losing its identity—in which case it would be no longer *that thing* which has changed? (See 'Back to the Presocratics', sections viii and ix.)

I believe that Heraclitus' great message was linked with his discovery of this exciting problem; and I believe that his discovery led to Parmenides' solution that change, indeed, is logically impossible for any thing—for any being; and later to the closely related theory of Leucippus and Democritus that things do not, indeed, change intrinsically, although they change their positions in the void.

The solution of this problem which, following Plato, Aristotle, *and* the fragments, I attribute to Heraclitus, is as follows: there are no (unchanging) things; what appears to us as a thing is a process. In reality a material thing is like a flame; for a flame *seems* to be a material thing, but it is not: it is a process; it is in flux; matter passes through it; it is like a river.

Thus all the apparently more or less stable things are really in flux; and some of them—those which indeed appear stable—are in *invisible* flux. (Thus Heraclitus' philosophy prepares the way for the Parmenidean distinction between appearance and reality.)

In order to appear as a stable thing, the process (which is the reality behind the thing) has to be regular, law-like, '*measured*': the lamp which holds a stable flame has to supply to it a definite measure of oil. It

[5] K. Reinhardt, *Parmenides*, 2nd edn., 1959, p. 220. I cannot mention this book without expressing my boundless admiration for it, even though I feel that I must reluctantly disagree with its fundamental doctrine: that Parmenides not only originated his problem independently of Heraclitus, but preceded Heraclitus, to whom he handed on his problem. I believe, however, that Reinhardt has given overwhelming reasons for the view that one of these two philosophers depends upon the other. I may perhaps say that my attempt to 'locate', as it were, Heraclitus' problem, may be regarded as an attempt to answer Reinhardt's challenge quoted in the text. (See also section vi of ch. 2, above.)

seems not unlikely that the idea of a measured or law-like process was developed by Heraclitus from suggestions of the Milesians, especially of Anaximander, about the significance of the cosmic periodic changes (such as day and night, perhaps also the tides, the waxing and waning of the moon, and especially the seasons of the year). These regularities might well have contributed to the idea that the apparent stability of things, and even of the cosmos, can be explained as a '*measured*' *process*—a process ruled by law.

(1) The first of the two main points on which I criticized Kirk's views on Heraclitus is this. Kirk suggested that Heraclitus did not believe, and that it was against common sense to believe, 'that a rock or a bronze cauldron . . . was invariably undergoing invisible changes'. Kirk's lengthy discussion (pp. 334 ff.) of my criticism ultimately arrives at a point about which he says:

> At this point the argument becomes somewhat rarefied. I agree, though, that it remains theoretically possible that certain *invisible* changes of our experience, for example the gradual rusting of iron, cited by Popper, struck Heraclitus so forcefully that they persuaded him to assert that all things which were not in visible change were in invisible change. I do not think however, that the extant fragments suggest that this was the case (p. 336).

I do not think that the argument need in any sense become rarefied; and there are many extant fragments which suggest the theory which I attribute to Heraclitus. Yet before referring to these I must repeat a question which I raised in my address: if, as Kirk and Raven agree, fire is, as it were, the structural model or the prototype (or the 'archetypal form' as they have it) of matter, what else can this mean but that material things are like flames, and therefore processes?

I do not, of course, assert that Heraclitus used an abstract term like 'process'. But I conjecture that he did apply his theory *not only* to matter in the abstract, or to 'the world order *as a whole*' (as Kirk says on p. 335), but also to concrete, single *things*; and these things, then, must be compared to concrete, single flames.

As to the extant fragments in support of this view and of my interpretation in general, there are first the fragments about the sun. It seems to me pretty clear that Heraclitus regarded the sun as a *thing*, or perhaps even as a new thing every day; see DK, B 6 which says[6] 'The sun is new every day', though this may perhaps only mean that it is, like a lamp, re-kindled every day: 'Were there no sun, it would be night in spite of the other stars' says B 99. (See also B 26 and my remark above concerning lamps and measures and compare B 94.) Or take B 125: 'If not stirred, the barley-brew decomposes.' Thus movement, process, is essential to the continued existence of the thing which otherwise ceases to exist. Or take B 51: 'What struggles with itself becomes committed to itself: there is a link or harmony due to recoil [or: to the turning back of the strings] and tension, as in the bow and the lyre.' It is the tension, the active force, the inherent strife (a process), which makes bow and lyre what they are, and only as long as the tension is kept up, only as long as the strife of their parts goes on, do they continue to be what they are.

Admittedly, Heraclitus likes generalizations and abstractions; and so he proceeds at once to a generalization which may well be intended as one on a cosmic scale, as in B 8: 'The opposites agree, and from discord results the best harmony.' (See also B 10.) But this does not mean that he loses sight of the single things, the bow, the lyre, the lamp, the flame, the river (B 12, 49*a*). 'Upon those who step into the same rivers, different and again different waters flow . . . We step into the same rivers, and we do not step [into the same rivers]. We are, and we are not.'

Yet before becoming symbols of the cosmic processes, the rivers are concrete rivers, and beyond that, symbols of other concrete things, including ourselves. And although 'we are, and we are not' (which, incidentally, Kirk and Raven prefer not to attribute to Heraclitus) is, in a sense, a sweeping and perhaps cosmic generalization and abstraction, it is no doubt also meant as a very concrete appeal to every man: it is a Heraclitean *memento mori*, like so many other fragments which remind us that life becomes death, and death becomes life. (Compare for example B 88, 20, 21, 26, 27, 62, 77.)

If B 49*a* moves towards something like a generalization, B 90 moves from the general and cosmic idea of a consuming (and dying) fire to

[6] Cp. Diels-Kranz. For B 51 see G. Vlastos, *AJP* **76**, 1955, pp. 348 ff.

the particular: 'Everything is an exchange for fire and fire for everything just as wares for gold and gold for wares.'

Thus when Kirk now asks (p. 336): 'Can we then say that the conclusion that all things separately are in permanent flux is necessarily entailed by any course of reasoning followed by Heraclitus?', then the answer is an emphatic 'yes', as far as we can speak at all of anything as being 'necessarily entailed' by a 'course of reasoning' in a field where everything must remain to some extent conjecture and interpretation.

Thus take for example B 126, 'What is cold becomes hot and what is hot becomes cold; what is moist becomes dry and what is dry becomes moist.' This may well have a cosmic significance: it may refer to the seasons, and to cosmic change. But how can it be doubted (especially if we attribute to Heraclitus 'common sense', whatever this may mean[7]) that it applies to concrete, individual things and their changes—and incidentally, to ourselves and our souls? (Cp. B 36, 77, 117, 118.)

But things are not only in flux—they are *invisibly* in flux. So we read in B 88: 'It is always one and the same: what is alive and what is dead; what is awake and what is asleep; what is young and what is old. For the one turns into the other and the other turns back into the one.' Thus our children age—as we know, invisibly; yet the parents also turn—somehow—into their children. (See also B 20, 21, 26, 62 and 90.) Or take B 103: 'In a circle, the beginning and the end are the same.' (The identity of opposites; opposites invisibly merging into each other; see also B 54, 65, 67, 126.)

That Heraclitus notices that these processes may indeed be invisible, and that he therefore felt that sight, and observation, were deceptive, may be seen from B 46: '. . . sight is deceptive.' B 54: 'Invisible

[7] Kirk, it seems, has misunderstood my criticism of his appeal to 'common sense'. I criticized the view that there was in these matters a straightforward standard of common sense to which the historiographer could appeal, and I suggested (but only suggested) that my interpretation of Heraclitus may attribute to him perhaps as much, or more, common sense, than Kirk's interpretation. (Besides, I also suggested that Heraclitus was the last man on earth whose sayings were to be measured by somebody else's standard of common sense.) And is not the invisible change in Ovid's '*gutta cavat*' common sense? (Alan Musgrave has called my attention to an elaborate argument for invisible change in Lucretius, *De rer. nat.*, i, 265–321, which may have been Ovid's source.)

harmony is stronger than visible.' (See also B 8 and 51.) B 123: 'Nature loves to hide.' (See also B 56 and 113.)

I have not the slightest doubt that any one or all of these fragments may be explained away. But they do seem to me to support what it is reasonable to assume in any case, and what is, in addition, supported by Plato *and* Aristotle. (And though the evidence of the latter has become suspect, especially in view of the great work of Harold Cherniss, nobody thinks—and least of all Harold Cherniss—that Aristotle's evidence has been completely discredited, including that which is supported by Plato or by the 'fragments'.)

(2) The last point of my reply, and my second and main point about Heraclitus, concerns the general summary of his philosophy which can be found in Kirk and Raven on p. 214 under the heading '*Conclusion*'.

I quoted part of this conclusion in my address, and said that I found the doctrine attributed by Kirk and Raven to Heraclitus 'absurd'; and in order to make quite clear *what* I regarded as 'absurd', I used italics. I repeat here my quotation from Kirk and Raven, with the italics as previously used by me. What I found 'absurd' is the allegedly Heraclitean doctrine 'that natural changes of all kinds [and thus presumably also earthquakes and great fires] are regular *and balanced and that the cause of this balance is fire, the common constituent of things that was also termed their Logos*'. (See p. 198, above.)

I did not object to anybody's attributing to Heraclitus the doctrine that change is ruled by law, or perhaps the more doubtful doctrine that the rule, or regularity, was their 'Logos'; or the doctrine that 'the common constituent of things was fire'. What I felt to be absurd were the doctrines (*a*) that *all* changes (or '*changes of all kinds*') are 'balanced' in the sense in which many important changes and processes such as the fire in a lamp, or the cosmic seasons, may well be called 'balanced'; (*b*) that fire is '*the cause of this balance*'; and (*c*) that the common constituent of things—that is, fire—'was also termed their Logos'.

Moreover, I could find no traces of these doctrines in Heraclitus' fragments, nor in any of the ancient sources, such as Plato or Aristotle.

Where then is the source of this summary or 'conclusion'—that is to say, the source of the three points (*a*), (*b*), and (*c*) which express Kirk's general view of Heraclitus' philosophy, and which colour so much of his interpretation of the fragments?

Reading the chapter on Heraclitus in Kirk and Raven again, I could find only *one* hint: the doctrines to which I object are first formulated on p. 200, with reference to the fragment which they number 223. (See also p. 434.) Now Kirk and Raven's 223 is the same fragment as DK, B 64: 'It is the thunderbolt which steers all things.'

Why should this fragment make Kirk ascribe to Heraclitus the doctrines (*a*), (*b*), and (*c*)? Is it not quite satisfactorily explained if we remember that the thunderbolt is the instrument of Zeus? For according to Heraclitus, DK, B 32 = KR, 231, 'One thing—the only one that is wise—wants, and does not want, to be called by the name of Zeus.' (This seems quite sufficient to explain DK, B 64. There is no necessity to connect it with DK, B 41 = KR, 230, though this could only further strengthen my interpretation.)

But Kirk and Raven interpret on pages 200 and 434, the fragment 'It is the thunderbolt which steers all things' more elaborately: first by identifying the thunderbolt with fire; secondly by attributing to fire a 'directive capacity'; thirdly by suggesting that fire 'reflects divinity'; and fourthly by suggesting its identification with the Logos.

What is the source of this somewhat over-elaborate interpretation of a short and simple fragment? I could find no trace of it in any of the ancient sources—the fragments themselves, or Plato, or Aristotle. The only trace I could find was an interpretation of Hippolytus, whom Kirk and Raven describe on p. 2 of their book as 'a theologian in Rome in the third century A.D.' (almost six centuries junior to Plato) who 'attacked Christian heresies by claiming them to be revivals of pagan philosophy'. It seems that Hippolytus, himself a schismatic Bishop, not only claimed that the Noetian heresy 'was a revival of Heraclitus' theory', but that he also contributed by his attacks to the extermination of the heresy.

Hippolytus is also the source of B 64, the beautiful fragment about the thunderbolt. He quotes it, apparently, because he wants to interpret it as closely related to the Noetian heresy. In this attempt he identifies the thunderbolt first with fire; next with eternal or divine fire, endowed with a providential 'directive capacity' (as Kirk and Raven have it), and thirdly with prudence or reason (Kirk and Raven have 'Logos'); and ultimately he interprets the Heraclitean fire as '*the cause* of the cosmic housekeeping', or of the 'directorship' or the 'economic government' that keeps the balance of the world. (Kirk and Raven have it that fire is 'the cause of this balance'.)

(The third of these identifications of Hippolytus might indeed have had a basis in the text: Karl Reinhardt, in an article in *Hermes*, **77**, 1942, conjectures that there was a lost fragment, alluded to in Hippolytus, which read '*pur phronimon*' or '*pur Phronoun*'. I am unable to evaluate the force of Reinhardt's arguments though to me they do not appear very compelling. But the alleged lost fragment itself would fit perfectly well into my interpretation: since I interpret Heraclitus to mean that we—our souls—are flames, 'thinking fire' or 'fire as a thought process' would of course fit very well. But only a Christian—or heretical Christian—interpretation could render it 'fire is providence'; and as to the 'cause' of Hippolytus, Reinhardt says explicitly that this is not Heraclitean. Fire as the cause of world-balance would come in—if at all—only through a Conflagration on the Day of Judgment, as the balance of justice; yet Kirk does not accept that this Conflagration is part of the teaching of Heraclitus.)

Thus the doctrine whose attribution to Heraclitus I found so unacceptable appears to be Kirk's interpretation of an interpretation through which Hippolytus may have tried to establish the semi-Christian character of Heraclitus' teaching—perhaps, as Karl Reinhardt suggests, in an attempt to fasten upon the Noetians heretic doctrines of pagan origin, such as the doctrine that fire is endowed with providential or divine powers.

Though Hippolytus may perhaps be a good source when he cites Heraclitus, he clearly cannot be taken very seriously when he interprets Heraclitus.

Considering its doubtful source, it is far from surprising that I could not make any sense of the quoted final summary or 'conclusion' of Kirk and Raven. I still feel that the doctrine there ascribed by Kirk and Raven to Heraclitus is absurd—especially the words which were italicized by me; and I am sure that I am not alone in this feeling. Yet Kirk now writes (on p. 338), referring to the passage in my address where I discuss his 'conclusion' and say that it is 'absurd': 'Popper is indeed isolated when he asserts that such an interpretation of Heraclitus is "absurd".' But when we look more closely at Kirk's present interpretation, we find that he has almost conceded my point: he now omits almost all the words which I put in italics because they seemed absurd to me (and in addition the words 'changes *of all kinds*'); and he omits, more especially, the statement that *the cause of the balance is fire* (and '*that was also termed their Logos*').

For Kirk now writes on p. 338, suggesting that *this* is the 'interpretation of Heraclitus' which I described as absurd: 'Heraclitus accepted change in all its manifest presence and inevitability, but claimed that the unity of the world-order was not thereby prejudiced: it was preserved through the logos which operates in all natural changes and ensures their ultimate equilibrium.'

I think that even this interpretation might perhaps be formulated more happily; but it is no longer absurd. On the contrary, it seems to agree, for example, with the interpretation which I myself gave in my *Open Society*, where I suggested that the 'logos' may be the law of change. Moreover, though I strongly object to describing fire (with Kirk and Raven, or with Hippolytus) as the *cause* of balance, I do not object to an interpretation which lays some stress on balance or on balanced change. Indeed, if the apparently stable material things are in reality processes like flames, then they must burn slowly, in a measured way. They will, like the flame of a lamp, or like that of the sun, 'not overstep their measure'; they will not get out of control, as a conflagration might. We may remember here that it is *movement*, a *process*, that keeps the barley brew from decomposing, separating, disintegrating; and that it is not every kind of movement that has this effect but, for example, a circular, and thus a *measured* movement. It is therefore the *measure* which may be called the cause of the balance of fire, of flames, and of things—of those processes and changes which appear as stable and as things at rest, and which are responsible for the preservation of things. The measure, the rule, the lawful change, the logos (but not the fire) is the cause of balance—including especially the balance of a fire when it is under control, such as a balanced flame or the sun or the moon (or the soul).

It is clear that according to this view most of the balanced change must necessarily be invisible: this kind of balanced or lawful change must be inferred by reasoning, by the reconstruction of the tale, the story, of how things do happen. (Perhaps this is why it is called the logos.)

This may well have been the way that led Heraclitus to his new epistemology, with its implicit distinction between reality and appearance, and its distrust of sense experience. This distrust, together with the doubts of Xenophanes, may later have helped Parmenides to arrive at his contrast between the 'well-rounded truth' (the invariant logos) on the one hand, and the delusive opinion, the erring thought of the mortals, on the other. Thus came about the first clear contrast between an intellectualism or rationalism, which Parmenides upheld; and an empiricism or sensualism, which he not only attacked, but which he was the first to formulate. For he taught (B 6 : 5) that the muddled horde of erring mortals, always in two minds about things, with erring thoughts (B 6 : 6) in their hearts, mistake sensations for knowledge; and that they take being and not-being for the same and yet for not the same. And against them he contended (B 7):

> Never shall it prevail that things that are not are existing.
> Keep back your thought from this way of enquiry; don't let experience,
> Much-tried habit, constrain you; and do not let wander your blinded
> Eye, or your deafenèd ear, or even your tongue, along this way!
> But by reason alone decide on the often contested
> Argument which I have here expounded to you as a disproof.

This is Parmenides' intellectualism or rationalism. He contrasts it with the sensualism of those erring mortals who hold the conventional and erroneous opinion that there is light and night, sound and silence, hot and cold; that their eyes mingle with light and night, and that their limbs mingle with hot and cold and so become themselves hot and cold; that this mixture determines *the physical state or 'nature'* of their sense organs or their limbs; and that this mixture or nature turns into thought. This doctrine according to which *there is nothing in the erring intellect* (the 'erring thought' or 'erring knowledge' of B 6 : 6) *which was not previously in the erring sense-organs* is stated by Parmenides as follows (B 16):[8]

[8] The significance of this passage, and my translation of it (which should also be compared with Empedocles B 108), are more fully discussed in *Addendum* 8 at the end of this book; see especially sections 6–10.

> For as, at any one time, is the much-erring sense-organs' mixture,
> So does knowledge appear in men. For these two are the same thing:
> That which thinks, and the mixture which makes up the sense-organs' nature.
> What in this mixture prevails becomes thought, in each man and all.

This anti-sensualist theory of knowledge soon afterwards turned, practically unchanged, into a pro-sensualist theory which extolled the sense-organs (disparaged by Parmenides) as more or less authoritative sources of knowledge.

The whole of this story is somewhat idealized and, of course, conjectural. I merely try to show how epistemological and logical problems and theories might have arisen in the course of a critical debate of cosmological problems and theories.

It almost seems more than a conjecture that something like this did happen.

6

A NOTE ON BERKELEY AS PRECURSOR OF MACH AND EINSTEIN

I had only a very vague idea who Bishop Berkeley was, but was thankful to him for having defended us from an incontrovertible first premise.

SAMUEL BUTLER

I

In this short contribution to the 200th anniversary of Berkeley's death, I wish to give a list of those ideas of Berkeley's in the field of the philosophy of physics which have a strikingly new look. They are mainly ideas which were rediscovered and reintroduced into the discussion of modern physics by Ernst Mach and Heinrich Hertz, and by a number of philosophers and physicists, some of them influenced by

First published in The British Journal for the Philosophy of Science, **4**, 1953.

Mach, such as Bertrand Russell, Philip Frank, Richard von Mises, Moritz Schlick,[1] Werner Heisenberg and others.

I may say at once that I do not agree with most of these positivistic views. I admire Berkeley without agreeing with him. But criticism of Berkeley is not the purpose of this note, and will be confined to some very brief and incomplete remarks in section v.[2]

Berkeley wrote only one work, *De Motu*, devoted exclusively to the philosophy of physical science; but there are passages in many of his other works in which similar ideas and supplementary ones are represented.[3]

The core of Berkeley's ideas on the philosophy of science is in his *criticism of Newton's dynamics*. (Newton's mathematics were criticized by Berkeley in *The Analyst* and its two sequels.) Berkeley was full of admiration for Newton, and no doubt realized that there could have been no worthier object for his criticism.

II

The following twenty-one theses are not always expressed in Berkeley's terminology; their order is not connected with the order in which they appear in Berkeley's writings, or in which they might be presented in a systematic treatment of Berkeley's thought.

For a motto, I open my list with a quotation from Berkeley (DM, 29).

(1) *'To utter a word and mean nothing by it is unworthy of a philosopher.'*

(2) The meaning of a word is the idea or the sense-quality with which it is associated (as its name). Thus the words 'absolute space'

[1] Schlick, under the influence of Wittgenstein, suggested an instrumentalist interpretation of universal laws which was practically equivalent to Berkeley's 'mathematical hypotheses'; see *Naturwissenschaften*, **19**, 1931, pp. 151 and 156. For further references see footnote 23 to section iv of ch. 3, above.

[2] I have since developed these ideas more fully in ch. 3, above; especially section 4.

[3] Apart from *DM* (= *De Motu*, 1721) I shall quote *TV* (= *Essay towards a New Theory of Vision*, 1709); *Pr* (= *Treatise concerning the Principles of Human Knowledge*, 1710); *HP* (= *Three Dialogues between Hylas and Philonous*, 1713); *Alc* (= *Alciphron*, 1732); *An* (= *The Analyst*, 1734); and *S* (= *Siris*, 1744). As far as I know, there does not exist an English translation of *DM* which succeeds in making clear what Berkeley meant to say; and the Editor of the latest edition of the *Works* even goes out of his way to belittle the significance of this highly original and in many ways unique essay.

and 'absolute time' are without any empirical (or operational) meaning; Newton's doctrine of absolute space and absolute time must therefore be rejected as a physical theory. (Cf. *Pr*, 97, 99, 116; *DM*, 53, 55, 62; *An*, 50, Qu. 8; *S*, 271: 'Concerning absolute space, that phantom of the mechanical and geometrical philosophers, it may suffice to observe that it is neither perceived by our sense, nor proved by our reason . . .'; *DM*, 64: 'for . . . the purpose of the philosophers of mechanics . . . it suffices to replace their "absolute space" by a relative space determined by the heavens of the fixed stars. . . . Motion and rest defined by this relative space can be conveniently used instead of the absolutes. . . .')

(3) The same holds for the word 'absolute motion'. The principle that all motion is relative can be established by appealing to the meaning of 'motion', or to operationalist arguments. (Cf. *Pr* as above, 58, 112, 115 'To denominate a body "moved" it is requisite . . . that it changes its distance or situation with regard to some other body . . .'; *DM*, 63: 'No motion can be discerned or measured, except with the help of sensible things'; *DM*, 62: '. . . the motion of a stone in a sling or of water in a whirled bucket cannot be called truly circular motion . . . by those who define [motion] with the help of absolute space. . . .')

(4) The words 'gravity' and 'force' are misused in physics; to introduce force as the cause or 'principle' of motion (or of an acceleration) is to introduce 'an occult quality' (*DM*, 1–4, and especially 5, 10, 11, 17, 22, 28; *Alc*, vii, 9). More precisely, we should say 'an occult metaphysical substance'; for the term 'occult quality' is a misnomer, in so far as 'quality' should more properly be reserved for observable or observed qualities—qualities which are given to our senses, and which, of course, are never 'occult'. (*An*, 50, Qu. 9; and especially *DM*, 6: 'It is plain, then, that it is useless to assume that the principle of motion is gravity or force; for how could this principle be known any more clearly through [its identification with] what is commonly called an *occult quality*? That which is itself occult explains nothing; not to mention that an unknown acting cause might more properly be called a [metaphysical] *substance* rather than a *quality*.')

(5) In view of these considerations Newton's theory cannot be accepted as an explanation which is truly *causal*, i.e. based on true natural causes. The view that gravity causally explains the motion of

bodies (that of the planets, of free-falling bodies, etc.), must be discarded, says Berkeley, and also the view that Newton discovered that gravity or attraction is 'an essential quality' (Pr, 106) whose inherence in the essence or nature of bodies explains the laws of their motion (S, 234; see also S, 246, last sentence). *But it must be admitted says Berkeley that Newton's theory leads to the correct results* (DM, 39, 41). To understand this, 'it is of the greatest importance . . . to distinguish between *mathematical hypotheses* and the *natures* [*or essences*] *of things*[4] . . . If we observe this distinction, then all the famous theorems of mechanical philosophy which . . . make it possible to subject the world system [i.e. the solar system] to human calculations, may be preserved; and at the same time, the study of motion will be freed of a thousand pointless trivialities and subtleties, and from [meaningless] abstract ideas' (DM, 66).

(6) In physics (mechanical philosophy) there is no causal explanation (cf. S, 231), i.e. no explanation based upon the discovery of the hidden nature or essence of things (Pr, 25). '. . . real efficient causes of the motion . . . of bodies do not in any way belong to the field of mechanics or of experimental science. Nor can they throw any light on these . . .' (DM, 41).

(7) The reason is, simply, that physical things have no secret or hidden, 'true or real nature', no 'real essence', no 'internal qualities' (Pr, 101).

(8) There is nothing physical *behind* the physical bodies, no occult physical reality. *Everything is surface*, as it were; physical bodies are nothing but their qualities. *Their appearance is their reality* (Pr, 87, 88).

(9) The province of the scientist (of the 'mechanical philosopher') is the discovery, 'by experiment and reasoning' (S, 234), of *Laws of Nature*, that is to say, of the regularities and uniformities of natural phenomena.

(10) The Laws of Nature are, in fact, regularities or similarities or analogies (Pr, 105) in the perceived motions of physical bodies (S, 234) '. . . these we learn from experience' (Pr, 30); they are observed, or inferred from observations (Pr, 30, 62; S, 228, 264).

(11) 'Once the Laws of Nature have been formed, it becomes the task of the philosopher to show of each phenomenon that it is in

[4] Concerning the equivalence of '*natures*' and '*essences*' see my *Open Society*, ch. 5, section vi.

conformity with these laws, that is, necessarily follows from these principles.' (*DM*, 37; cf. *Pr*, 107; and *S*, 231: 'their [i.e. the "mechanical philosophers'"] province being . . . to account for particular phenomena by reducing them under, and showing their conformity to, such general rules.')

(12) This process *may* be called, if we like, 'explanation' (even 'causal explanation'), so long as we distinguish it clearly from the truly causal (i.e. metaphysical) explanation based upon the true nature or essence of things. *S*, 231; *DM*, 37: 'A thing may be said to be mechanically explained if it is reduced to those most simple and universal principles' (i.e. 'the primary laws of motion which have been proved by experiments . . .' *DM*, 36) 'and proved, by accurate reasoning, to be in agreement and connection with them . . . This means to *explain* and solve the phenomena, and to assign them their *cause* . . .' This terminology is admissible (cf. *DM*, 71) but it must not mislead us. We must always clearly distinguish (cf. *DM*, 72) between an 'essentialist'[5] explanation which appeals to the nature of things and a 'descriptive' explanation which appeals to a Law of Nature, i.e. to the description of an observed regularity. Of these two kinds of explanation only the latter is admissible in physical science.

(13) From both of these we must now distinguish a third kind of 'explanation'—an explanation which appeals to *mathematical hypotheses*. A mathematical hypothesis may be described as a procedure for calculating certain results. It is a mere formalism, a mathematical tool or instrument, comparable to a calculating machine. It is judged merely by its efficiency. It may not only be admissible, it may be useful and it may be admirable, yet it is *not science*: even if it produces the correct results, it is only a trick, 'a knack' (*An*, 50, Qu. 35). And, as opposed to the explanation by essences (which, in mechanics, are simply false) and to that by laws of nature (which, if the laws 'have been proved by experiment', are simply true), the question of the *truth* of a mathematical hypothesis does not arise—only that of its *usefulness as a calculating tool*.

(14) Now, those principles of the Newtonian theory which 'have been proved by experiment'—those of the laws of motion which

[5] The term 'essentialist' (and 'essentialism') is not Berkeley's but was introduced by me in *The Poverty of Historicism*, and in *The Open Society and Its Enemies*.

simply describe the observable regularities of the motion of bodies—are true. But the part of the theory involving the concepts which have been criticized above—absolute space, absolute motion, force, attraction, gravity—is not true, since these are 'mathematical hypotheses'. As such, however, they should not be rejected, if they work well (as in the case of force, attraction, gravity). Absolute space and absolute motion have to be rejected because they do not work (they are to be replaced by the system of fixed stars, and motion relative to it). '"Force", "gravity", "attraction",[6] and words such as these are useful for purposes of reasoning and for computations of motions and of moving bodies; but they do not help us to understand the simple nature of motion itself, nor do they serve to designate so many distinct qualities. . . . As far as attraction is concerned it is clear that it was not introduced by Newton as a true physical quality but merely as a mathematical hypothesis' (*DM*, 17).[7]

(15) Properly understood, a mathematical hypothesis does not claim that anything exists in nature which corresponds to it—neither to the words or terms with which it operates, nor to the functional dependencies which it appears to assert. It erects, as it were, a fictitious mathematical world behind that of appearance, but without the claim that this world exists. 'But what is said of forces residing in bodies, whether attracting or repelling, is to be regarded only as a mathematical hypothesis, and not as anything really existing in nature' (*S*, 234; cf. *DM*, 18, 39 and especially *Alc*, vii, 9, *An*, 50, Qu. 35). It claims only that from its assumptions the correct consequences can be drawn. But it can easily be misinterpreted as claiming more, as claiming to describe a real world behind the world of appearance. No such world *could* be described, however; for the description would necessarily be meaningless.

(16) It can be seen from this that the same appearances *may* be successfully calculated from more than one mathematical hypothesis, and that two mathematical hypotheses which yield the same results concerning the calculated appearances may not only differ, but even

[6] The italics in the Latin original function here as quotation marks.

[7] This was more or less Newton's own opinion; cp. Newton's letters to Bentley, 17th January, and especially 25th February 1692–3, and section 3 of ch. 3, above.

contradict each other (especially if they are misinterpreted as describing a world of essences behind the world of appearances); nevertheless, there may be nothing to choose between them. 'The foremost of men proffer . . . many different doctrines, and even opposite doctrines, and yet their conclusions [i.e. their calculated results] attain the truth . . . Newton and Torricelli seem to disagree with one another, . . . but the thing is well enough explained by both. For all forces attributed to bodies are merely mathematical hypotheses . . . ; thus the same thing may be explained in different ways' (*DM*, 67).

(17) Berkeley's analysis of Newton's theory thus yields the following results: We must distinguish

(*a*) Observations of concrete, particular things.
(*b*) Laws of Nature, which are either observations of regularities, or which are proved ('*comprobatae*', *DM*, 36; this may perhaps mean here 'supported' or 'corroborated'; see *DM*, 31) by experiments, or discovered 'by a diligent observation of the phenomena' (*Pr*, 107).
(*c*) Mathematical hypotheses, which are not based on observation but whose consequences agree with the phenomena (or 'save the phenomena', as the Platonists said).
(*d*) Essentialist or metaphysical causal explanations, which have no place in physical science.

Of these four, (*a*) and (*b*) are based on observation, and can, from experience, be known to be true; (*c*) is not based on observation and has only an instrumental significance—thus more than one instrument may do the trick (cf. (16), above); and (*d*) is known to be false whenever it constructs a world of essences behind the world of appearances. Consequently (*c*) is also known to be false whenever it is interpreted in the sense of (*d*).

(18) These results clearly apply to cases other than Newtonian theory, for example to atomism (corpuscular theory). In so far as this theory attempts to explain the world of appearances by constructing an invisible world of 'inward essences' (*Pr*, 102) behind the world of appearances, it must be rejected. (Cf. *Pr*, 50; *An*, 50, Qu. 56; *S*, 232, 235.)

(19) The work of the scientist leads to something that may be called 'explanation', but it is hardly of great value for *understanding* the thing explained, since the attainable explanation is not one based upon an insight into the nature of things. But it is of practical importance. It enables us to make both *applications* and *predictions*. '. . . laws of nature or motions direct us how to act, and teach us what to expect' (*S*, 234; cf. Pr, 62). Prediction is based merely upon regular sequence (not upon causal sequence—at least not in the essentialist sense). A sudden darkness at noon may be a 'prognostic' indicator, a warning 'sign', a 'mark' of the coming downpour; nobody takes it as its cause. Now *all* observed regularities are of this nature even though 'prognostics' or 'signs' are usually mistaken for true causes (*TV*, 147; Pr, 44, 65, 108; S, 252–4; *Alc*, iv, 14, 15).

(20) A general practical result—which I propose to call 'Berkeley's razor'—of this analysis of physics allows us *a priori* to eliminate from physical science all essentialist explanations. If they have a mathematical and a predictive content they may be admitted *qua* mathematical hypotheses (while their essentialist interpretation is eliminated). If not, they may be ruled out altogether. This razor is sharper than Ockham's: *all* entities are ruled out except those which are perceived.

(21) The ultimate argument for these views, the reason why occult substances and qualities, physical forces, structures of corpuscles, absolute space, and absolute motion, etc. are eliminated, is this: we know that there are no entities such as these because we know that the words professedly designating them must be meaningless. *To have a meaning, a word must stand for an 'idea'*; that is to say, for a perception, or the memory of a perception; in Hume's terminology, for an impression or its reflection in our memory. (It may also stand for a 'notion', such as God; but the words belonging to physical science cannot stand for 'notions'.) Now the words here in question do not stand for ideas. 'Those who assert that active force, action, and the principle of motion are in reality inherent in the bodies, maintain a doctrine that is based upon no experience, and support it by obscure and general terms, and so do not themselves understand what they want to say' (DM, 31).

III

Everybody who reads this list of twenty-one theses must be struck by their modernity. They are surprisingly similar, especially in the criticism of Newton, to the philosophy of physics which Ernst Mach taught for many years in the conviction that it was new and revolutionary; in which he was followed by, for example, Joseph Petzold; and which had an immense influence on modern physics, especially on the Theory of Relativity. There is only one difference: Mach's 'principle of the economy of thought' (*Denkoekonomie*) goes beyond what I have called 'Berkeley's razor', in so far as it allows us not only to discard certain 'metaphysical elements', but also to distinguish in some cases between various competing hypotheses (of the kind called by Berkeley 'mathematical') with respect to their *simplicity*. (Cf. (16) above.) There is also a striking similarity to Hertz's *Principles of Mechanics* (1894), in which he tried to eliminate the concept of 'force', and to Wittgenstein's *Tractatus*.

What is perhaps most striking is that Berkeley and Mach, both great admirers of Newton, criticize the ideas of absolute time, absolute space, and absolute motion, on very similar lines. Mach's criticism, exactly like Berkeley's, culminates in the suggestion that all arguments for Newton's absolute space (like Foucault's pendulum, the rotating bucket of water, the effect of centrifugal forces upon the shape of the earth) fail because these movements are relative to the system of the fixed stars.

To show the significance of this anticipation of Mach's criticism, I may cite two passages, one from Mach and one from Einstein. Mach wrote (in the 7th edition of the *Mechanics*, 1912, ch. ii, section 6, § 11) of the reception of his criticism of *absolute motion*, propounded in earlier editions of his *Mechanics*: 'Thirty years ago the view that the notion of "absolute motion" is meaningless, without any empirical content, and scientifically without use, was generally felt to be very strange. Today this view is upheld by many well-known investigators.' And Einstein said in his obituary notice for Mach ('Nachruf auf Mach', *Physikalische Zeitschr.*, 1916), referring to this view of Mach's: 'It is not improbable that Mach would have found the Theory of Relativity if, at a time when his mind was still young, the problem of the constancy of velocity of light had agitated the physicists.' This remark of Einstein's is no doubt more than generous.[8] Of the bright light it throws upon Mach some reflection must fall upon Berkeley.[9]

IV

A few words may be said about the relation of Berkeley's philosophy of science to his metaphysics. It is very different indeed from Mach's.

While the positivist Mach was an enemy of all traditional, that is non-positivistic, metaphysics, and especially of all theology, Berkeley was a Christian theologian, and intensely interested in Christian apologetics. While Mach and Berkeley agreed that such words as 'absolute time', 'absolute space' and 'absolute motion' are meaningless and therefore to be eliminated from science, Mach surely would not have agreed with Berkeley on the reason why physics cannot treat of real causes. Berkeley believed in causes, even in 'true' or 'real' causes; but all true or real causes were to him 'efficient or final causes' (S, 231), and therefore *spiritual* and utterly beyond physics (cf. HP, ii). He also believed in true or real causal *explanation* (S, 231) or, as I may perhaps call it, in 'ultimate explanation'. This, for him, was God.

All appearances are truly caused by God, and explained through God's intervention. This for Berkeley is the simple reason why physics can only describe regularities, and why it cannot find true causes.

It would be a mistake, however, to think that the similarity between Berkeley and Mach is by these differences shown to be only superficial. On the contrary, Berkeley and Mach are both convinced that there is no physical world (of primary qualities, or of atoms; cf. Pr, 50; S, 232, 235) behind the world of physical appearances (Pr, 87, 88). Both believed in a form of the doctrine nowadays called phenomenalism—the view that physical things are bundles, or complexes, or constructs of phenomenal *qualities*, of particular experienced colours, noises, etc.;

[8] Mach survived Einstein's Special Theory of Relativity by more than eleven years, at least eight of which were very active years; but he remained strongly opposed to it; and though he alluded to it in the preface to the last (seventh) German edition (1912) of the *Mechanik* published during his lifetime, the allusion was by way of compliment to the opponent of Einstein, Hugo Dingler: Einstein's name and that of the theory were not mentioned.

[9] This is not the place to discuss other predecessors of Mach, such as Leibniz.

Mach calls them 'complexes of elements'. The difference is that for Berkeley, these are directly caused by God. For Mach, they are just there. While Berkeley says that there can be nothing physical behind the physical phenomena, Mach suggests that there is nothing at all behind them.

V

The great historical importance of Berkeley lies, I believe, in his protest against essentialist explanations in science. Newton himself did not interpret his theory in an essentialist sense; he himself did not believe that he had discovered the fact that physical bodies, by their nature, are not only extended but endowed with a force of attraction (radiating from them, and proportional to the amount of matter in them). But soon after him the essentialist interpretation of his theory became the ruling one, and remained so till the days of Mach.

In our own day essentialism has been dethroned; a Berkeleian or Machian positivism or instrumentalism has, after all these years, become fashionable.

Yet there is clearly a third possibility—a 'third view' (as I call it).

Essentialism is, I believe, untenable. It implies the idea of an *ultimate* explanation, for an essentialist explanation is neither in need of, nor capable of, further explanation. (If it is in the nature of a body to attract others, then there is no need to ask for an explanation of this fact, and no possibility of finding such an explanation.) Yet we know, at least since Einstein, that explanation may be pushed, unexpectedly, further and further.

But although we must reject essentialism, this does not mean that we have to accept positivism; for we may accept the 'third view'.

I shall not here discuss the positivist dogma of meaning, since I have done so elsewhere. I shall make only six observations. (i) One can work with something like a world 'behind' the world of appearance without committing oneself to essentialism (especially if one assumes that we can never know whether there may not be a further world behind that world). To put it less vaguely, one can work with the idea of hierarchical levels of explanatory hypotheses. There are comparatively low level ones (somewhat like what Berkeley had in mind when he spoke of

'Laws of Nature'); higher ones such as Kepler's laws, still higher ones such as Newton's theory, and, next, Relativity. (ii) These theories are not mathematical hypotheses, that is, *nothing but* instruments for the prediction of appearances. Their function goes very much further; for (iii) there is no pure appearance or pure observation: what Berkeley had in mind when he spoke of these things was always the result of interpretation, and (iv) it had therefore a theoretical or hypothetical admixture. (v) New theories, moreover, may lead to re-interpretation of old appearances, and in this way change the world of appearances. (vi) The multiplicity of explanatory theories which Berkeley noted (see Section ii (16), above) is used, wherever possible, to construct, for any two competing theories, conditions in which they yield different observable results, so that we can make a crucial test to decide between them, winning in this way new experience.

A main point of this third view is that science aims at *true* theories, even though we can never be sure that any particular theory is true; and that science *may* progress (and know that it does so) by inventing theories which compared with earlier ones may be described as better approximations to what is true.

So we can now admit, without becoming essentialist, that in science we always try *to explain the known by the unknown*; the observed (and observable) by the unobserved (and, perhaps, unobservable). At the same time we can now admit, without becoming instrumentalist, what Berkeley said of the nature of hypotheses in the following passage (S, 228), which shows both his weakness—his failure to realize the conjectural character of all science, including what he calls the 'laws of nature'—and his strength, his admirable understanding of the logical structure of hypothetical explanation.

'It is one thing', Berkeley writes, 'to arrive at general laws of nature from a contemplation of the phenomena; and another to frame an hypothesis, and from thence deduce the phenomena. Those who suppose epicycles, and by them explain the motions and appearances of the planets, may not therefore be thought to have discovered principles true in fact and nature. And, albeit we may from the premises infer a conclusion, it will not follow that we can argue reciprocally, and from the conclusion infer the premises. For instance, supposing an elastic fluid, whose constituent minute particles are equidistant from each other, and of equal densities and diameters, and recede one from another with a centrifugal force which is inversely as the distance of the centres; and admitting that from such supposition it must follow that the density and elastic force of such fluid are in the inverse proportion of the space it occupies when compressed by any force; yet we cannot reciprocally infer that a fluid endowed with this property must therefore consist of such supposed equal particles.'

7

KANT'S CRITIQUE AND COSMOLOGY

One hundred and fifty years ago Immanuel Kant died, having spent the eighty years of his life in the Prussian provincial town of Königsberg. For years his retirement had been complete,[1] and his friends intended a quiet burial. But this son of an artisan was buried like a king. When the rumour of his death spread through the town the people flocked to his house demanding to see him. On the day of the funeral the life of the town was at a standstill. The coffin was followed by thousands, while the bells of all the churches tolled. Nothing like this had ever before happened in Königsberg, say the chroniclers.[2]

It is difficult to account for this astonishing upsurge of popular feeling. Was it due solely to Kant's reputation as a great philosopher and a good man? It seems to me that there was more in it than this;

A broadcast given on the eve of the hundred and fiftieth anniversary of Kant's death. First published (without the footnotes) under the title 'Immanuel Kant: Philosopher of the Enlightenment' in The Listener, **51**, 1954.

[1] Six years before Kant's death, Pörschke reports (see his letter to Fichte of 2nd July 1798) that owing to Kant's retired way of life, he was being forgotten even in Königsberg.

[2] C.E.A.Ch. Wasianski, *Immanuel Kant in seinen letzten Lebensjahren* (from *Ueber Immanuel Kant. Dritter Band*, Königsberg, bei Nicolovius, 1804). 'The public newspapers and a special publication have made Kant's funeral known in all its circumstances.'

and I suggest that in the year 1804, under the absolute monarchy of Frederick William, those bells tolling for Kant carried an echo of the American and French revolutions—of the ideas of 1776 and 1789. I suggest that to his countrymen Kant had become an embodiment of these ideas.[3] They came to show their gratitude to a teacher of the Rights of Man, of equality before the law, of world citizenship, of peace on earth, and, perhaps most important, of emancipation through knowledge.[4]

1. KANT AND THE ENLIGHTENMENT

Most of these ideas had reached the Continent from England through a book published in 1733, Voltaire's *Letters Concerning the English Nation*. In this book Voltaire contrasts English constitutional government with Continental absolute monarchy; English religious toleration with the attitude of the Roman Church; and the explanatory power of Newton's cosmology and of Locke's analytic empiricism with the dogmatism of Descartes. Voltaire's book was burnt; but its publication marks the beginning of a philosophical movement—a movement whose peculiar mood of intellectual aggressiveness was little understood in England, where there was no occasion for it.

Sixty years after Kant's death these same English ideas were being presented to the English as a 'shallow and pretentious intellectualism': and ironically enough the English word 'Enlightenment', which was then used to name the movement started by Voltaire, is still beset by this connotation of shallowness and pretentiousness; this, at least, is what the *Oxford English Dictionary* tells us.[5] I need hardly add

[3] Kant's sympathies with the ideas of 1776 and 1789 were well known, for he used to express them in public. (Cf. Motherby's eye-witness report on Kant's first meeting with Green in R. B. Jachmann, *Immanuel Kant geschildert in Briefen—Ueber Immanuel Kant, Zweiter Band*, Königsberg bei Nicolovius, 1804; eighth letter; pp. 54 f. of the edition of 1902).

[4] I say 'most important' because Kant's well-deserved rise from near poverty to fame and comparatively easy circumstances helped to create on the Continent the idea of emancipation through self-education, in this form hardly known in England where the 'self-made man' was the uncultured upstart. On the Continent, the educated had been for a long time the middle classes, while in England they were the upper classes.

[5] The *O.E.D.* says (some of the italics are mine): 'Enlightenment . . . 2. Sometimes used [after the German *Aufklärung, Aufklärerei*] to designate the spirit and the aims of the French

that no such connotation is intended when I use the word 'Enlightenment'.

Kant believed in the Enlightenment. He was its last great defender. I realize that this is not the usual view. While I see Kant as the defender of the Enlightenment, he is more often taken as the founder of the school which destroyed it—of the Romantic School of Fichte, Schelling, and Hegel. I contend that these two interpretations are incompatible.

Fichte, and later Hegel, tried to appropriate Kant as the founder of their school. But Kant lived long enough to reject the persistent advances of Fichte, who proclaimed himself Kant's successor and heir. In *A Public Declaration Concerning Fichte*,[6] which is too little known, Kant wrote: 'May God protect us from our friends. . . . For there are fraudulent and perfidious so-called friends who are scheming for our ruin while speaking the language of good-will.' It was only after Kant's death, when he could no longer protest, that this world-citizen was successfully pressed into the service of the nationalistic Romantic School, in spite of all his warnings against romanticism, sentimental enthusiasm, and *Schwärmerei*. But let us see how Kant himself describes the idea of the Enlightenment:[7]

> Enlightenment is the emancipation of man from a state of self-imposed tutelage . . . of incapacity to use his own intelligence without external guidance. Such a state of tutelage I call 'self-imposed' if it is due, not to lack of intelligence, but to lack of courage or determination to use one's own intelligence without the help of a leader. *Sapere aude*! Dare to use your own intelligence! This is the battle-cry of the Enlightenment.

philosophers of the 18th c., of *others whom it is intended to associate with them in the implied charge of shallow and pretentious intellectualism*, unreasonable contempt of tradition and authority, etc.' The *O.E.D.* does not mention that 'Aufklärung' is a translation of the French '*éclaircissement*', and that it does not have these connotations in German, while '*Aufklärerei*' (or '*Aufkläricht*') are disparaging neologisms invented and exclusively used by the Romantics, the enemies of the Enlightenment. The *O.E.D.* quotes J. H. Stirling, *The Secret of Hegel*, 1865, and Caird, *The Philosophy of Kant*, 1889, as users of the word in sense 2.

[6] The date of this Declaration is 1799. Cf. *WWC* (i.e. *Immanuel Kant's Werke*, ed. Ernst Cassirer, *et al.*), vol. VIII, pp. 515 f., and my *Open Society*, note 58 to ch. 12 (4th edn. 1962, vol. II, p. 313).

[7] *What is Enlightenment* (1785); *WWC*, IV, p. 169.

Kant is saying something very personal here. It is part of his own history. Brought up in near poverty, in the narrow outlook of Pietism—a severe German version of Puritanism—his own life was a story of emancipation through knowledge. In later years he used to look back with horror to what he called[8] 'the slavery of childhood', his period of tutelage. One might well say that the dominant theme of his whole life was the struggle for spiritual freedom.

2. KANT'S NEWTONIAN COSMOLOGY

A decisive role in this struggle was played by Newton's theory, which had been made known on the Continent by Voltaire. The cosmology of Copernicus and Newton became the powerful and exciting inspiration of Kant's intellectual life. His first important book,[9] *The Theory of the Heavens*, has the interesting sub-title: *An Essay on the Constitution and the Mechanical Origin of the Universe, Treated According to Newtonian Principles*. It is one of the greatest contributions ever made to cosmology and cosmogony. It contains the first formulation not only of what is now called the 'Kant-Laplace hypothesis' of the origin of the solar system, but also, anticipating Jeans, an application of this idea to the 'Milky Way' (which Thomas Wright had interpreted as a stellar system five years earlier). But all this is excelled by Kant's identification of the nebulae as other 'Milky Ways'—distant stellar systems similar to our own.

It was the cosmological problem, as Kant explains in one of his letters,[10] which led him to his theory of knowledge, and to his *Critique of*

[8] See T. G. von Hippel's Biography (Gotha, 1801, pp. 78 f.). See also the letter to Kant from D. Ruhnken (one of Kant's schoolfellows in the Pietist Frederickan College), in Latin, of 10th March 1771, in which he speaks of the 'stern yet not regrettable discipline *of the fanatics*' who had educated them.

[9] Published in 1755. The full principal title might be translated: *General Natural History [of the Heavens] and Theory of the Heavens*. The words '*General Natural History*' are used to indicate that the work is a contribution to the theory of the *evolution* of stellar systems.

[10] To C. Garve, 21st September 1798. 'My starting point was not an investigation into the existence of God, but the antinomy of pure reason: "The world has a beginning: it has no beginning", etc. down to the fourth . . .' (Here comes a place where Kant, apparently, mixes up his third and fourth antinomies.) 'It was these [antinomies] which first stirred me from my dogmatic slumber and drove me to the critique of reason . . . , in order to resolve the scandal of the apparent contradiction of reason with itself.'

Pure Reason. He was concerned with the knotty problem (which has to be faced by every cosmologist) of the finitude or infinity of the universe, with respect to both space and time. As far as space is concerned a fascinating solution has been suggested since, by Einstein, in the form of a world which is both finite and without limits. This solution cuts right through the Kantian knot, but it uses more powerful means than those available to Kant and his contemporaries. As far as time is concerned no equally promising solution of Kant's difficulties has been offered up to now.

3. THE CRITIQUE AND THE COSMOLOGICAL PROBLEM

Kant tells us[11] that he came upon the central problem of his *Critique* when considering whether the universe had a beginning in time or not. He found to his dismay that he could produce seemingly valid proofs for both of these possibilities. The two proofs[12] are interesting; it needs concentration to follow them, but they are not long, and not hard to understand.

For the first proof we start by analysing the idea of an infinite sequence of years (or days, or any other equal and finite intervals of time). Such an infinite sequence of years must be a sequence which goes on and on and never comes to an end. It can never be completed: a completed or an elapsed infinity of years is a contradiction in terms. Now in his first proof Kant simply argues that the world must have a beginning in time since otherwise, at this present moment, an infinite number of years must have elapsed; which is impossible. This concludes the first proof.

For the second proof we start by analysing the idea of a completely empty time—the time before there was a world. Such an empty time, in which there is nothing whatever, must be a time none of whose time-intervals is differentiated from any other by its temporal relation to things and events, since things and events simply do not exist at all.

[11] See the foregoing note. Cf. also Leibniz's correspondence with Clarke (*Philos. Bibl.*, edited by Kirchmann, **107**, pp. 134 f., 147 f., 188 ff.), and Kant's *Reflexionen zur Kritischen Philosophie*, edited by B. Erdmann; esp. No. 4.

[12] See *Critique of Pure Reason* (2nd edn.), 454 ff.

Now take the last interval of the empty time—the one immediately before the world begins. Clearly, this interval is differentiated from all earlier intervals since it is characterized by its close temporal relation to an event—the beginning of the world; yet the same interval is supposed to be empty, which is a contradiction in terms. Now in his second proof Kant simply argues that the world cannot have a beginning in time since otherwise there would be a time-interval—the moment immediately before the world began—which is empty and yet characterized by its immediate temporal relation to an event in the world; which is impossible.

We have here a clash between two proofs. Such a clash Kant called an 'antinomy'. I shall not trouble you with the other antinomies in which Kant found himself entangled, such as those concerning the limits of the universe in space.

4. SPACE AND TIME

What lesson did Kant draw from these bewildering antinomies? He concluded[13] that our ideas of space and time are inapplicable to the universe as a whole. We can, of course, apply the ideas of space and time to ordinary physical things and physical events. But space and time themselves are neither things nor events: they cannot even be observed: they are more elusive. They are a kind of framework for things and events: something like a system of pigeon-holes, or a filing system, for observations. Space and time are not part of the real empirical world of things and events, but rather part of our mental outfit, our apparatus for grasping this world. Their proper use is as instruments of observation: in observing any event we locate it, as a rule, immediately and intuitively in an order of space and time. Thus space and time may be described as a frame of reference which is not based upon experience but intuitively used in experience, and properly applicable to experience. This is why we get into trouble if we misapply the ideas of space and time by using them in a field which transcends all possible experience—as we did in our two proofs about the universe as a whole.

[13] Op. cit., 518 ff. 'The Doctrine of Transcendental Idealism as the Key to the Solution of the Cosmological Dialectic.'

To the view which I have just outlined Kant chose to give the ugly and doubly misleading name 'Transcendental Idealism'. He soon regretted this choice,[14] for it made people believe that he was an idealist in the sense of denying the reality of physical things: that he declared physical things to be mere ideas. Kant hastened to explain that he had only denied that space and time are empirical and real—empirical and real in the sense in which physical things and events are empirical and real. But in vain did he protest. His difficult style sealed his fate: he was to be revered as the father of German Idealism. I suggest that it is time to put this right. Kant always insisted[15] that the physical things in space and time are real. And as to the wild and obscure metaphysical speculations of the German Idealists, the very title of Kant's *Critique* was chosen to announce a critical attack upon all such speculative reasoning. For what the *Critique* criticizes is pure reason; it criticizes and attacks all reasoning about the world that is 'pure' in the sense of being untainted by sense experience. Kant attacked pure reason by showing that pure reasoning about the world must always entangle us in antinomies. Stimulated by Hume, Kant wrote his *Critique* in order to establish[16] that the limits of sense experience are the limits of all sound reasoning about the world.

[14] *Prolegomena* (1783), *Appendix*: 'Specimen of a Judgment on the *Critique* Anticipating its Investigation'. See also the *Critique*, 2nd edn. (1787; the first edition had been published in 1781), pp. 274–9, 'The Refutation of Idealism', and the last footnote to the Preface of the *Critique of Practical Reason*.

[15] See the passages mentioned in the foregoing note.

[16] See Kant's letter to M. Herz, of 21st February 1772, in which he gives, as a tentative title of what became the first *Critique*, 'The Limits of Sense Experience and of Reason'. See also the *Critique of Pure Reason* (2nd edn.), pp. 738 f. (italics mine): '*There is no need for a critique of reason* in its empirical use; for its principles are continuously submitted to tests, being tested by the touchstone of experience. Similarly, there is no need for it within the field of mathematics, where its conceptions must be presented at once in pure intuition [of space and time] . . . But in a field in which reason is constrained neither by sense-experience nor by pure intuition to follow a visible track—namely, in the field of its transcendental use . . . —*there is much need to discipline reason, so that its tendency to overstep the narrow limits of possible experience may be subdued* . . .'

5. KANT'S COPERNICAN REVOLUTION

Kant's faith in his theory of space and time as an intuitive frame of reference was confirmed when he found in it a key to the solution of a second problem. This was the problem of the validity of Newtonian theory in whose absolute and unquestionable truth he believed,[17] in common with all contemporary physicists. It was inconceivable, he felt, that this exact mathematical theory should be nothing but the result of accumulated observations. But what else could be its basis? Kant approached this problem by first considering the status of geometry. Euclid's geometry is not based upon observation, he said, but upon our intuition of spatial relations. Newtonian science is in a similar position. Although confirmed by observations it is the result not of these observations but of our own ways of thinking, of our attempts to order our sense-data, to understand them, and to digest them intellectually. It is not these sense-data but our own intellect, the organization of the digestive system of our mind, which is responsible for our theories. Nature as we know it, with its order and with its laws, is thus largely a product of the assimilating and ordering activities of our mind. In Kant's own striking formulation of this view,[18] 'Our intellect does not draw its laws from nature, but imposes its laws upon nature'.

This formula sums up an idea which Kant himself proudly calls his 'Copernican Revolution'. As Kant puts it, Copernicus,[19] finding that no progress was being made with the theory of the revolving heavens, broke the deadlock by turning the tables, as it were: he assumed that it is not the heavens which revolve while we the observers stand still, but that we the observers revolve while the heavens stand still. In a similar

[17] See, for example, Kant's *Metaphysical Foundations of Natural Science* (1786), containing the *a priori* demonstration of Newtonian mechanics. See also the end of the penultimate paragraph of the *Critique of Practical Reason*. I have tried to show elsewhere (chapter 2 of this volume) that some of the greatest difficulties in Kant are due to the tacit assumption that Newtonian Science is demonstrably true (that it is *epistēmē*), and that, with the realization that this is not so, one of the most fundamental problems of the *Critique* dissolves. See also ch. 8, below.

[18] See *Prolegomena*, end of section 37. Kant's footnote referring to Crusius is interesting: it suggests that Kant had an inkling of the analogy between what he called his 'Copernican Revolution' and his principle of autonomy in ethics.

[19] My text here is a free translation from the *Critique of Pure Reason*, 2nd edn., pp. xvi f.

way, Kant says, the problem of scientific knowledge is to be solved—the problem how an exact science, such as Newtonian theory, is possible, and how it could ever have been found. We must give up the view that we are passive observers, waiting for nature to impress its regularity upon us. Instead we must adopt the view that in digesting our sense-data we actively impress the order and the laws of our intellect upon them. Our cosmos bears the imprint of our minds.

By emphasizing the role played by the observer, the investigator, the theorist, Kant made an indelible impression not only upon philosophy but also upon physics and cosmology. There is a Kantian climate of thought without which Einstein's theories or Bohr's are hardly conceivable; and Eddington might be said to be more of a Kantian, in some respects, than Kant himself. Even those who, like myself, cannot follow Kant all the way can accept his view that the experimenter must not wait till it pleases nature to reveal her secrets, but that he must question her.[20] He must cross-examine nature in the light of his doubts, his conjectures, his theories, his ideas, and his inspirations. Here, I believe, is a wonderful philosophical find. It makes it possible to look upon science, whether theoretical or experimental, as a human creation, and to look upon its history as part of the history of ideas, on a level with the history of art or of literature.

There is a second and even more interesting meaning inherent in Kant's version of the Copernican Revolution, a meaning which may perhaps indicate an ambivalence in his attitude towards it. For Kant's Copernican Revolution solves a human problem to which Copernicus' own revolution gave rise. Copernicus deprived man of his central position in the physical universe. Kant's Copernican Revolution takes the sting out of this. He shows us not only that our location in the physical universe is irrelevant, but also that in a sense our universe may well be said to turn about us; for it is we who produce, at least in part, the order we find in it; it is we who create our knowledge of it. We are discoverers: and discovery is a creative art.

[20] *Op. cit.*, pp. xii f.; cf. especially the passage: 'The physicists . . . realized that they . . . had to compel Nature to reply to their questions, rather than let themselves be tied to her apron-strings, as it were.'

6. THE DOCTRINE OF AUTONOMY

From Kant the cosmologist, the philosopher of knowledge and of science, I now turn to Kant the moralist. I do not know whether it has been noticed before that the fundamental idea of Kant's ethics amounts to another Copernican Revolution, analogous in every respect to the one I have described. For Kant makes man the lawgiver of morality just as he makes him the lawgiver of nature. And in doing so he gives back to man his central place both in his moral and in his physical universe. Kant humanized ethics, as he had humanized science.

Kant's Copernican Revolution in the field of ethics[21] is contained in his doctrine of autonomy—the doctrine that we cannot accept the command of an authority, however exalted, as the ultimate basis of ethics. For whenever we are faced with a command by an authority, it is *our* responsibility to judge whether this command is moral or immoral. The authority may have power to enforce its commands, and we may be powerless to resist. But unless we are physically prevented from choosing the responsibility remains ours. It is our decision whether to obey a command, whether to accept authority.

Kant boldly carries this revolution into the field of religion. Here is a striking passage:[22]

> Much as my words may startle you, you must not condemn me for saying: every man creates his God. From the moral point of view . . . you even *have* to create your God, in order to worship in Him your creator. For in whatever way . . . the Deity should be made known to

[21] See the *Grundlegung zur Met. d. Sitten*, 2nd section (*WWC*, pp. 291 ff., especially 299 ff.): 'The Autonomy of the Will as the Highest Principle of Morality', and the 3rd section (*WWC*, pp. 305 ff.).

[22] This is a free translation (although as close as is compatible with lucidity, I believe) from a passage contained in the footnote to the Fourth Chapter, Part II, § 1, of *Religion within the Limits of Pure Reason* (2nd edn., 1794 = *WWC*, vi, p. 318; the passage is not in the 1st edn., 1793. See also the Introduction to the present volume, note 9). The passage is foreshadowed by the following: 'We ourselves judge revelation by the moral law' (*Lectures on Ethics by Immanuel Kant*, translated by L. Infield 1930; the translation of the passage is corrected by P. A. Schilpp, *Kant's Pre-Critical Ethics*, 1938, p. 166, note 63). Just before Kant says of the moral law that 'our own reason is capable of revealing it to us'.

> you, and even . . . if He should reveal Himself to you: it is you . . . who must judge whether you are permitted [by your conscience] to believe in Him, and to worship Him.

Kant's ethical theory is not confined to the statement that a man's conscience is his moral authority. He also tries to tell us what our conscience may demand from us. Of this, the moral law, he gives several formulations. One of them is:[23] 'Always regard every man as an end in himself, and never use him merely as a means to your ends.' The spirit of Kant's ethics may well be summed up in these words: dare to be free; and respect the freedom of others.

Upon the basis of these ethics Kant erected his most important theory of the state,[24] and his theory of international law. He demanded[25] a league of nations, or a federal union of states, which ultimately was to proclaim and to maintain eternal peace on earth.

I have tried to sketch in broad outline Kant's philosophy of man and his world, and its two main inspirations—Newtonian cosmology, and the ethics of freedom; the two inspirations to which Kant referred when he spoke[26] of the starry heavens above us and the moral law within us.

Stepping back further to get a still more distant view of Kant's historical role, we may compare him with Socrates. Both were accused of perverting the state religion, and of corrupting the minds of the young. Both denied the charge; and both stood up for freedom of thought. Freedom meant more to them than absence of constraint; it was for both a way of life.

From Socrates' apology and from his death there sprang a new idea of a free man: the idea of a man whose spirit cannot be subdued; of a man who is free because he is self-sufficient; who is not in need of

[23] See the *Grundlegung*, 2nd section (*WWC*, iv, p. 287). My translation is, again, free.

[24] See, especially, Kant's various formulations to the effect that the principle of the just state is to establish equality in those limitations of the freedom of its citizens which are unavoidable if *the freedom of each should coexist with the freedom of all* (e.g. *Critique of Pure Reason*, 2nd edn., p. 373).

[25] *On Peace Eternal* (1795).

[26] At the 'Conclusion' of the *Critique of Practical Reason*; see especially the end of the penultimate paragraph, referred to in note 17 above.

constraint because he is able to rule himself, and to accept freely the rule of law.

To this Socratic idea of self-sufficiency, which forms part of our western heritage, Kant has given a new meaning in the fields of both knowledge and morals. And he has added to it further the idea of a community of free men—of all men. For he has shown that every man is free; not because he is born free, but because he is born with the burden of responsibility for free decision.

8

ON THE STATUS OF SCIENCE AND OF METAPHYSICS

1. KANT AND THE LOGIC OF EXPERIENCE

In this talk I do not propose to speak of ordinary everyday experience. I intend, rather, to use the word 'experience' in the sense in which we use it when we say that science is based on experience. Since, however, experience in science is after all no more than an extension of ordinary everyday experience what I shall have to say will apply, by and large, to everyday experience also.

In order not to get lost in abstractions I intend to discuss the logical status of a specific empirical science—Newtonian dynamics. I do not, however, presuppose any knowledge of physics on the part of my audience.

One of the things a philosopher may do, and one of those that may rank among his highest achievements, is to see a *riddle*, a *problem*, or a *paradox*, not previously seen by anyone else. This is an even greater achievement than resolving the riddle. The philosopher who

Two Radio Talks written for the Free Radio-University, Berlin; first published in Ratio, **1**, 1958, pp. 97–115.

first sees and understands a new problem disturbs our laziness and complacency. He does to us what Hume did for Kant: he rouses us from our 'dogmatic slumber'. He opens out a new horizon before us.

The first philosopher clearly to apprehend *the riddle of natural science* was Kant. I do not know of any philosopher, either before or since, who has been so profoundly stirred by it.

When Kant talked of 'natural science' he almost invariably had Isaac Newton's celestial mechanics in mind. Kant himself made important contributions to Newtonian physics and he was one of the greatest cosmologists of all time. His two principal cosmological works are the *Natural History and Theory of the Heavens* (1755) and the *Metaphysical Foundations of Natural Science* (1786). Both themes were (in Kant's own words) 'treated according to Newtonian Principles'.[1]

Like almost all of his contemporaries who were knowledgeable in this field, Kant believed in the *truth* of Newton's celestial mechanics. The almost universal belief that Newton's theory *must* be true was not only understandable but seemed to be well-founded. Never had there been a better theory, nor one more severely tested. Newton's theory not only accurately predicted the orbits of all the planets, including their deviations from Kepler's ellipses, but also the orbits of all their satellites. Moreover, its few simple principles supplied at the same time a celestial mechanics and a terrestrial mechanics.

Here was a universally valid system of the world that described the laws of cosmic motion in the simplest and clearest way possible—and with absolute accuracy. Its principles were as simple and precise as geometry itself—as Euclid's supreme achievement, that unsurpassed model of all science. Newton had indeed propounded a kind of cosmic geometry consisting of Euclid supplemented by a theory (which too could be represented geometrically) of the motion of mass-points under the influence of forces. It added, apart from the concept of time, only two essentially new concepts to Euclidean geometry: the concept of mass or of a material *mass-point*, and the even more important

[1] Also of great importance is the Latin *Physical Monadology* of 1756 in which Kant anticipated the main idea of Boscovič; but in his work of 1786 Kant repudiated the theory of matter propounded in his *Monadology*.

concept of a directed *force* (*vis* in Latin and *dynamis* in Greek from which the name 'dynamics' for Newton's theory is derived).

Here then was a science of the cosmos, of nature; and, it was claimed, a science based upon experience. It was a deductive science, exactly like geometry. Yet Newton himself asserted that he had wrested its functional principles from experience *by induction*. In other words, Newton asserted that the truth of his *theory* could be logically derived from the truth of certain *observation-statements*. Although he did not describe these observation-statements precisely it is nevertheless clear that he must have been referring to Kepler's laws, the laws of the elliptic motions of the planets. And we can still find prominent physicists who maintain that Kepler's laws can be derived inductively from observation-statements, and that Newton's principles can in turn be derived, entirely or almost entirely, from Kepler's laws.

It was one of Kant's greatest achievements that, roused by Hume, he recognized that this contention was paradoxical. Kant saw more clearly than anyone before or since how absurd it was to assume that Newton's theory could be derived from observations. Since this important insight of Kant's is falling into oblivion, partly because of his own contributions towards a solution of the problem he had discovered, I will now present and discuss it in detail.

The assertion that Newton's theory was derived from observation will be criticized here on three counts:

First, the assertion is *intuitively not credible*, especially when we compare the character of the theory with the character of observation-statements.

Secondly, the assertion is *historically false*.

Thirdly, the assertion is *logically false*: it is a logically impossible assertion.

Let us examine the first point—that it is intuitively not credible that *observations* can show Newton's mechanics to be true.

To see this we merely have to remember how utterly Newtonian theory differs from any observation-statement. In the first place observations are *always* inexact, while the theory makes absolutely exact assertions. Moreover, it was one of the triumphs of Newtonian theory that it stood up to subsequent observations which as regards precision went far beyond what could be attained in Newton's own time. Now it is incredible that more precise statements, let alone the absolutely precise statements of the theory itself, could be logically derived from less exact or inexact ones.[2] But even if we forget all about the question of precision we should realize that an observation is always made under very special conditions, and that each observed situation is always a highly specific situation. The theory, on the other hand, claims to apply in all possible circumstances—not only to the planets Mars or Jupiter, or even to the satellites in the solar system, but to *all* planetary motion and to *all* solar systems. Indeed, its claims go far beyond all this. For example the theory makes assertions about gravitational pressure inside the stars, assertions which even today have never been tested by observation. Moreover, observations are always *concrete*, while theory is *abstract*. For example we never observe mass points but rather extended planets. This may perhaps not be so very important; but what is of the utmost importance is that we can never—I repeat, never—*observe* anything like Newtonian *forces*. Admittedly, since forces are so defined that they may be measured by measuring accelerations, we can indeed *measure* forces; and we may at times measure a force not by measuring an acceleration, but for instance with the help of a spring balance. *Yet in all these measurements, without exception, we always presuppose the truth of Newtonian dynamics*. Without the prior assumption of a dynamical theory it is simply impossible to measure forces. But forces, and changes of forces, are among the most important things of which the theory treats. Thus we may assert that at least some of the objects of which the theory treats are abstract and unobservable objects. For all these reasons it is intuitively not credible that the theory should be logically derivable from observations.

This result would not be affected even if it were possible so to reformulate Newton's theory that any reference to forces was avoided. Nor would it be affected by a dismissal of force as a mere fiction, or perhaps as a purely theoretical construction which serves only as a tool or instrument for prediction. Because the thesis which we are questioning says that Newton's theory can be shown to be true by observation. And our objection was that we can only observe *concrete things*, while

[2] A similar consideration may be found in Bertrand Russell's *The Analysis of Mind*, 1922, pp. 95 f.

theory, and particularly Newtonian forces, are *abstract*. These difficulties are in no way mitigated if we make the theory even more abstract by eliminating the notion of force or by unmasking it as a mere auxiliary construction.

So much for my first point.

My second point was that it is historically false to believe that Newton's dynamics was derived from observation. Though this belief is widespread, it is nevertheless a belief in a historical myth—or, if you like, a bold distortion of history. To show this I shall briefly refer to the part played by the three most important precursors of Newton in this field: Nicolaus Copernicus, Tycho Brahe, and Johannes Kepler.

Copernicus studied in Bologna under the Platonist Novara; and Copernicus' idea of placing the sun rather than the earth in the centre of the universe was not the result of new observations but of a *new interpretation* of old and well-known facts in the light of semi-religious Platonic and Neo-Platonic ideas. The crucial idea can be traced back to the sixth book of Plato's *Republic*, where we can read that the sun plays the same role in the realm of visible things as does the idea of the good in the realm of ideas. Now the idea of the good is the highest in the hierarchy of Platonic ideas. Accordingly the sun, which endows visible things with their visibility, vitality, growth and progress, is the highest in the hierarchy of the visible things in nature.

This passage in the *Republic* is of outstanding importance among the passages upon which Neo-Platonic philosophy—particularly Christian Neo-Platonic philosophy—was based.

Now if the sun was to be given pride of place, if the sun merited a divine status in the hierarchy of visible things, then it was hardly possible for it to revolve about the earth. The only fitting place for so exalted a star was the centre of the universe.[3] So the earth was bound to revolve about the sun.

This Platonic idea, then, forms the historical background of the Copernican revolution. It does not start with observations, but with a

[3] Cp. Aristotle, *De Caelo*, 293b1–5, where the doctrine that the centre of the universe is 'precious' and therefore to be occupied by a central fire is criticized and ascribed to the 'Pythagoreans' (which perhaps means his rivals, the successors of Plato who stayed in the Academy).

religious or mythological idea. Such beautiful but wild ideas have often been put forward by great thinkers, and just as often by cranks. But Copernicus, for one, was not a crank. He was highly critical of his own mystical intuitions, which he rigorously examined in the light of astronomical observations reinterpreted with the aid of the new idea. He rightly considered these observations to be extremely important. Yet looked at from a historical or genetical point of view observations were not the source of his idea. The idea came first, and it was indispensable for the interpretation of the observations: they had to be interpreted in its light.

Johannes Kepler—the pupil and assistant of Tycho Brahe, to whom that great teacher left his unpublished observations—was a Copernican. Like Plato himself, Kepler, though always a critical thinker, was steeped in astrological lore; and he also was like Plato deeply influenced by the number-mysticism of the Pythagoreans. What he hoped to discover, what he searched for throughout his life, was the arithmetical law underlying the structure of the world, the law upon which rested the construction of the circles of Copernicus' solar system, and upon which, in particular, their relative distances from the sun were based. He never found what he sought. He did not find in Tycho's observations the hoped-for confirmation of his belief that Mars revolved about the sun in a perfectly circular orbit with uniform velocity. On the contrary, he discovered in Tycho's observations a *refutation* of the circle hypothesis. Thus he discarded the circle hypothesis; and having tried in vain various other solutions, he hit upon the next best thing: the hypothesis of the ellipse. And he found that the observations could be made to agree with the new hypothesis—although only under the assumption, at first far from welcome, that Mars did not travel with uniform velocity.

Historically, therefore, Kepler's laws were not the result of observations. What happened was that Kepler tried in vain to interpret Tycho's observations by means of his original circle hypothesis. The observations *refuted* this hypothesis, and so he tried the next best solutions—the oval, and the ellipse. The observations still did not prove that the hypothesis of an ellipse was correct, but they could now be *explained* by means of this hypothesis: they could be reconciled with it.

Moreover, Kepler's laws partly support, and are partly inspired by, his belief in a cause, a power, emanating like light rays from the sun and influencing, steering, or causing the movement of the planets, including the earth. But the view that there is an influx or '*Influence*' from the stars reaching the earth was at the time considered as the fundamental tenet of astrology as opposed to Aristotelian rationalism. Here we have an important dividing line which separated two schools of thought: for example, Galileo (himself a great critic of Aristotle), or Descartes or Boyle or Newton, belonged to the (Aristotelian) rationalist tradition. This is why Galileo remained sceptical of Kepler's views and also why he was unable to accept any theory of the tides which explained them by the 'influence' of the moon, so that he felt compelled to develop a non-lunar theory which explained the tides merely by the motion of the earth. This is also why Newton was so reluctant to accept his own theory of attraction (or Robert Hooke's) and why he was never quite reconciled to it. And this is why the French Cartesians were so long unwilling to accept Newton's theory. But in the end the originally astrological view proved so successful that it was accepted by all rationalists and its disreputable origin was forgotten.[4]

Such, from an historical and genetical point of view, were the main antecedents of Newton's theory. Our story shows that as a matter of historical fact the theory was not derived from observations.

Kant realized much of this; and he also appreciated the fact that *even physical experiments* are not, genetically, prior to theories—no more than are astronomical observations. They too simply represent crucial questions which man poses to nature with the help of theories—just as Kepler asked nature whether his circle hypothesis was true. Thus Kant wrote in the preface to the 2nd edition of the *Critique of Pure Reason*:

> When Galileo let his globes run down an inclined plane with a gravity which he had chosen himself; when Torricelli caused the air to sustain a weight which he had calculated beforehand to be equal to that of a

[4] I think that Arthur Koestler's criticism of Galileo, in his remarkable book *The Sleep-walkers*, suffers from the fact that he does not take into account the schism described here. Galileo was as justified in trying to see whether he could not solve the problems within the rationalist framework as was Kepler in his attempts to solve them within the astrological framework. For the influence of astrological ideas see also note 4 to ch. 1 of the present volume.

> column of water of known height; . . . then a light dawned upon all natural philosophers. They learnt that our reason can understand only *what it creates according to its own design: that we must compel Nature to answer our questions*, rather than cling to Nature's apron strings and allow her to guide us. *For purely accidental observations, made without any plan having been thought out in advance, cannot be connected by a . . . law—which is what reason is searching for.*[5]

This quotation from Kant shows how well he understood that we ourselves must confront nature with hypotheses and demand a reply to our questions; and that, lacking such hypotheses, we can only make haphazard observations which follow no plan and which can therefore never lead us to a natural law. In other words, Kant saw with perfect clarity that the history of science had refuted the Baconian myth that we must begin with observations in order to derive our theories from them. And Kant also realized very clearly that behind this historical fact lay a logical fact; that there were logical reasons why this kind of thing did not occur in the history of science: that it was logically impossible to derive theories from observations.

My third point—the contention that it is logically impossible to derive Newton's theory from observations—follows immediately from Hume's critique of the validity of inductive inferences, as pointed out by Kant. Hume's decisive point may be put as follows:

Take a class consisting of any number of true observation-statements and designate it by the letter K. The statements in the class K will describe actual observations, i.e. *past* observations: thus K will be any class whatsoever of true statements about observations actually made in the past. Since we have assumed that K consists only of true statements, all statements in the class K must also be self-consistent statements, and, furthermore, all statements belonging to the class K must be *compatible with one another*. Now take a further observation-statement which we shall designate by the letter B. We assume B describes some *future, logically possible observation*; for example, B may say that an eclipse of the sun will be observed tomorrow. Since eclipses of the sun have already been observed, we can be certain that a statement B, asserting

[5] The original has no italics.

that there will be an eclipse of the sun tomorrow, is a statement which, on purely logical grounds, is *possible*; that is to say, our B is self-consistent. Now Hume shows the following: if B is a self-consistent statement about some possible future observation, and K any class of true statements about some past observations, then B can *always* be conjoined with K without contradiction; or, in other words, if we add a statement B about a possible future observation to statements in K we can *never* arrive at a logical contradiction. Hume's finding can also be formulated as follows: *no logically possible future observation can ever contradict the class of past observations*.

Let us now add to Hume's simple finding a theorem of pure logic, namely: whenever a statement B can be conjoined without contradiction to a class of statements K, then it can also be conjoined without contradiction to any class of statements that consists of statements of K together with any statement that can be *derived from* K.

And so we have proved our point: if Newton's theory could be derived from a class K of true observation-statements, then no future observation B could possibly contradict Newton's theory and the observations K.

Yet it is known, on the other hand, that from Newton's theory and past observations we can logically derive a statement that tells us whether or not there will be an eclipse of the sun tomorrow. Now if this derived statement tells us that tomorrow there will be no eclipse of the sun, then our B is clearly *incompatible* with Newton's theory and the class K. From this and our previous results it follows logically that it is impossible to assume that Newton's theory can be derived from observations.

Thus we have proved our third point. And we can now see the whole riddle of experience—the paradox of the empirical sciences, as discovered by Kant:

Newton's dynamics goes essentially beyond all observations. It is universal, exact and abstract; it arose historically out of myths; and we can show by purely logical means that it is not derivable from observation-statements.

Kant also showed that what holds for Newtonian theory must hold for *everyday experience*, though not, perhaps, to quite the same extent: that everyday experience, too, goes far beyond all observation. Everyday experience too must *interpret* observation; for without theoretical interpretation, observation remains blind—uninformative. Everyday experience constantly operates with abstract ideas, such as that of cause and effect, and so it cannot be derived from observations.

In order to *solve* the riddle of experience, and to explain how natural science and experience are at all possible, Kant constructed his *theory of experience and of natural science*. I admire this theory as a truly heroic attempt to solve the paradox of experience, yet I believe that it answers a false question, and hence that it is *in part* irrelevant. Kant, the great discoverer of the riddle of experience, was in error about one important point. But his error, I hasten to add, was quite unavoidable, and it detracts in no way from his magnificent achievement.

What was this error? As I have said, Kant, like almost all philosophers and epistemologists right into the twentieth century, was convinced that Newton's theory was *true*. This conviction was inescapable. Newton's theory had made the most astonishing and exact predictions, all of which had proved strikingly correct. Only ignorant men could doubt its truth. How little we may reproach Kant for his belief is best shown by the fact that even Henri Poincaré, the greatest mathematician, physicist and philosopher of his generation, who died shortly before the First World War, believed like Kant that Newton's theory was true and irrefutable. Poincaré was one of the few scientists who felt about Kant's paradox almost as strongly as Kant himself; and though he proposed a solution which differed somewhat from Kant's, it was only a variant of it. The important point, however, is that he fully shared Kant's error, as I have called it. It was an unavoidable error—unavoidable, that is, before Einstein.

Even those who do not accept Einstein's theory of gravitation ought to admit that his was an achievement of truly epoch-making significance. For his theory established at least that Newton's theory, no matter whether true or false, was certainly *not the only possible* system of celestial mechanics that could explain the phenomena in a simple and convincing way. For the first time in more than 200 years Newton's theory became *problematical*. It had become, during these two centuries, a dangerous *dogma*—a dogma of almost stupefying power. I have no objection to those who oppose Einstein's theory on scientific grounds. But even Einstein's opponents, like his greatest admirers, ought to be grateful to him for having freed physics of the paralysing belief in the incontestable truth of Newton's theory. Thanks to Einstein we now look upon this theory as a hypothesis (or a system of hypotheses)—perhaps the most magnificent and the most important hypothesis in the history of science, and certainly an astonishing approximation to the truth.[6]

Now if, unlike Kant, we consider Newton's theory as a hypothesis whose truth is problematic, then we must radically alter Kant's problem. No wonder then that his solution no longer suits the new post-Einsteinian formulation of the problem, and that it must be amended accordingly.

Kant's solution of the problem is well known. He assumed, correctly I think, that *the world as we know it is our interpretation of the observable facts in the light of theories that we ourselves invent*. As Kant puts it: 'Our intellect does not draw its laws from nature . . . but imposes them upon nature.' While I regard this formulation of Kant's as essentially correct, I feel that it is a little too radical, and I should therefore like to put it in the following modified form: 'Our intellect does not draw its laws from nature, but tries—with varying degrees of success—to impose upon nature laws which it freely invents.' The difference is this. Kant's formulation not only implies that our reason attempts to impose laws upon nature, but also that it is invariably successful in this. For Kant believed that Newton's laws were successfully imposed upon nature by us: that we were bound to interpret nature by means of these laws; from which he concluded that they must be true *a priori*. This is how Kant saw these matters; and Poincaré saw them in a similar way.

Yet we know since Einstein that very different theories and very different interpretations are also possible, and that they may even be superior to Newton's. Thus reason is capable of more than one interpretation. Nor can it impose its interpretation upon nature once and for all time. Reason works by trial and error. We invent our myths and our theories and we try them out: we try to see how far they take us. And we improve our theories if we can. The better theory is the one that

[6] See Einstein's own formulation in his Herbert Spencer lecture 'On the Method of Theoretical Physics', 1933, p. 11, where he writes: 'It was the general Theory of Relativity which showed . . . that it was possible for us, using basic principles, very far removed from those of Newton, to do justice to the entire range of the data of experience . . .'

has the greater explanatory power: that explains more; that explains with greater precision; and that allows us to make better predictions.

Since Kant believed that it was his task to explain the uniqueness and the truth of Newton's theory, he was led to the belief that this theory followed inescapably and with logical necessity from the laws of our understanding. The modification of Kant's solution which I propose, in accordance with the Einsteinian revolution, frees us from this compulsion. In this way, theories are seen to be the *free* creations of our own minds, the result of an almost poetic intuition, of an attempt to understand intuitively the laws of nature. But we no longer try to force our creations upon nature. On the contrary, we question nature, as Kant taught us to do; and we try to elicit from her *negative* answers concerning the truth of our theories: we do not try to prove or to *verify* them, but we test them by trying to *disprove* or to falsify them, to *refute* them.

In this way the freedom and boldness of our theoretical creations can be controlled and tempered by self-criticism, and by the severest tests we can design. It is here, through our critical methods of testing, that scientific rigour and logic enter into empirical science.

We have seen that theories cannot be logically derived from observations. They can, however, clash with observations: they can contradict observations. This fact makes it possible to infer from observations that a theory is *false*. The possibility of refuting theories by observations is the basis of all empirical tests. For the test of a theory is, like every rigorous examination, always an attempt to show that the candidate is mistaken—that is, that the theory entails a false assertion. From a logical point of view, all empirical tests are therefore *attempted refutations*.

In conclusion I should like to say that ever since Laplace attempts have been made to attribute to our theories instead of truth at least *a high degree of probability*. I regard these attempts as misconceived. All we can ever hope to say of a theory is that it explains this or that; that it has been tested severely, and that it has stood up to all our tests. We may also compare, say, two theories in order to see which of them has stood up better to our severest tests—or in other words, which of them is *better corroborated* by the results of our tests. But it can be shown by purely mathematical means that *degree of corroboration can never be equated with mathematical probability*. It can even be shown that all theories, including the best, have the same probability, namely zero. But the degree to which they are corroborated (which, in theory at least, can be found out with the help of the calculus of probability) may approach very closely to unity, i.e. its maximum, though the probability of the theory is zero. That an appeal to probability is incapable of solving the riddle of experience is a conclusion first reached long ago by David Hume.

Thus logical analysis shows that experience does not consist in the mechanical accumulation of observations. Experience is creative. It is the result of free, bold and creative interpretations, controlled by severe criticism and severe tests.

2. THE PROBLEM OF THE IRREFUTABILITY OF PHILOSOPHICAL THEORIES

In order to avoid right from the start the danger of getting lost in generalities, it might be best to explain at once, with the help of five examples, what I mean by a *philosophical theory*.

A typical example of a philosophical theory is Kant's doctrine of *determinism*, with respect to the world of experience. Though Kant was an indeterminist at heart, he said in the *Critique of Practical Reason*[7] that full knowledge of our psychological and physiological conditions and of our environment would make it possible to predict our future behaviour with the same certainty with which we can predict an eclipse of the sun or of the moon.

In more general terms, one could formulate the determinist doctrine as follows.

The future of the empirical world (or of the phenomenal world) is completely predetermined by its present state, down to its smallest detail.

Another philosophic theory is *idealism*, for example, Berkeley's or Schopenhauer's; we may perhaps express it here by the following thesis: 'The empirical world is my idea', or '*The world is my dream*'.

A third philosophic theory—and one that is very important today—is epistemological *irrationalism*, which might be explained as follows.

Since we know from Kant that human reason is incapable of grasping, or knowing, the world of things in themselves, we must either give up hope of ever knowing it, or else try to know it otherwise than by

[7] *Kritik der praktischen Vernunft*, 4th to 6th edn., p. 172; *Works*, ed. Cassirer, vol. v, p. 108.

means of our reason; and since we cannot and will not give up this hope, we can only use irrational or supra-rational means, such as instinct, poetic inspiration, moods, or emotions.

This, irrationalists claim, is possible because in the last analysis we are ourselves such things-in-themselves; thus if we can manage somehow to obtain an intimate and immediate knowledge of ourselves, we can thereby find out what things-in-themselves are like.

This simple argument of irrationalism is highly characteristic of most nineteenth-century post-Kantian philosophers; for example of the ingenious Schopenhauer, who in this way discovered that since we, as things-in-themselves, are *will*, will must be the thing-in-itself. The world, as a thing-in-itself, is *will*, while the world as phenomenon is an *idea*. Strangely enough this obsolescent philosophy, dressed up in new clothes, has once again become the latest fashion, although, or perhaps just because, its striking similarity to old post-Kantian ideas has remained hidden (so far as anything may remain hidden under the emperor's new clothes). Schopenhauer's philosophy is nowadays propounded in obscure and impressive language, and his self-revealing intuition that man, as a thing-in-itself, is ultimately *will*, has now given place to the self-revealing intuition that man may so utterly bore himself that his very boredom proves that the thing-in-itself is Nothing—that it is Nothingness, Emptiness-in-itself. I do not wish to deny a certain measure of originality to this existentialist variant of Schopenhauer's philosophy: its originality is proved by the fact that Schopenhauer could never have thought so poorly of his powers of self-entertainment. What he discovered in himself was will, activity, tension, excitement—roughly the opposite of what some existentialists discovered: the utter boredom of the bore-in-himself bored by himself. Yet Schopenhauer is no longer the fashion: the great fashion of our post-Kantian and post-rationalist era is what Nietzsche ('haunted by premonitions, and suspicious of his own progeny') rightly called 'European nihilism'.[8]

[8] Cf. Julius Kraft, *Von Husserl zu Heidegger*, 2nd edn., 1957, e.g. pp. 103 f., 136 f. and particularly p. 130, where Kraft writes: 'Thus it is hard to understand how existentialism could ever have been considered to be something new in philosophy, from an epistemological point of view.' Cf. also the stimulating paper by H. Tint, in the *Proc. Aristot. Society* 1956–7, pp. 253 ff.

Yet all this is only by the way. We now have before us a list of five philosophical theories.

First, determinism: the future is contained in the present, inasmuch as it is fully determined by the present.

Second, idealism: the world is my dream.

Third, irrationalism: we have irrational or supra-rational experiences in which we experience ourselves as things-in-themselves; and so we have some kind of knowledge of things-in-themselves.

Fourth, voluntarism: in our own volitions we know ourselves as wills. The thing-in-itself is the will.

Fifth, nihilism: in our boredom we know ourselves as nothings. The thing-in-itself is Nothingness.

So much for our list. I have chosen my examples in such a way that I can say of any one of these five theories, after careful consideration, that I am convinced that it is false. To put it more precisely; I am *first* of all an indeterminist, *secondly* a realist, *thirdly* a rationalist. As regards my fourth and fifth examples, I gladly admit—with Kant and other critical rationalists—that we cannot possess anything like full knowledge of the real world with its infinite richness and beauty. Neither physics nor any other science can help us to this end. Yet I am sure the voluntarist formula, 'The world is will', cannot help us either. And as to our nihilists and existentialists who bore themselves (and perhaps others), I can only pity them. They must be blind and deaf, poor things, for they speak of the world like a blind man of Perugino's colours or a deaf man of Mozart's music.

Why then have I made a point of selecting for my examples a number of philosophical theories that I believe to be false? Because I hope in this way to put more clearly the problem contained in the following important statement.

Although I consider each one of these five theories to be *false*, I am nevertheless convinced that each of them is *irrefutable*.

Listening to this statement you may well wonder how I can possibly hold a theory to be *false and irrefutable* at one and the same time—I who claim to be a rationalist. For how can a rationalist say of a theory that it is false and irrefutable? Is he not bound, as a rationalist, to refute a theory before he asserts that it is false? And conversely, is he not bound to admit that if a theory is irrefutable, it is true?

With these questions I have at last arrived at our problem.

The last question can be answered fairly simply. There have been thinkers who believed that the truth of a theory may be inferred from its irrefutability. Yet this is an obvious mistake, considering that there may be two incompatible theories which are equally irrefutable—for example, determinism and its opposite, indeterminism. Now since two incompatible theories cannot both be true, we see from the fact that both theories are irrefutable that irrefutability cannot entail truth.

To infer the truth of a theory from its irrefutability is therefore inadmissible, no matter how we interpret irrefutability. For normally 'irrefutability' would be used in the following two senses:

The first is a purely logical sense: we may use 'irrefutable' to mean the same as 'irrefutable by purely logical means'. But this would mean the same as 'consistent'. Now it is quite obvious that the truth of a theory cannot possibly be inferred from its consistency.

The second sense of 'irrefutable' refers to refutations that make use not only of logical (or analytic) but also of empirical (or synthetic) assumptions; in other words, it admits empirical refutations. In this second sense, 'irrefutable' means the same as 'not empirically refutable', or more precisely 'compatible with any possible empirical statement' or 'compatible with every possible experience'.

Now both the logical and the empirical irrefutability of a statement or a theory can easily be reconciled with its falsehood. In the case of logical irrefutability this is clear from the fact that every empirical statement and its negation must both be *logically* irrefutable. For example, the two statements, 'Today is Monday', and, 'Today is not Monday', are both logically irrefutable; but from this it follows immediately that there exist false statements which are logically irrefutable.

With empirical irrefutability the situation is a little different. The simplest examples of empirically irrefutable statements are so-called strict or pure existential statements. Here is an example of a strict or pure existential statement. 'There exists a pearl which is ten times larger than the next largest pearl.' If in this statement we restrict the words 'There exists' to some finite region in space and time, then it may of course become a refutable statement. For example, the following statement is obviously empirically refutable: 'At this moment and in this box here there exist at least two pearls one of which is ten times larger than the next largest pearl in this box.' But then this statement is no longer a strict or pure existential statement; rather it is a *restricted* existential statement. A strict or pure existential statement applies to the whole universe, and it is irrefutable simply because there can be no method by which it could be refuted. For even if we were able to search our entire universe, the strict or pure existential statement would not be refuted by our failure to discover the required pearl, seeing that it might always be hiding in a place where we are not looking.

Examples of empirically irrefutable existential statements which are of greater interest are the following.

'There exists a completely effective cure for cancer, or, more precisely, there is a chemical compound which can be taken without ill effect, and which cures cancer.' Needless to say, this statement must not be interpreted as meaning that such a chemical compound is actually *known* or that it will be discovered within a given time.

Similar examples are: 'There exists a cure for any infectious disease', and, 'There exists a Latin formula which, if pronounced in proper ritual manner, cures all diseases.'

Here we have an empirically irrefutable statement that few of us would hold to be true. The statement is irrefutable because it is obviously impossible to try out every conceivable Latin formula in combination with every conceivable manner of pronouncing it. Thus there always remains the logical possibility that there might be, after all, a magical Latin formula with the power of curing all diseases.

Even so, we are justified in believing that this irrefutable existential statement is false. We certainly cannot *prove* its falsehood; but everything we know about diseases tells against its being true. In other words, though we cannot establish its falsity, the conjecture that there is no such magical Latin formula is much more reasonable than the irrefutable conjecture that such a formula does exist.

I need hardly add that through almost 2,000 years learned men have believed in the truth of an existential statement very much like this one; this is why they persisted in their search for the philosopher's stone. Their failure to find it does not prove anything—precisely because existential propositions are irrefutable.

Thus the logical or empirical irrefutability of a theory is certainly not a sufficient reason for holding the theory to be true, and hence I have vindicated my right to believe, at the same time, that these five philosophical theories are irrefutable, and that they are false.

Some twenty-five years ago I proposed to distinguish empirical or scientific theories from non-empirical or non-scientific ones precisely by defining the empirical theories as the refutable ones and the non-empirical theories as the irrefutable ones. My reasons for this proposal were as follows. Every serious test of a theory is an attempt to refute it. Testability is therefore the same as refutability, or falsifiability. And since we should call 'empirical' or 'scientific' only such theories as can be empirically tested, we may conclude that it is the possibility of an empirical refutation which distinguishes empirical or scientific theories.

If this 'criterion of refutability' is accepted, then we see at once that *philosophical* theories, or metaphysical theories, will be *irrefutable by definition*.

My assertion that our five philosophical theories are irrefutable may now sound almost trivial. At the same time it will have become obvious that though I am a rationalist I am in no way obliged to refute these theories before being entitled to call them 'false'. And this brings us to the crux of our problem:

If philosophical theories are all irrefutable, how can we ever distinguish between true and false philosophical theories?

This is the serious problem which arises from the *irrefutability of philosophical theories*.

In order to state the problem more clearly, I should like to reformulate it as follows.

We may distinguish here between three types of theory.

First, logical and mathematical theories.

Second, empirical and scientific theories.

Third, philosophical or metaphysical theories.

How can we, in each of these groups, distinguish between true and false theories?

Regarding the first group the answer is obvious. Whenever we find a mathematical theory of which we do not know whether it is true or false we test it, first superficially and then more severely, by trying to refute it. If we are unsuccessful we then try to prove it or to refute its negation. If we fail again, doubts as to the truth of the theory may have cropped up again, and we shall again try to refute it, and so on, until we either reach a decision or else shelve the problem as too difficult for us.

The situation could also be described as follows. Our task is the testing, the critical examination, of two (or more) rival theories. We solve it by trying to refute them—either the one or the other—until we come to a decision. In mathematics (but only in mathematics) such decisions are generally *final*: invalid proofs that escape detection are rare.

If we now look at the empirical sciences, we find that we follow, as a rule, fundamentally the same procedure. Once again we test our theories: we examine them critically, we try to refute them. The only important difference is that now we can also make use of empirical arguments in our critical examinations. But these empirical arguments occur only together with other critical considerations. Critical thought as such remains our main instrument. Observations are used only if they fit into our critical discussion.

Now if we apply these considerations to philosophical theories, our problem can be reformulated as follows:

Is it possible to examine irrefutable philosophical theories *critically*? If so, what can a critical discussion of a theory consist of, if not of *attempts to refute the theory*?

In other words, is it possible to assess an irrefutable theory rationally—which is to say, critically? And what reasonable argument can we adduce for and against a theory which we know to be neither demonstrable nor refutable?

In order to illustrate these various formulations of our problem by examples, we may first refer again to the problem of determinism. Kant knew perfectly well that we are unable to predict the future actions of a human being as accurately as we can predict an eclipse. But he explained the difference by assuming that we know far less about the present conditions of a man—about his wishes and fears, his feelings and his motives—than about the present state of the solar system. Now this assumption contains, implicitly, the following hypothesis:

'*There exists* a true description of the present state of this man which would suffice (in conjunction with true natural laws) for the prediction of his future actions.'

This is of course again a purely existential statement, and it is thus irrefutable. Can we, in spite of this fact, discuss Kant's argument rationally and critically?

As a second example we may consider the thesis: 'The world is my dream.' Though this thesis is clearly irrefutable, few will believe in its truth. But can we discuss it rationally and critically? Is not its irrefutability an insurmountable obstacle to any critical discussion?

As to Kant's doctrine of determinism, it might perhaps be thought that the critical discussion of it might begin by saying to him: 'My dear Kant, it simply is not enough to assert that *there exists* a true description that is sufficiently detailed to enable us to predict the future. What you must do is tell us exactly what this description would consist of, so that we may test your theory empirically.' This speech, however, would be tantamount to the assumption that philosophical—that is, irrefutable—theories can never be discussed and that a responsible thinker *is bound* to replace them by empirically testable theories, in order to make a rational discussion possible.

I hope that our *problem* has by now become sufficiently clear; so I will now proceed to *propose a solution* of it.

My solution is this: if a philosophical theory were no more than an isolated assertion about the world, flung at us with an implied 'take it or leave it' and without a hint of any connection with anything else, then it would indeed be beyond discussion. But the same might be said of an empirical theory also. Should anybody present us with Newton's equations, or even with his arguments, without explaining to us first what the problems were which his theory was meant to solve, then we should not be able to discuss its truth rationally—no more than the truth of the *Book of Revelation*. Without any knowledge of the results of Galileo and Kepler, of the problems that were resolved by these results, and of Newton's problem of explaining Galileo's and Kepler's solutions by a unified theory, we should find Newton's theory just as much beyond discussion as any metaphysical theory. In other words every *rational* theory, no matter whether scientific or philosophical, is rational in so far as it tries to *solve certain problems*. A theory is comprehensible and reasonable only in its relation to a given *problem-situation*, and it can be rationally discussed only by discussing this relation.

Now if we look upon a theory as a proposed solution to a set of problems, then the theory immediately lends itself to critical discussion—even if it is non-empirical and irrefutable. For we can now ask questions such as, Does it solve the problem? Does it solve it better than other theories? Has it perhaps merely shifted the problem? Is the solution simple? Is it fruitful? Does it perhaps contradict other philosophical theories needed for solving other problems?

Questions of this kind show that a critical discussion even of irrefutable theories may well be possible.

Once again let me refer to a specific example: the idealism of Berkeley or Hume (which I have replaced by the simplified formula 'The world is my dream'). It is notable that these authors were far from wishing to offer us an extravagant, an incredible theory. This may be seen from Berkeley's repeated insistence that his theories were really in agreement with sound common sense.[9] Now if we try to understand the *problem situation* which induced them to propound this theory, then we find that Berkeley and Hume believed that all our knowledge was reducible to *sense-impressions* and to associations between *memory-images*. This assumption led these two philosophers to adopt idealism; and in the case of Hume, in particular, very unwillingly. Hume was an idealist only because he failed in his attempt to reduce realism to *sense-impressions*.

It is therefore perfectly *reasonable* to criticize Hume's idealism by pointing out that his sensualistic theory of knowledge and of learning was in any case inadequate, and that there are less inadequate theories of learning which have no unwanted idealistic consequences.

In a similar way we could now proceed to discuss Kant's determinism rationally and critically. Kant was in his fundamental intention an indeterminist: even though he believed in determinism with respect to the phenomenal world as an unavoidable consequence of Newton's theory, he never doubted that man, as a moral being, was not

[9] It may also be seen from Hume's frank admission that 'whatever may be the reader's opinion at this present moment, . . . an hour hence he will be persuaded there is both an external and internal world' (*Treatise*, I, IV, end of section ii; Selby-Bigge, p. 218).

determined. Kant never succeeded in solving the resulting conflict between his theoretical and practical philosophy in a way that satisfied himself completely, and he despaired of ever finding a real solution.

In the setting of this *problem-situation* it becomes possible to criticize Kant's determinism. We may ask, for example, whether it really follows from Newton's theory. Let us conjecture for a moment that it does not. I do not doubt that a clear proof of the truth of this conjecture would have persuaded Kant to renounce his doctrine of determinism—even though this doctrine happens to be irrefutable and even though he would not, for this very reason, have been logically compelled to renounce it.

Similarly with irrationalism. It first entered rational philosophy with Hume—and those who have read Hume, that calm analyst, cannot doubt that this was not what he intended. Irrationalism was the unintended consequence of Hume's conviction that *we do in fact learn* by Baconian induction coupled with Hume's logical proof that *it is impossible to justify induction rationally*. 'So much the worse for rational justification' was a conclusion which Hume, of necessity, was compelled to draw from this situation. He accepted this irrational conclusion with the integrity characteristic of the real rationalist who does not shrink from an unpleasant conclusion if it seems to him unavoidable.

Yet in this case it was not unavoidable, though it seemed to be so to Hume. We are not in fact the Baconian induction machines that Hume believed us to be. Habit or custom does not play the role in the process of learning which Hume assigned to it. And so Hume's problem dissolves and with it his irrationalist conclusions.

The situation of post-Kantian irrationalism is somewhat similar. Schopenhauer in particular was genuinely opposed to irrationalism. He wrote with only *one* desire: to be understood; and he wrote more lucidly than any other German philosopher. His striving to be understood made him one of the few great masters of the German language.

Yet Schopenhauer's problems were those of Kant's metaphysics—the problem of determinism in the phenomenal world, the problem of the thing-in-itself, and the problem of our own membership of a world of things-in-themselves. He solved these problems—*problems transcending all possible experience*—in his typically rational manner. But the solution was *bound* to be irrational. For Schopenhauer was a Kantian and as such he believed in the Kantian limits of reason: he believed that the limits of human reason coincided with *the limits of possible experience*.

But here again there are other possible solutions. Kant's problems can and must be revised; and the direction that this revision should take is indicated by his fundamental idea of critical, or self-critical, rationalism. The discovery of a philosophical problem can be something final; it is made once, and for all time. But the solution of a philosophical problem is never final. It cannot be based upon a final proof or upon a final refutation: this is a consequence of the irrefutability of philosophical theories. Nor can the solution be based upon the magical formulae of inspired (or bored) philosophical prophets. Yet it may be based upon the conscientious and critical examination of a problem-situation and its underlying assumptions, and of the various possible ways of resolving it.

9

WHY ARE THE CALCULI OF LOGIC AND ARITHMETIC APPLICABLE TO REALITY?

Professor Ryle has confined his contribution[1] to the applicability of the rules of logic, or more precisely, to the logical rules of inference. I intend to follow him in this, and only later to extend the discussion to the applicability of logical and arithmetical calculi. The distinction I have just made between the *logical rules of inference* and the so-called *logical calculi* (such as the propositional calculus or the class calculus or the calculus of relations) needs, however, some clarification, and I shall discuss the distinction, as well as the connection between the rules of inference and the calculi, in section i, before taking up the two main problems before us: that of the applicability of the rules of inference (in section ii), and that of the applicability of the logical calculi (in section viii).

This was the third paper of a symposium held at the Joint Session of the Mind Association and the Aristotelian Society at Manchester in 1946. It was published in the Proceedings of the Aristotelian Society, Supplementary volume **20**. *The first symposiast was Professor Gilbert Ryle. Dr C. Lewy was the second symposiast, but his contribution came too late to be discussed in my paper, whose first paragraph is here omitted.*

[1] Professor Ryle's contribution to this discussion is summarized in my paper so far as is necessary to the understanding of my paper.

I shall allude to, and make use of, some ideas from Professor Ryle's paper, and also from his Presidential Address to the Aristotelian Society, *Knowing How and Knowing That* (1945).[2]

I

Let us consider a simple example of an argument or of reasoning, formulated in some language, say in ordinary English. The argument will consist of a series of statements. We may assume, perhaps, that somebody argues: 'Rachel is the mother of Richard. Richard is the father of Robert. The mother of the father is the paternal grandmother. Thus, Rachel is the paternal grandmother of Robert.'

The word 'thus' in the last sentence may be taken as an indication that the speaker claims that his argument is conclusive, or valid; or in other words, that the last statement (the conclusion) has been validly drawn from the three foregoing statements (the premises). In this claim, he may be right or wrong. If he is usually right in claims of this kind, then we can say that he *knows how* to argue. And he may know how to argue without being able to explain to us in words the rules of the procedure which he observes (in common with others who know how to argue); just as a pianist may know how to play well without being able to explain the rules of procedure that underlie a good performance. If a man knows how to argue without always being aware of the rules of procedure, then we usually say that he argues or reasons 'intuitively'. And if we now read through the above argument, then we may be able to say, intuitively, that the argument is valid. There is little doubt that most of us reason, as a rule, intuitively, in the sense indicated. The formulation and discussion of the rules of procedure that underlie ordinary intuitive arguments is a rather specialized and sophisticated sort of inquiry; it is a business peculiar to the logician. While every reasonably intelligent man knows how to argue—provided the arguments do not become too complicated—there are few who can formulate the rules which underlie these performances and which we may call 'rules of inference'; there are few who *know that* (and fewer perhaps who know *why*) a certain rule of inference is valid.

[2] Cp. Aristotle, *An. Post.*, ii, 19; 100a, 8.

The particular rule of inference which underlies the argument stated above can be formulated, making use of variables and a few other artificial symbols, by a scheme like this:[3]

From three premises of the form:

'x R y'
'y S z'
'R 'S = T'

———

a conclusion may be drawn of the form: 'x T z'

Here, for 'x', 'y', and 'z', any proper name of individuals may be substituted, and for 'R', 'S', and 'T' any names of relations between individuals; for 'x R y', etc., any statement asserting that R holds between x and y, etc.; for 'R 'S' any name of a relation holding between x and z if, and only if, there exists a y such that x R y and y S z; and '=' expresses here equality of extension between relations.

It should be noted that this rule of inference makes assertions about *statements of a certain kind or form*. This fact distinguishes it clearly from a formula of a calculus (in this case, the calculus of relations) such as:

'For all R, S, and T; and for all x, y, and z: if x R y and y S z and R 'S = T, then x T z.'

This formula, undoubtedly, has some similarity to our rule of inference; in fact, it is that statement (in the calculus of relations) which corresponds to our rule of inference. But it is not the same: *it asserts something conditionally about all relations and individuals of a certain kind*, while the rule of inference asserts something, unconditionally, about all statements of a certain kind—namely that every statement of a certain form is deducible, unconditionally, from a set of statements of another form.

In a similar way, we should distinguish, for instance, between the rule of inference (called 'Barbara') of traditional logic:

'M *a* P'
'S *a* M'
———
'S *a* P'

[3] The best method, I believe, of formulating such schemata is one that uses Quine's 'quasi-quotation'; but I shall not introduce Quine's notation here.

and the formula of the calculus of classes 'If M *a* P and S *a* M, then S *a* P', (or in slightly more modern writing: 'If $c \subset b$ and $a \subset c$, then $a \subset b$'); or between the rule of inference—which is called the 'principle of inference of propositional logic', or the *modus ponendo ponens*:

$$\frac{\begin{array}{l} p \\ \text{If } p \text{ then } q \end{array}}{q}$$

and the formula of the calculus of propositions: 'If p, and if p then q, then q'.

The fact that, to every well-known rule of inference, there corresponds a logically true hypothetical or conditional formula of some well-known calculus—a 'logician's hypothetical', as Professor Ryle calls these formulae—has led to confusion between *rules of inference* and the corresponding conditional formulae. But there are important differences. (1) Rules of inference are always *statements about statements*, or about classes of statements (they are 'meta-linguistic'); but the formulae of the calculi are not. (2) The rules of inference are unconditional statements about deducibility; but the corresponding formulae of the calculi are conditional or hypothetical 'If . . . then . . .' statements, which do not mention deducibility or inference, or premises or conclusions. (3) A rule of inference, after substitution of constants for the variables, asserts something *about* a certain argument—an 'observance' of the rule—namely, that this argument is valid; but the corresponding formula, after substitution, yields a *logical truism*, i.e. a statement such as 'All tables are tables', although in hypothetical form, as for example, 'If it is a table, then it is a table' or 'If all men are mortal, and all Greeks are men, then all Greeks are mortal'. (4) The rules of inference are *never* used as premises in those arguments which are formulated in accordance with them; but the corresponding formulae *are* used in this way. In fact, one of the main motives in constructing logical calculi is this: by using the 'logician's hypotheticals' (i.e. those hypothetical truisms which correspond to a certain rule of inference) *as a premise*, we can dispense with the corresponding rule of inference. By this method we can eliminate all the different rules of inference—except *one*, the above-mentioned 'principle of inference' (or *two*, if we make use of the 'principle of substitution', which, however, can be avoided). In other words, the method of building up a logical calculus is a method of systematically reducing a vast number of rules of inference to one (or two). The place of all the others is taken by formulae of the calculus; which has the advantage that all these formulae—an infinite number, in fact—can be, in turn, systematically inferred (using the 'principle of inference') from a very few formulae.

We have indicated that for each of the well-known rules of inference there exists an asserted (or demonstrable) formula in a well-known logical calculus. The converse is not true in general (though it is true for hypothetical formulae). For example, to the formula '*p* or non-*p*'; or to 'non-(*p* and non-*p*)'; and to many others which are not hypothetical, there exists no corresponding rule of inference.

Thus rules of inference and formulae of logical calculi have to be carefully distinguished. This need not, however, prevent us from *interpreting* a certain sub-set of these formulae—the 'logician's hypotheticals'—as rules of inference. In fact, our assertion that to every such hypothetical formula there corresponds a rule of inference justifies such an interpretation.

II

After these somewhat technical preliminaries, we now turn to Professor Ryle's treatment of the question: 'Why are rules of inference applicable to reality?' This question forms an important part of our original problem, for we have just seen that a certain sub-set of the formulae of the logical calculi (viz., those which Professor Ryle calls 'the logician's hypotheticals') can be interpreted as rules of inference.

Professor Ryle's central thesis, if I understand him rightly, is this. The rules of logic, or more precisely, the rules of inference, are rules of procedure. This means that they apply to certain procedures, rather than to things or facts. They do not apply to reality, if by 'reality' we mean the things and facts described, for example, by scientists and historians. They do not 'apply' in the sense in which a description, say of a man, may apply to—or fit—either the man described or some other man; or in the sense in which a descriptive theory, for example of nuclear resonance absorption, may apply to—or fit—the atoms of uranium. Logical rules, rather, apply to the procedure of drawing inferences, comparable to the way in which the rules of the highway code apply to the procedure of riding a bicycle or driving a car. Logical rules can be observed or contravened, and to apply them does not mean to make them fit, but means to *observe* them, to act in accordance with them. If the question 'Why are the rules of logic applicable to reality?' is mistakenly intended to mean 'Why do the rules of logic fit the things and the facts of our world?' then the answer would be that the question assumes that they can, and do, fit the facts, whereas it is not possible to predicate of the rules of logic that they are 'fitting the facts of the world' or 'not fitting the facts of the world'. This is not possible any more than it is possible to predicate such a thing of the highway code or of the rules of chess.

Thus it seems that our problem disappears. Those who wonder why the rules of inference apply to the world, vainly trying to imagine what an illogical world would be like, are the victims of an ambiguity. Rules of inference are procedural rules or rules of performance, so that they cannot 'apply' in the sense of 'fit' but only in the sense of being observed. Thus a world in which they do not apply would not be an illogical world, but a world peopled by illogical men.

This analysis (which is Professor Ryle's) seems to me true and important, and it may well indicate the direction in which a solution of our problem can be found. But I do not feel satisfied that in itself it offers a solution.

The position appears to me in this way. Professor Ryle's analysis shows that one way of interpreting the problem reduces it to nonsense, or to a pseudo-problem. Now I have for many years made it a personal rule of procedure not to be easily satisfied with the reduction of problems to pseudo-problems. Whenever somebody succeeds in reducing a problem to a pseudo-problem, I always ask myself whether one could not find another interpretation of the original problem—an interpretation which shows, if possible, that apart from the pseudo-problem there is also a real problem behind the original problem. I have found in many cases that this rule of procedure was fruitful and successful. I fully admit that an analysis which attempts to reduce the original problem to a pseudo-problem may often be extremely valuable; it may show that there was a danger of muddled thinking, and it may often help us to find the real problem. But it does not settle it. All this is the case here too, I believe.

III

I accept Professor Ryle's view that the rules of logic (or of inference) are rules of procedure, and, as he himself indicates, that they may be considered as good or useful or helpful rules of procedure. And now I suggest that the problem 'Why are the rules of logic applicable to reality?' might be interpreted to mean 'Why are the rules of logic good, or useful, or helpful rules of procedure?'

That this interpretation is justifiable can hardly be denied. The man who applies the rules of logic, in the sense that he acts according to them, or, as Professor Ryle says, observes them, does so probably because he finds them useful in practice. But this means, ultimately, that he finds them useful in dealing with real situations, i.e. with reality. If we ask, 'Why are these rules useful?', we ask something very similar to the question 'Why are they applicable?' and the similarity is sufficient, I believe, for claiming that this may very well be what the original questioner had in mind. On the other hand, there is no doubt any longer that our question ceases to be a pseudo-problem.

IV

I believe that our question can be answered comparatively easily. The man who finds observance of the rules of logic useful is, we have seen, a man who draws inferences. That is to say, he obtains from some statements or descriptions of facts, called 'premises', other statements or descriptions of facts, called 'conclusions'. And he finds the procedure useful because he finds that, whenever he observes the rules of logic, whether consciously or intuitively, the conclusion will be *true*, provided the premises were *true*. In other words, he will be able to obtain reliable (and possibly valuable) indirect information, provided his original information was reliable and valuable.

If this is correct, then we must substitute for our question 'Why are the rules of logic good rules of procedure?' another question, namely,

'What is the explanation of the fact that the logical rules of inference always lead to true conclusions, provided the premises are true?'

V

I believe this question, too, can be answered comparatively easily. Having learned to speak, and to use our language for the purpose of describing facts, we soon become more or less conversant with the procedure called 'reasoning' or 'arguing', that is to say, with the intuitive procedure of obtaining some kind of secondary information which was not explicitly stated in our original information. Part of this intuitive procedure can be analysed in terms of rules of inference. The formulation of these rules is the principal task of logic.

Accordingly we may lay it down that a logician's rule of inference is, by definition, a good or 'valid' rule of inference if, and only if, its observance ensures that we obtain true conclusions, provided our premises are true. And if we succeed in finding an observance of a suggested rule which allows us to obtain a false conclusion from true premises—I call this a 'counter example'—then we are satisfied that this rule was invalid. In other words, we call a *rule of inference* 'valid' if, and only if, no counter example to this rule exists; and we may be able to establish that none exists. Similarly, we call an *observance* of a rule of inference—that is to say an inference—'valid', if, and only if, no counter example exists to the observed rule.

Thus a 'good' or 'valid' rule of inference is useful because no counter example can be found, i.e. because we can rely on it as a rule of procedure that leads from true descriptions of facts to true descriptions of facts. But since we can say of a true description that it fits the facts, 'applying' in the sense of 'fitting' does enter into our analysis in some indirect way, after all. For we can say that rules of inference apply to facts in so far as every observance of them which starts with a fitting description of facts can be relied on to lead to a description which likewise fits the facts.

It is perhaps not without interest that the fundamental importance of the principle that a valid inference from true premises invariably leads to true conclusions has been discussed at some length by Aristotle (*Anal. Prior.*, II, 1–4).

VI

In order to see whether this result is of any use I shall try to apply it to a criticism of the three main views of the nature of logic. The views I have in mind are

(A) The rules of logic are laws of thought.

(A1) They are natural laws of thought—they describe how we actually do think; and we cannot think otherwise.

(A2) They are normative laws—they tell us how we ought to think.

(B) The rules of logic are the most general laws of nature—they are descriptive laws holding for any object whatsoever.

(C) The rules of logic are laws of certain descriptive languages—of the use of words and especially of sentences.

The reason why (A1) has been so widely held is, I believe, the fact that there is something compelling and inescapable about logical rules—at least about the simple ones. They are said to hold good because we are compelled to think in accordance with them—because a state of affairs for which they do not hold good is inconceivable. But an argument that proceeds from inconceivability, is, like other appeals to self-evidence, always suspect. The fact that a rule, or a proposition, appears to be true, convincing, compelling, self-evident, or what not, is obviously no sufficient reason why it should be true, although the opposite may well be the case—its truth may be the reason why it appears to us to be true, or convincing. In other words, if the laws of logic hold for all objects, i.e. if (B) is correct, then their compelling character would be clear and reasonable; otherwise we may perhaps feel compelled to think in this way merely because we have a neurotic compulsion. In this way, our criticism of (A1) leads to (B).

But another criticism of (A1) leads to (A2); namely, the observation that we do not always reason in accordance with the laws of logic, but that we sometimes commit what is usually called a 'fallacy'. (A2) asserts that we ought to avoid such breaches of the rules of logic. But why? Is it immoral? Certainly not. 'Alice in Wonderland' is not immoral. Is it stupid? Hardly. Obviously, we ought to avoid breaches of the rules of logic if, and only if, we are interested in formulating or deriving statements which are true, that is to say, which are true descriptions of facts. This consideration, again, leads us to (B).

But (B)—a position which has been held by men like Bertrand Russell, Morris Cohen, and Ferdinand Gonseth—seems to me not altogether satisfactory. First, because the rules of inference, as we have emphasized with Professor Ryle, are rules of procedure rather than descriptive statements; secondly, because an important class of logically true formulae (viz., precisely those which Professor Ryle would call the logician's hypotheticals) can be interpreted as, or correspond to, rules of inference, and because these, as we have shown, following Professor Ryle, do not *apply* to facts in the sense in which a fitting description does. Thirdly, because any theory which does not allow for the radical difference between the status of a physical truism (such as 'All rocks are heavy') and a logical truism (such as 'All rocks are rocks' or perhaps 'Either all rocks are heavy or some rocks are not heavy') must be unsatisfactory. We feel that such a logically true proposition is true not because it describes the behaviour of all possible facts but simply because it does not take the risk of being falsified by any fact; it does not exclude any possible fact, and it therefore does not assert anything whatsoever of any fact at all. But we need not go here into the problem of the status of these logical truisms. For whatever their status may be, logic is not primarily the doctrine of logical truisms; it is, primarily, the doctrine of valid inference.

The position (C) has been criticized—rightly, I think—as unsatisfactory so long as it was bound up with the view that by a *language* we can, for the purpose of logic, understand a 'mere symbolism', i.e. a symbolism apart from any 'meaning' (whatever this may mean). I do not think that this view can be upheld. And our definition of a valid inference would most certainly not be applicable to such a mere symbolism since this definition makes use of the term 'truth'; for we could not say of a 'mere symbolism' (which is void of meaning) that it contains true or false statements. We should therefore have no inference in our sense, and no rules of inference; and as a consequence, we should have no answer to our question why the rules of logic are valid or good or useful.

But if we mean by a language a symbolism that allows us to make *true* statements (and in respect of which we can explain, as was first done by Tarski, what we mean when we say of a certain statement that it is true) then, I believe, the objections which so far have been raised against (C) lose most of their force. A valid rule of inference with regard to such a semantic language system would be a rule to which, in the language in question, no counter example can be found, because no counter example exists.

It may be said in passing that these rules of inference need not necessarily have that 'formal' character which we know from our logical studies; their character will depend, rather, on the character of the semantic language system under investigation. (Examples of semantic language systems have been analysed by Tarski and Carnap.) Yet for languages similar to those usually considered by logicians, the rules of inference will be of that 'formal' character to which we are accustomed.

VII

As indicated by my last remarks, the rules of procedure which we are discussing, i.e. the rules of inference, are, *to a certain extent*, always relative to a language system. But they all have this in common: their observance leads from true premises to true conclusions. Thus there cannot be alternative logics in the sense that their rules of inference lead from true premises to conclusions which are not true, simply because we have defined the term 'rule of inference' in such a way that this is impossible. (This does not exclude the possibility of considering the rules of inference as special cases of more general rules, for example, of rules which allow us to attach to certain quasi-conclusions a certain 'probability', provided that certain quasi-premises are true.) Yet there can be alternative logics in the sense that they formulate alternative systems of rules of inference with respect to more or less widely different languages—languages which differ in what we call their 'logical structure'.

We may take, for example, the language of categorical propositions (subject-predicate statements), for which the traditional system of categorical syllogisms formulates the rules of inference. The logical structure of this language is characterized by the fact that it contains only a small number of logical signs—signs for the copula and its negation, for universality and particularity, and perhaps for the complementation (or negation) of its so-called 'terms'. If we now consider the argument formulated in section I, third paragraph, then we see that all three premises as well as the conclusion can be formulated in the language of categorical propositions. Nevertheless, if so formulated, it is impossible to formulate a valid rule of inference which exhibits the general form of this argument; and accordingly, it is no longer possible to defend the validity of this argument, once it has been couched in the language of categorical propositions. Once we have fused the words 'mother of Richard' into one term—the predicate of our first premise—we cannot separate them again. The logical structure of this language is too poor to exhibit the fact that this predicate contains, in some way or other, the subject of the second premise, and part of the subject of the third premise. Similar remarks hold for the other two premises and for the conclusion. Accordingly, if we try to formulate the rule of inference, we get something like

'*A* is *b*'
'*C* is *d*'
'*All e are f*'
―――――――
'*A* is *g*'

(Here '*A*' and '*C*' stand for 'Rachel' and 'Richard', '*b*' for 'mother of Richard', '*d*' for 'father of Robert', '*e*' for 'mothers of fathers', '*f*' for 'paternal grandmothers', and '*g*' for 'paternal grandmother of Robert'.) This rule, of course, is invalid since we can produce in the language of categorical propositions as many counter examples as we like. Thus a language, even though it may be rich enough for describing all the facts we wish to describe, may not permit the formulation of the rules of inference needed to cover all the cases in which we can safely pass from true premises to true conclusions.

VIII

These last considerations may be used for extending our analysis to the problem of the applicability of the calculi of logic and arithmetic; for we must not forget that so far (following Professor Ryle) we have discussed only the applicability of rules of inference.

I believe that the construction of so-called 'logical calculi' can be said to be due, mainly, to the desire to build up languages with regard to which *all* those inferences which we intuitively *know how* to draw can be 'formalized', that is to say, shown to be drawn in accordance with a very few explicit, and valid, rules of inference. (These rules of inference, as rules of procedure, speak *about* the language or calculus we are investigating. They are, therefore, not to be stated *in* the calculus under investigation, but in the so-called 'meta-language' of this calculus, i.e. in the language which we use when discussing this calculus.) Syllogistic logic, for example, can be said to have been an attempt to construct such a language, and many of its adherents still believe that it was successful and that *all* inferences which are really valid are formalized in its figures and moods. (We have seen that this is not the case.) Other systems have been built up, with similar aims (for example *Principia Mathematica*), and have succeeded in formalizing practically all valid rules of inference as observed not only in ordinary discourse but also in mathematical arguments. One is tempted to describe the task of constructing a language or calculus such that we can formalize *all* valid rules of inference (partly with the help of the logical formulae of the calculus itself, and partly with the help of a few rules of inference pertaining to this calculus) as the *prima facie* fundamental problem of logic. Now there is good reason to believe that this problem is insoluble, at least if we do not admit, for the purpose of formalizing relatively simple intuitive inferences, procedures of an entirely different character (such as inferences drawn from an infinite class of premises). The position appears to be this: although it is possible, for *any* given valid intuitive inference, to construct some language permitting the formalization of this inference, it is not possible to construct one language permitting the formalization of *all* valid intuitive inferences. This interesting situation which was first discussed, to my knowledge, by Tarski, with reference to investigations by Gödel, bears on our problem in so far as it shows that the applicability of every calculus (in the sense of its suitability as a language with regard to which every valid intuitive inference can be formulated) breaks down at some stage or other.

I shall now turn to our problem of applicability, this time, however, confined to the logical calculi, or more precisely, to the asserted formulae (the axioms and theorems) of the logical calculi, rather than to the rules of inference. Why are these calculi—which may contain arithmetic—applicable to reality?

I shall try to answer this question in the form of three statements.

(*a*) These calculi as a rule are semantical systems,[4] that is to say, languages designed with the intention of being used for the description of certain facts. If it turns out that they serve this purpose then we need not be surprised.

(*b*) They *may* be so designed that they do not serve the purpose; this can be seen from the fact that certain calculi—for example, the arithmetic of natural numbers, or that of real numbers—are helpful in describing certain kinds of fact, but not other kinds.

(*c*) In so far as a calculus is applied to reality, it loses the character of a *logical* calculus and becomes a descriptive theory *which may be empirically refutable*; and in so far as it is treated as irrefutable, i.e. as a system of *logically true* formulae, rather than a descriptive scientific theory, it is not applied to reality.

A remark that bears on (*a*) will be found in section ix. In the present section, only (*b*) and (*c*) will be briefly discussed.

As to (*b*), we may note that the calculus of natural numbers is used in order to count billiard balls, or pennies, or crocodiles, while the calculus of real numbers provides a framework for the measurement or the calculation of continuous magnitudes such as geometrical distances or velocities. (This is especially clear in Brouwer's theory of the real numbers.) We should not say that we have, for instance, 3.6, or perhaps π, crocodiles in our zoo. In order to count crocodiles, we make use of the calculus of natural numbers. But in order to determine the latitude of our zoo, or its distance from Greenwich, we may have to make use of π. The belief that any one of the calculi of arithmetic is applicable to any reality (a belief that seems to underlie the problem which was set to our symposium) is therefore hardly tenable.

As to (*c*), if we consider a proposition such as '$2 + 2 = 4$', then it may be applied—for example to apples—in different senses, of which I shall discuss only two. In the first of these senses, the statement '2

[4] I am using this term in a slightly wider sense than Carnap; for I do not see why a calculus designed to have an (L-true) interpretation in a certain semantical system cannot itself be simply described or interpreted as a formalized semantical system.

apples + 2 apples = 4 apples' is taken to be irrefutable and logically true. But it does not describe any fact involving apples—any more than the statement 'All apples are apples' does. Like this latter statement, it is a logical truism; and the only difference is that it is based, not on the definition of the signs 'All' and 'are', but on certain definitions of the signs '2', '4', '+', and '='. (These definitions may be either explicit or implicit.) We might say in this case that the application is not real but only apparent; that we do not here describe any reality, but only assert that a certain way of describing reality is equivalent to another way.

More important is the application in the second sense. In this sense, '2 + 2 = 4' may be taken to mean that, if somebody has put two apples in a certain basket, and then again two, and has not taken any apples out of the basket, there will be four in it. In this interpretation the statement '2 + 2 = 4' helps us to calculate, i.e. to describe certain physical facts, and the symbol '+' stands for a physical manipulation—for physically adding certain things to other things. (We see here that it is sometimes possible to interpret an apparently logical symbol descriptively.[5]) But in this interpretation the statement '2 + 2 = 4' becomes a physical theory, rather than a logical one; and as a consequence, we cannot be sure whether it remains universally true. As a matter of fact, it does not. It may hold for apples, but it hardly holds for rabbits. If you put 2 + 2 rabbits in a basket, you *may* soon find 7 or 8 in it. Nor is it applicable to such things as drops. If you put 2 + 2 drops into a dry flask, you will never get four out of it. In other words, if you wonder what a world would look like in which '2 + 2 = 4' is not applicable, it is easy to satisfy your curiosity. A couple of rabbits of different sexes or a few drops of water may serve as a model for such a world. If you answer that these examples are not fair because something has happened to the rabbits and to the drops, and because the equation '2 + 2 = 4' only applies to objects to which nothing happens, then my answer is that, if you interpret it in this way, then it does not hold for 'reality' (for in 'reality' something happens all the time) but only for an abstract world of distinct objects in which nothing happens. To the extent, it is clear, to which our real world resembles such an abstract world, for

[5] This bears on some fundamental problems discussed by Tarski in his *Logic, Semantics, Metamathematics* (ch. 16) and by Carnap in his *Introduction to Semantics*.

example, to the extent to which our apples do not rot, or rot only very slowly, or to which our rabbits or crocodiles do not happen to breed; to the extent, in other words, to which physical conditions resemble the pure logical or arithmetical operation of addition, to the same extent, of course, does arithmetic remain applicable. But this statement is trivial.

An analogous statement may be made about the addition of measurements. That any two straight sticks which, if placed side by side, are each of the length *a*, will, if placed end to end, be together of the length 2*a*, is by no means logically necessary. We can easily imagine a world in which sticks do behave according to the rules of perspective, i.e. exactly as they appear to behave in the visual field and on photographic plates—a world in which they shrink if moved away from a certain centre (e.g. that of the lens). In fact, for the purpose of the addition of certain measurable quantities—velocities—we do seem to live in such a world. According to special relativity, the ordinary calculus of addition is inapplicable to the measurement of velocities[6] (i.e. it leads to false results); it has to be replaced by a different one. Of course, it is possible to reject the claim that the ordinary calculus of addition of velocities is inapplicable, and to resist, *on principle*, any demand that it should be changed. Such a principle would be tantamount to saying that velocities must necessarily be added in the ordinary way, or in other words, to claiming, implicitly, that they are to be *defined* as obeying the ordinary laws of addition. But in this case, of course, velocities can no longer be defined by empirical measurements (for we cannot define the same concept in two different ways) and our calculus no longer applies to empirical reality.

Professor Ryle has helped us to approach the problem from the angle of an analysis of the word 'applicable'. My last remarks may be taken as a complementary attempt to tackle the problem by analysing the word 'reality' (and also the problem of the distinction between the logical and the descriptive use of symbols). For I believe that whenever we are

[6] Remark on the Problem of the Addition of Velocities. If a train moves with the velocity V_1 (relative to the surface of the Earth), and if the ticket inspector moves in the train (towards the front of the train) with the velocity V_2, then, according to Einstein, his velocity relative to the surface of the Earth is indeed somewhat less than $V_1 + V_2$.

doubtful whether or not our statements deal with the real world, we can decide it by asking ourselves whether or not we are ready to accept an empirical refutation. If we are determined, on principle, to defend our statements in the face of refutations (such as are provided by rabbits or drops or velocities), we are not speaking about reality. Only if we are ready to accept refutations are we speaking about reality. In Professor Ryle's language, we should have to say: Only if we *know how* to abide by a refutation do we *know how* to speak about reality. If we wish to formulate this readiness or 'knowledge how', then we have to do it again with the help of a rule of procedure. It is clear that only a performance rule can help us here, for *speaking about reality* is a performance.[7]

IX

My last remarks—on (*c*)—indicate the direction in which, perhaps, an answer may be found to what I hold to be the most important aspect of our many-sided problem. Yet I do not wish to conclude this paper without making it quite clear that I believe that the problem can be taken further. Why, we could ask, are we at all successful in speaking about reality? Is it not true that reality must have a definite structure in order that we can speak about it? Could we not conceive of a reality which would be like a thick fog—and nothing else, no solids, no movement? Or perhaps like a fog with certain changes in it—rather indefinite changes of light, for example? Of course, by my very attempt to describe this world I have shown that it *can* be described in our language, but this is not to say that *any* such world could be so described.

I do not think that, in this form, the question is a very serious one, but I also do not think that it should be too quickly dismissed. In fact, I believe that we are all most intimately acquainted with a world that cannot be properly described by our language which has developed mainly as an instrument for describing and dealing with our physical environment—more precisely, with physical bodies of medium size in moderately slow motion. The indescribable world I have in mind is, of

[7] With these questions, cp. my *L.Sc.D.*

course, the world I have 'in my mind'—the world which most psychologists (except the behaviourists) attempt to describe, somewhat unsuccessfully, with the help of what is nothing but a host of metaphors taken from the languages of physics, of biology, and of social life.

But whatever the world to be described may be like, and whatever may be the languages we use, and their logical structure, there is one thing we can be sure of: as long as our interest in describing the world does not change, we shall be interested in *true descriptions*, and in *inferences*—that is to say, in operations which lead from true premises to true conclusions: in the application of logic to reality. On the other hand, there is certainly no reason to believe that our ordinary languages are the best means for the description of any world. On the contrary, they are probably not even the best possible means for a finer description of our own physical world. The development of mathematics, which is a somewhat artificial development of certain parts of our ordinary languages, shows that with new linguistic means new kinds of facts can be described. In a language possessing, say, five numerals and the word 'many', even the simple fact that in field *A* there are 6 more sheep than in field B cannot be stated. The use of an arithmetical calculus permits us to describe relations which simply could not be described without it.

There are, however, further and possibly deeper problems concerning the relations between the means of description and the described facts. These relations are rarely seen in the right way. The same philosophers who oppose a naïve realism with regard to things are often naïve realists with regard to facts. While perhaps believing that things are logical constructs (which, I am satisfied, is a mistaken view) they believe that facts are part of the world in a sense similar to that in which processes or things may be said to be parts of the world; that the world consists of facts in a sense in which it may be said to consist of (four dimensional) processes or of (three dimensional) things. They believe that, just as certain nouns are names of things, sentences are names of facts. And they sometimes even believe that sentences are something like pictures of facts, or that they are projections of facts.[8]

[8] I had in mind Wittgenstein in the *Tractatus*. Note that this paper was written in 1946.

But all this is mistaken. The fact that there is no elephant in this room is not one of the processes or parts of the world; nor is the fact that a hailstorm in Newfoundland occurred exactly 111 years after a tree collapsed in the New Zealand bush. Facts are something like a common product of language and reality; they are reality pinned down by descriptive statements. They are like abstracts from a book, made in a language which is different from that of the original, and determined not only by the original book but nearly as much by the principles of selection and by other methods of abstracting, and by the means of which the new language disposes. New linguistic means not only help us to describe new kinds of facts. In a way, they even help us to create these new kinds of (perfectly objective) facts, new kinds of states of affairs. In a certain sense, these facts obviously existed before the new means were created which were indispensable for their description; I say, 'obviously' because a calculation, for example, of the movements of the planet Mercury of 100 years ago, carried out today with the help of the calculus of the theory of relativity, may certainly be a true description of the facts concerned, even though the theory was not yet invented when these facts occurred. But in another sense we might say that these facts do not exist *as facts* before they are singled out from the continuum of events and pinned down by statements—the theories which describe them. These questions, however, although closely connected with our problem, must be left for another discussion. I have mentioned them only in order to make clear that even should the solutions I have proposed be more or less correct, there would still be open problems left in this field.

10

TRUTH, RATIONALITY, AND THE GROWTH OF SCIENTIFIC KNOWLEDGE

1. THE GROWTH OF KNOWLEDGE: THEORIES AND PROBLEMS

I

My aim in this lecture is to stress the significance of one particular aspect of science—its need to grow, or, if you like, its need to progress. I do not have in mind here the practical or social significance of this need. What I wish to discuss is rather its intellectual significance. I assert that continued growth is essential to the rational and empirical character of scientific knowledge; that if science ceases to grow it must lose that character. It is the way of its growth which makes science rational and empirical; the way, that is, in which scientists discriminate between available theories and choose the better one or (in the absence

This lecture was never delivered, or published before. It was prepared for the International Congress for the Philosophy of Science in Stanford, August 1960, but because of its length only a small part of it could be presented there. Another part formed my Presidential Address to the British Society for the Philosophy of Science, delivered in January 1961. I believe that the lecture contains (especially in parts 3 to 5) some essential further developments of the ideas of my Logic of Scientific Discovery.

of a satisfactory theory) the way they give reasons for rejecting all the available theories, thereby suggesting some of the conditions with which a satisfactory theory should comply.

You will have noticed from this formulation that it is not the accumulation of observations which I have in mind when I speak of the growth of scientific knowledge, but the repeated overthrow of scientific theories and their replacement by better or more satisfactory ones. This, incidentally, is a procedure which might be found worthy of attention even by those who see the most important aspect of the growth of scientific knowledge in new experiments and in new observations. For our critical examination of our theories leads us to attempts to test and to overthrow them; and these lead us further to experiments and observations of a kind which nobody would ever have dreamt of without the stimulus and guidance both of our theories and of our criticism of them. For indeed, the most interesting experiments and observations were carefully designed by us in order to *test* our theories, especially our new theories.

In this paper, then, I wish to stress the significance of this aspect of science and to solve some of the problems, old as well as new, which are connected with the notions of scientific progress and of discrimination among competing theories. The new problems I wish to discuss are mainly those connected with the notions of objective truth, and of getting nearer to the truth—notions that seem to me of great help in analysing the growth of knowledge.

Although I shall confine my discussion to the growth of knowledge in science, my remarks are applicable without much change, I believe, to the growth of pre-scientific knowledge also—that is to say, to the general way in which men, and even animals, acquire new factual knowledge about the world. The method of learning by trial and error—of learning from our mistakes—seems to be fundamentally the same whether it is practised by lower or by higher animals, by chimpanzees or by men of science. My interest is not merely in the theory of scientific knowledge, but rather in the theory of knowledge in general. Yet the study of the growth of scientific knowledge is, I believe, the most fruitful way of studying the growth of knowledge in general. For the growth of scientific knowledge may be said to be the growth of ordinary human knowledge *writ large* (as I have pointed out in the 1958 Preface to my *Logic of Scientific Discovery*).

But is there any danger that our need to progress will go unsatisfied, and that the growth of scientific knowledge will come to an end? In particular, is there any danger that the advance of science will come to an end because science has completed its task? I hardly think so, thanks to the infinity of our ignorance. Among the real dangers to the progress of science is not the likelihood of its being completed, but such things as lack of imagination (sometimes a consequence of lack of real interest); or a misplaced faith in formalization and precision (which will be discussed below in section v); or authoritarianism in one or another of its many forms.

Since I have used the word 'progress' several times, I had better make quite sure, at this point, that I am not mistaken for a believer in a historical law of progress. Indeed I have before now struck various blows against the belief in a law of progress,[1] and I hold that even science is not subject to the operation of anything resembling such a law. The history of science, like the history of all human ideas, is a history of irresponsible dreams, of obstinacy, and of error. But science is one of the very few human activities—perhaps the only one—in which errors are systematically criticized and fairly often, in time, corrected. This is why we can say that, in science, we often learn from our mistakes, and why we can speak clearly and sensibly about making progress there. In most other fields of human endeavour there is change, but rarely progress (unless we adopt a very narrow view of our possible aims in life); for almost every gain is balanced, or more than balanced, by some loss. And in most fields we do not even know how to evaluate change.

Within the field of science we have, however, a *criterion of progress*: even before a theory has ever undergone an empirical test we may be able to say whether, provided it passes certain specified tests, it would be an improvement on other theories with which we are acquainted. This is my first thesis.

To put it a little differently, I assert that we *know* what a good

[1] See especially my *Poverty of Historicism* (2nd edn., 1960), and ch. 16 of the present volume.

scientific theory should be like, and—even before it has been tested—what kind of theory would be better still, provided it passes certain crucial tests. And it is this (meta-scientific) knowledge which makes it possible to speak of progress in science, and of a rational choice between theories.

II

Thus it is my first thesis that we can know of a theory, even before it has been tested, that if it passes certain tests it will be better than some other theory.

My first thesis implies that we have a criterion of relative *potential* satisfactoriness, or of *potential* progressiveness, which can be applied to a theory even before we know whether or not it will turn out, by the passing of some crucial tests, to be satisfactory *in fact*.

This criterion of relative potential satisfactoriness (which I formulated some time ago,[2] and which, incidentally, allows us to grade theories according to their degree of relative potential satisfactoriness) is extremely simple and intuitive. It characterizes as preferable the theory which tells us more; that is to say, the theory which contains the greater amount of empirical information or *content*; which is logically stronger; which has the greater explanatory and predictive power; and which can therefore be *more severely tested* by comparing predicted facts with observations. In short, we prefer an interesting, daring, and highly informative theory to a trivial one.

All these properties which, it thus appears, we desire in a theory can be shown to amount to one and the same thing: to a higher degree of empirical *content* or of testability.

[2] See the discussion of degrees of testability, empirical content, corroborability, and corroboration in my *L.Sc.D.*, especially sections 31 to 46; 82 to 85; new appendix *ix; also the discussion of degrees of explanatory power in this appendix, and especially the comparison of Einstein's and Newton's theories (in note 7 on p. 401). In what follows, I shall sometimes refer to testability, etc., as the 'criterion of progress', without going into the more detailed distinctions discussed in my book.

III

My study of the *content* of a theory (or of any statement whatsoever) was based on the simple and obvious idea that the informative content of the *conjunction*, *ab*, of any two statements, *a* and *b*, will always be greater than, or at least equal to, that of any of its components.

Let *a* be the statement 'It will rain on Friday'; *b* the statement 'It will be fine on Saturday'; and *ab* the statement 'It will rain on Friday and it will be fine on Saturday': it is then obvious that the informative content of this last statement, the conjunction *ab*, will exceed that of its component *a* and also that of its component *b*. And it will also be obvious that the probability of *ab* (or, what is the same, the probability that *ab* will be true) will be smaller than that of either of its components.

Writing Ct(*a*) for 'the content of the statement *a*', and Ct(*ab*) for 'the content of the conjunction *a* and *b*', we have

$$\text{(1)} \qquad Ct(a) \leqslant Ct(ab) \geqslant Ct(b).$$

This contrasts with the corresponding law of the calculus of probability,

$$\text{(2)} \qquad p(a) \geqslant p(ab) \leqslant p(b),$$

where the inequality signs of (1) are inverted. Together these two laws, (1) and (2), state that with increasing content, probability decreases, and *vice versa*; or in other words, that content increases with increasing improbability. (This analysis is of course in full agreement with the general idea of the logical *content* of a statement as the class of *all those statements which are logically entailed* by it. We may also say that a statement *a* is logically stronger than a statement *b* if its content is greater than that of *b*—that is to say, if it entails more than *b* does.)

This trivial fact has the following inescapable consequences: if growth of knowledge means that we operate with theories of increasing content, it must also mean that we operate with theories of decreasing probability (in the sense of the calculus of probability). Thus if our aim is the advancement or growth of knowledge, then a high probability (in the sense of the calculus of probability) cannot possibly be our aim as well: *these two aims are incompatible.*

I found this trivial though fundamental result about thirty years ago, and I have been preaching it ever since. Yet the prejudice that a high probability must be something highly desirable is so deeply ingrained that my trivial result is still held by many to be 'paradoxical'.[3] Despite this simple result the idea that a high degree of probability (in the sense of the calculus of probability) must be something highly desirable seems to be so obvious to most people that they are not prepared to consider it critically. Dr Bruce Brooke-Wavell has therefore suggested to me that I should stop talking in this context of 'probability' and should base my arguments on a 'calculus of content' and of 'relative content'; or in other words, that I should not speak about science aiming at improbability, but merely say that it aims at maximum content. I have given much thought to this suggestion, but I do not think that it would help; a head-on collision with the widely accepted and deeply ingrained probabilistic prejudice seems unavoidable if the matter is really to be cleared up. Even if, as would be easy enough, I were to base my own theory upon the calculus of content, or of logical strength, it would still be necessary to explain that the probability calculus, in its ('logical') application to propositions or statements, is nothing but a *calculus of the logical weakness or lack of content of these statements* (either of absolute logical weakness or of relative logical weakness). Perhaps a head-on collision would be avoidable if people were not so generally inclined to assume uncritically that a high probability must be an aim of science, and that, therefore, the theory of induction must explain to us how we can attain a high degree of probability for our theories. (And it then becomes necessary to point out that there is something else—a 'truthlikeness' or 'verisimilitude'—with a calculus totally different from the calculus of probability with which it seems to have been confused.)

To avoid these simple results, all kinds of more or less sophisticated

[3] See for example J. C. Harsanyi, 'Popper's Improbability Criterion for the Choice of Scientific Hypotheses', *Philosophy*, **35**, 1960, pp. 332 ff. Incidentally, I do not propose any 'criterion' for the choice of scientific hypotheses: every choice remains a risky guess. Moreover, the theoretician's choice is the hypothesis most worthy of *further critical discussion* (rather than of *acceptance*).

theories have been designed. I believe I have shown that none of them is successful. But what is more important, they are quite unnecessary. One merely has to recognize that the property which we cherish in theories and which we may perhaps call 'verisimilitude' or 'truthlikeness' (see section XI below) is *not* a probability *in the sense of the calculus of probability* of which (2) is an inescapable theorem.

It should be noted that the problem before us is not a problem of words. I do not mind what you call 'probability', and I do not mind if you call those degrees for which the so-called 'calculus of probability' holds by any other name. I personally think that it is most convenient to reserve the term 'probability' for whatever may satisfy the well-known rules of this calculus (which Laplace, Keynes, Jeffreys and many others have formulated, and for which I have given various formal axiom systems). If (and only if) we accept this terminology, then there can be no doubt that the absolute probability of a statement *a* is simply the *degree of its logical weakness, or lack of informative content*, and that the relative probability of a statement *a*, given a statement *b*, is simply the degree of the relative weakness, or the relative *lack* of *new* informative content in statement *a*, assuming that we are already in possession of the information *b*.

Thus if we aim, in science, at a high informative content—if the growth of knowledge means that we know more, that we know *a* and *b*, rather than *a* alone, and that the content of our theories thus increases—then we have to admit that we also aim at a low probability, in the sense of the calculus of probability.

And since a low probability means a high probability of being falsified, it follows that a high degree of falsifiability, or refutability, or testability, is one of the aims of science—in fact, precisely the same aim as a high informative content.

The criterion of potential satisfactoriness is thus testability, or improbability: only a highly testable or improbable theory is worth testing, and is actually (and not merely potentially) satisfactory if it withstands severe tests—especially those tests to which we could point as crucial for the theory before they were ever undertaken.

It is possible in many cases to compare the severity of tests objectively. It is even possible, if we find it worth while, to define a measure of the severity of tests. (See the *Addenda* to this volume.) By the same method we can define the explanatory power and the degree of corroboration of a theory.[4]

IV

The thesis that the criterion here proposed actually dominates the progress of science can easily be illustrated with the help of historical examples. The theories of Kepler and Galileo were unified and superseded by Newton's logically stronger and better testable theory, and similarly Fresnel's and Faraday's by Maxwell's. Newton's theory, and Maxwell's, in their turn, were unified and superseded by Einstein's. In each such case the progress was towards a more informative and therefore logically less probable theory: towards a theory which was more severely testable because it made predictions which, in a purely logical sense, were more easily refutable.

A theory which is not in fact refuted by testing those new and bold and improbable predictions to which it gives rise can be said to be corroborated by these severe tests. I may remind you in this connection of Galle's discovery of Neptune, of Hertz's discovery of electromagnetic waves, of Eddington's eclipse observations, of Elsasser's interpretation of Davisson's maxima as interference fringes of de Broglie waves, and of Powell's observations of the first Yukawa mesons.

All these discoveries represent corroborations by severe tests—by predictions which were highly improbable in the light of our previous knowledge (previous to the theory which was tested and corroborated). Other important discoveries have also been made while testing a theory, though they did not lead to its corroboration but to its refutation. A recent and important case is the so-called refutation of parity. And Lavoisier's classical experiments which show that the volume of air decreases while a candle burns in a closed space, or that the weight of burning iron-filings increases, do not establish the oxygen theory of combustion; yet they tend to refute the phlogiston theory.

Lavoisier's experiments were carefully thought out; but even most so-called 'chance-discoveries' are fundamentally of the same logical structure. For these so-called 'chance-discoveries' are as a rule

[4] See especially appendix *ix to my *L.Sc.D.*

refutations of theories which were consciously or unconsciously held: they are made when some of our expectations (based upon these theories) are unexpectedly disappointed. Thus the catalytic property of mercury was discovered when it was accidentally found that in its presence a chemical reaction had been speeded up which had not been expected to be influenced by mercury. But neither Oersted's nor Röntgen's nor Becquerel's nor Fleming's discoveries was really accidental, even though they had accidental components: every one of these men was searching for an effect of the kind he found.

We can even say that some discoveries, such as Columbus' discovery of America, corroborate one theory (of the spherical earth) while refuting at the same time another (the theory of the size of the earth, and with it, of the nearest way to India); and that they were chance-discoveries to the extent to which they contradicted all expectations, and were not consciously undertaken as tests of those theories which they refuted.

V

The stress I am laying upon change in scientific knowledge, upon its growth, or its progressiveness, may to some extent be contrasted with the current ideal of science as an axiomatized deductive system. This ideal has been dominant in European epistemology from Euclid's Platonizing cosmology (for this is, I believe, what Euclid's *Elements* were really intended to be) to that of Newton, and further to the systems of Boscovič, Maxwell, Einstein, Bohr, Schrödinger, and Dirac. It is an epistemology that sees the final task and end of scientific activity in the construction of an axiomatized deductive system.

As opposed to this, I now believe that these most admirable deductive systems should be regarded as stepping stones rather than as ends:[5] as important stages on our way to richer, and better testable, scientific knowledge.

[5] I have been influenced in adopting this view by Dr J. Agassi who, in a discussion in 1956, convinced me that the attitude of looking upon the finished deductive systems as an end is a relic of the long domination of Newtonian ideas (and thus, I may add, of the Platonic, and Euclidean, tradition). For an even more radical view of Dr Agassi's see the last footnote to this chapter.

Regarded thus as means or stepping stones, they are certainly quite indispensable, for we are bound to develop our theories in the form of deductive systems. This is made unavoidable by the logical strength, by the great informative content, which we have to demand of our theories if they are to be better and better testable. The wealth of their consequences has to be unfolded deductively; for as a rule, a theory cannot be tested except by testing, one by one, some of its more remote consequences; consequences, that is, which cannot immediately be seen upon inspecting it intuitively.

Yet it is not the marvellous deductive unfolding of the system which makes a theory rational or empirical but the fact that we can examine it critically; that is to say, subject it to attempted refutations, including observational tests; and the fact that, in certain cases, a theory may be able to withstand those criticisms and those tests—among them tests under which its predecessors broke down, and sometimes even further and more severe tests. It is in the repeated and flexible choice of a new theory that the rationality of science lies, rather than in the deductive development of the theory.

Consequently there is little merit in formalizing and elaborating a deductive system (intended for use as an empirical science) beyond the requirements of the task of criticizing and testing it, and of comparing it critically with competitors. This critical comparison, though it has, admittedly, some minor conventional and arbitrary aspects, is largely non-conventional, thanks to our criterion of progress. It is this critical procedure which contains both the rational and the empirical elements of science. It contains those choices, those rejections, and those decisions, which show that we have learnt from our mistakes, and thereby added to our scientific knowledge.

VI

Yet perhaps even this picture of science—as a procedure whose rationality consists in the fact that we learn from our mistakes—is not quite good enough. It may still suggest that science progresses from theory to theory and that it consists of a sequence of better and better deductive systems. Yet what I really wish to suggest is that science should be visualized as *progressing from problems to problems*—to problems of ever increasing depth.

For a scientific theory—an explanatory theory—is, if anything, an attempt to solve a scientific problem, that is to say, a problem concerned or connected with the discovery of an explanation.[6]

Admittedly, our expectations, and thus our theories, may precede, historically, even our problems. *Yet science starts only with problems*. Problems crop up especially when we are disappointed in our expectations, or when our theories involve us in difficulties, in contradictions; and these may arise either within a theory, or between two different theories, or as the result of a clash between our theories and our observations. Moreover, it is only through a problem that we become conscious of holding a theory. It is the problem which challenges us to learn; to advance our knowledge; to experiment; and to observe.

Thus science starts from problems, and not from observations; though observations may give rise to a problem, especially if they are *unexpected*; that is to say, if they clash with our expectations or theories. The conscious task before the scientist is always the solution of a problem through the construction of a theory which solves the problem; for example, by explaining unexpected and unexplained observations. Yet every worthwhile new theory raises new problems; problems of reconciliation, problems of how to conduct new and previously unthought-of observational tests. And it is mainly through the new problems which it raises that it is fruitful.

Thus we may say that the most lasting contribution to the growth of scientific knowledge that a theory can make are the new problems which it raises, so that we are led back to the view of science and of the growth of knowledge as always starting from, and always ending with, problems—problems of an ever increasing depth, and an ever increasing fertility in suggesting new problems.

[6] Compare this and the following two paragraphs with my *Poverty of Historicism*, section 28, pp. 121 ff., and chs. 1 and 16 of this volume.

2. THE THEORY OF OBJECTIVE TRUTH: CORRESPONDENCE TO THE FACTS

VII

So far I have spoken about science, its progress, and its criterion of progress, without even mentioning *truth*. Perhaps surprisingly, this can be done without falling into pragmatism or instrumentalism: it is perfectly possible to argue in favour of the intuitive satisfactoriness of the criterion of progress in science without ever speaking about the truth of its theories. In fact, before I became acquainted with Tarski's theory of truth,[7] it appeared to me safer to discuss the criterion of progress without getting too deeply involved in the highly controversial problem connected with the use of the word 'true'.

My attitude at the time was this: although I accepted, as almost everybody does, the objective or absolute or correspondence theory of truth—truth as correspondence with the facts—I preferred to avoid the topic. For it appeared to me hopeless to try to understand clearly this strangely elusive idea of a correspondence between a statement (or a proposition) and a fact.

In order to recall why the situation appeared so hopeless we only have to remember, as one example among many, Wittgenstein's *Tractatus* with its surprisingly naïve picture theory, or projection theory, of truth. In his book a proposition was conceived as a picture or a projection (or a photograph) of the fact which it was intended to describe and as having the same structure (or 'form') as that fact; just as a gramophone record is indeed a picture or a projection of a sequence of sounds, and shares some of its structural properties.[8]

Another of these unavailing attempts to explain this correspondence was due to Schlick, who gave a beautifully clear and truly devastating criticism[9] of various correspondence theories—including the picture or projection theory—but who unfortunately produced in his turn another one which was no better. He interpreted the correspondence in question as a one-to-one correspondence between our designations and the designated objects, although counterexamples abound

[7] See my *L.Sc.D*, especially section 84, and my *Open Society*, especially pp. 369–74.

[8] Cp. Wittgenstein's *Tractatus*, especially 4.0141; also 2.161; 2.17; 2.223; 3.11.

[9] See especially pp. 56–7 of Schlick's remarkable *Allgemeine Erkenntnislehre*, 2nd edn., 1925.

(designations applying to many objects, objects designated by many designations) which refute this interpretation.

All this was changed by Tarski's theory of truth and of the correspondence of a statement with the facts. Tarski's greatest achievement, and the real significance of his theory for the philosophy of the empirical sciences, is that he rehabilitated the correspondence theory of absolute or objective truth which had become suspect. He vindicated the free use of the intuitive idea of truth as correspondence to the facts. (The view that his theory is applicable only to formalized languages is, I think, mistaken. It is applicable to any consistent and even to a 'natural' language, if only we learn from Tarski's analysis how to dodge its inconsistencies; which means, admittedly, the introduction of some 'artificiality'—or caution—into its use. See also *Addendum* 5, below.)

Although I may assume in this assembly some familiarity with Tarski's theory of truth, I may perhaps explain the way in which it can be regarded, from an intuitive point of view, as a simple elucidation of the idea of *correspondence to the facts*. I shall have to stress this almost trivial point because, in spite of its triviality, it will be crucial for my argument.

The highly intuitive character of Tarski's ideas seems to become more evident (as I have found in teaching) if we first decide explicitly to take 'truth' as a synonym for 'correspondence to the facts', and then (forgetting all about 'truth') *proceed to explain the idea of 'correspondence to the facts'*.

Thus we shall first consider the following two formulations, each of which states very simply (in a metalanguage) under what conditions a certain assertion (of an object language) corresponds to the facts.

(1) The statement, or the assertion, '*Snow is white*' corresponds to the facts if, and only if, snow is, indeed, white.

(2) The statement, or the assertion, '*Grass is red*' corresponds to the facts if, and only if, grass is, indeed, red.

These formulations (in which the word 'indeed' is only inserted for ease, and may be omitted) sound, of course, quite trivial. But it was left to Tarski to discover that, in spite of their apparent triviality, they contained the solution of the problem of explaining correspondence to the facts.

The decisive point is Tarski's discovery that, in order to speak of correspondence to the facts, as do (1) and (2), we must use a metalanguage in which we can *speak about two things: statements; and the facts to which these statements refer*. (Tarski calls such a metalanguage 'semantical'; a metalanguage in which we can speak about an object language but not about the facts to which it refers is called 'syntactical'.) Once the need for a (semantical) metalanguage is realized, everything becomes clear. (Note that while (3) '"John called" *is true*' is essentially a statement belonging to such a metalanguage, (4) 'It *is true that* John called' may belong to the same language as 'John called'. Thus the phrase 'It *is true that*'—which, like double negation, is logically redundant—differs widely from the metalinguistic predicate '*is true*'. The latter is needed for general remarks such as, 'If the conclusion is not true, the premises cannot all be true' or 'John once made a true statement'.)

I have said that Schlick's theory was mistaken, yet I think that certain comments he made (*loc. cit.*) about his own theory throw some light on Tarski's. For Schlick says that the problem of truth shared the fate of some others whose solutions were not easily seen because they were mistakenly supposed to lie on a very deep level, while actually they were fairly plain and, at first sight, unimpressive. Tarski's solution may well appear unimpressive at first sight. Yet its fertility and its power are impressive indeed.

VIII

Thanks to Tarski's work, the idea of objective or absolute truth—that is truth as correspondence to the facts—appears to be accepted today with confidence by all who understand it. The difficulties in understanding it seem to have two sources: first, the combination of an extremely simple intuitive idea with a certain amount of complexity in the execution of the technical programme to which it gives rise; secondly, the widespread but mistaken dogma that a satisfactory theory of truth should yield a criterion of *true belief*—of well-founded, or rational belief. Indeed, this dogma underlies the three rivals of the correspondence theory of truth—the coherence theory which mistakes consistency for truth, the evidence theory which mistakes 'known to be true' for 'true', and the pragmatic or instrumentalist theory which mistakes usefulness for truth. These are all subjective (or 'epistemic') theories of truth, in contradistinction to Tarski's objective (or 'metalogical') theory. They are subjective in the sense that *they all stem from the fundamental subjectivist position which can conceive of knowledge only as a special kind of mental state, or as a disposition, or as a special kind of belief*, characterized, for example, by its history or by its relation to other *beliefs*.

If we start from our subjective experience of believing, and thus look upon knowledge as a special kind of belief, then we may indeed have to look upon truth—that is, true knowledge—as some even more special kind of belief: as one that is well-founded or justified. This would mean that there should be some more or less effective criterion, if only a partial one, of well-foundedness; some symptom by which to differentiate the experience of a well-founded belief from other experiences of belief. It can be shown that all subjective theories of truth aim at such a criterion: they try to define truth in terms of the sources or origins of our beliefs,[10] or in terms of our operations of verification, or of some set of rules of acceptance, or simply in terms of the quality of our subjective convictions. They all say, more or less, that truth is what we are justified in believing or in accepting, in accordance with certain rules or criteria, of origins or sources of our knowledge, or of reliability, or stability, or success, or strength of conviction, or inability to think otherwise.

The theory of objective truth leads to a very different attitude. This may be seen from the fact that it allows us to make assertions such as the following: a theory may be true even though nobody believes it, and even though we have no reason to think that it is true; and another theory may be false even though we have comparatively good reasons for accepting it.

Clearly, these assertions would appear to be self-contradictory from the point of view of any subjective or epistemic theory of truth. But within the objective theory, they are not only consistent, but quite obviously true.

A similar assertion which the objective correspondence theory would make quite natural is this: even if we hit upon a true theory, we shall as a rule be merely guessing, and it may well be impossible for us to know that it is true.

[10] See my Introduction to this volume, 'On the Sources of Knowledge and of Ignorance'.

An assertion like this was made, apparently for the first time, by Xenophanes[11] who lived 2,500 years ago; which shows that the objective theory of truth is very old indeed—antedating Aristotle, who also held it. But only with Tarski's work has the suspicion been removed that the objective theory of truth as correspondence to the facts may be either self-contradictory (because of the paradox of the liar), or empty and redundant (as Ramsey suggested), or barren, or at the very least redundant, in the sense that we can do without it.

In my theory of scientific progress I might perhaps do without it, up to a point. Since Tarski, however, I no longer see any reason why I should try to avoid it. And if we wish to elucidate the difference between pure and applied science, between the search for knowledge and the search for power or for powerful instruments, then we cannot do without it. For the difference is that, in the search for knowledge, we are out to find true theories, or at least theories which are nearer than others to the truth—which correspond better to the facts; whereas in the search for powerful instruments we are, in many cases, quite well served by theories which are known to be false.[12]

So one great advantage of the theory of objective or absolute truth is that it allows us to say—with Xenophanes—that we search for truth, but may not know when we have found it; that we have no criterion of truth, but are nevertheless guided by the idea of truth as a *regulative principle* (as Kant or Peirce might have said); and that, though there are no general criteria by which we can recognize truth—except perhaps tautological truth—there are criteria of progress towards the truth (as I shall explain presently).

The status of truth in the objective sense, as correspondence to the facts, and its role as a regulative principle, may be compared to that of a mountain peak usually wrapped in clouds. A climber may not merely have difficulties in getting there—he may not know when he gets there, because he may be unable to distinguish, in the clouds, between the main summit and a subsidiary peak. Yet this does not affect the objective existence of the summit; and if the climber tells us 'I doubt whether I reached the actual summit', then he does, by implication,

[11] See my Introduction, p. 26, and ch. 5, pp. 152 f., above.

[12] See the discussion of the 'second view' (called 'instrumentalism') in ch. 3, above.

recognize the objective existence of the summit. The very idea of error or of doubt (in its normal straightforward sense) implies the idea of an objective truth which we may fail to reach.

Though it may be impossible for the climber ever to make sure that he has reached the summit, it will often be easy for him to realize that he has not reached it (or not yet reached it); for example, when he is turned back by an overhanging wall. Similarly, there will be cases when we are quite sure that we have not reached the truth. Thus while coherence, or consistency, is no criterion of truth, simply because even demonstrably consistent systems may be false in fact, incoherence or inconsistency do establish falsity; so, if we are lucky, we may discover the falsity of some of our theories.[13]

In 1944, when Tarski published the first English outline of his investigations into the theory of truth (which he had published in Poland in 1933), few philosophers would have dared to make assertions like those of Xenophanes; and it is interesting that the volume in which Tarski's paper was published also contained two subjectivist papers on truth.[14]

Though things have improved since then, subjectivism is still rampant in the philosophy of science, and especially in the field of probability theory. The subjectivist theory of probability, which interprets degrees of probability as degrees of rational belief, stems directly from the subjectivist approach to truth—especially from the coherence theory. Yet it is still embraced by philosophers who have accepted Tarski's theory of truth. At least some of them, I suspect, have turned to probability theory in the hope that it would give them what they had originally expected from a subjectivist or epistemological theory of the attainment of truth *through verification*; that is, a theory of rational and justifiable belief, based upon observed instances.[15]

It is an awkward point in all these subjectivist theories that they are irrefutable (in the sense that they can too easily evade any criticism). For it is always possible to uphold the view that everything we say

[13] See Alfred Tarski's somewhat popularized paper, 'The Semantic Conception of Truth', in *Philosophy and Phenom. Research*, **4**, 1943–4, pp. 341 ff. (Cp. especially section 21.)

[14] See the volume referred to in the preceding note, especially pp. 279 and 336.

[15] Cp. Carnap, *Logical Foundations of Probability*, 1950, p. 177, and my *L.Sc.D.*, especially section 84.

about the world, or everything we print about logarithms, should be replaced by a belief statement. Thus we may replace the statement 'Snow is white' by 'I believe that snow is white' or perhaps even by 'In the light of all the available evidence I believe that it is rational to believe that snow is white'. That we can (in a way) 'replace' assertions about the objective world by one of these subjectivist circumlocutions is trivial, although in the case of the assertions found in logarithm tables—which might well be produced by machines—somewhat unconvincing. (It may be mentioned in passing that the subjective interpretation of logical probability links these subjectivist replacements, exactly as in the case of the coherence theory of truth, with an approach which, on closer analysis, turns out to be essentially 'syntactic' rather than 'semantic'—though it may be presented within a 'semantical system'.)

It may be useful to sum up the relationships between the objective and subjective theories of scientific knowledge with the help of a little table:

OBJECTIVE OR LOGICAL OR ONTOLOGICAL THEORIES	SUBJECTIVE OR PSYCHOLOGICAL OR EPISTEMOLOGICAL THEORIES
truth as correspondence with the facts	*truth as property of our state of mind—or knowledge or belief*
objective probability (*inherent in the situation, and testable by statistical tests*)	*subjective probability* (*degree of rational belief based upon our total knowledge*)
objective randomness (*statistically testable*)	*lack of knowledge*
equiprobability (*physical or situational symmetry*)	*lack of knowledge*

In all these cases I am inclined to say not only that these two approaches should be distinguished, but also that the subjectivist approach should be discarded as a lapse, as based on a mistake—though perhaps a tempting mistake. There is, however, a similar table in which the epistemological (right hand) side is not based on a mistake.

truth	*conjecture*
testability	*empirical test*
explanatory or predictive power	*degree of corroboration (that is, report of the results of tests)*
'verisimilitude'	

3. TRUTH AND CONTENT: VERISIMILITUDE VERSUS PROBABILITY

IX

Like many other philosophers I am at times inclined to classify philosophers as belonging to two main groups—those with whom I disagree, and those who agree with me. I might call them the verificationists or the justificationist philosophers of knowledge or of belief, and the falsificationists or fallibilists or critical philosophers of conjectural knowledge. I may mention in passing a third group with whom I also disagree. They may be called the disappointed justificationists—the irrationalists and sceptics.

The members of the first group—the verificationists or justificationists—hold, roughly speaking, that whatever cannot be supported by positive reasons is unworthy of being believed, or even of being taken into serious consideration.

On the other hand, the members of the second group—the falsificationists or fallibilists—say, roughly speaking, that what cannot (at present) in principle be overthrown by criticism is (at present) unworthy of being seriously considered; while what can in principle be so overthrown and yet resists all our critical efforts to do so may quite possibly be false, but is at any rate not unworthy of being seriously considered and perhaps even of being believed—though only tentatively.

Verificationists, I admit, are eager to uphold the most important tradition of rationalism—the fight of reason against superstition and arbitrary authority. For they demand that we should accept a belief *only if it can be justified by positive evidence*; that is to say, *shown* to be true, or, at least, to be highly probable. In other words, they demand that we should accept a belief only if it can be *verified*, or probabilistically *confirmed*.

Falsificationists (the group of fallibilists to which I belong) believe—as most irrationalists also believe—that they have discovered logical arguments which show that the programme of the first group cannot be carried out: that we can never give positive reasons which justify the belief that a theory is true. But, unlike irrationalists, we falsificationists believe that we have also discovered a way to realize the old ideal of distinguishing rational science from various forms of superstition, in spite of the breakdown of the original inductivist or justificationist programme. We hold that this ideal can be realized, very simply, by recognizing that the rationality of science lies not in its habit of appealing to empirical evidence in support of its dogmas—astrologers do so too—but solely in the *critical approach*: in an attitude which, of course, involves the critical use, among other arguments, of empirical evidence (especially in refutations). For us, therefore, science has nothing to do with the quest for certainty or probability or reliability. We are not interested in establishing scientific theories as secure, or certain, or probable. Conscious of our fallibility we are only interested in criticizing them and testing them, hoping to find out where we are mistaken; of learning from our mistakes; and, if we are lucky, of proceeding to better theories.

Considering their views about the positive or negative function of argument in science, the first group—the justificationists—may be also nicknamed the 'positivists' and the second—the group to which I belong—the critics or the 'negativists'. These are, of course, mere nicknames. Yet they may perhaps suggest some of the reasons why some people believe that only the positivists or verificationists are seriously interested in truth and in the search for truth, while we, the critics or negativists, are flippant about the search for truth, and addicted to barren and destructive criticism and to the propounding of views which are clearly paradoxical.

This mistaken picture of our views seems to result largely from the adoption of a justificationist programme, and of the mistaken subjectivist approach to truth which I have described.

For the fact is that we too see science as the search for truth, and that, at least since Tarski, we are no longer afraid to say so. Indeed, it is only with respect to this aim, the discovery of truth, that we can say that though we are fallible, we hope to learn from our mistakes. It is only the idea of truth which allows us to speak sensibly of mistakes and of rational criticism, and which makes rational discussion possible—that is to say, critical discussion in search of mistakes with the serious purpose of eliminating as many of these mistakes as we can, in order to get nearer to the truth. Thus the very idea of error—and of fallibility—involves the idea of an objective truth as the standard of which we may fall short. (It is in this sense that the idea of truth is a *regulative* idea.)

Thus we accept the idea that the task of science is the search for truth, that is, for true theories (even though as Xenophanes pointed out we may never get them, or know them *as true* if we get them). Yet we also stress that *truth is not the only aim of science*. We want more than mere truth: what we look for is *interesting truth*—truth which is hard to come by. And in the natural sciences (as distinct from mathematics) what we look for is truth which has a high degree of explanatory power, in a sense which implies that it is logically improbable truth.

For it is clear, first of all, that we do not merely want truth—we want more truth, and new truth. We are not content with 'twice two equals four', even though it is true: we do not resort to reciting the multiplication table if we are faced with a difficult problem in topology or in physics. Mere truth is not enough; what we look for are *answers to our problems*. The point has been well put by the German humorist and poet Busch, of Max-and-Moritz fame, in a little nursery rhyme—I mean a rhyme for the epistemological nursery:[16]

> Twice two equals four: 'tis true,
> But too empty, and too trite.
> What I look for is a clue
> To some matters not so light.

Only if it is an answer to a problem—a difficult, a fertile problem, a problem of some depth—does a truth, or a conjecture about the truth, become relevant to science. This is so in pure mathematics, and it is so

[16] From W. Busch, *Schein und Sein* (first published posthumously in 1909; p. 28 of the *Insel* edition, 1952). My attention has been drawn to this rhyme by an essay on Busch as a philosopher which my late friend Julius Kraft contributed to the volume *Erziehung und Politik* (Essays for Minna Specht, 1960); see p. 262. My translation makes it perhaps more like a nursery rhyme than Busch intended.

in the natural sciences. And in the latter, we have something like a logical measure of the depth or significance of the problem in the increase of logical improbability or explanatory power of the proposed new answer, as compared with the best theory or conjecture previously proposed in the field. This logical measure is essentially the same thing which I have described above as the logical criterion of potential satisfactoriness and of progress.

My description of this situation might tempt some people to say that truth does not, after all, play a very big role with us negativists even as a regulative principle. There can be no doubt, they will say, that negativists (like myself) much prefer an attempt to solve an interesting problem by a bold conjecture, *even if it soon turns out to be false*, to any recital of a sequence of true but uninteresting assertions. Thus it does not seem, after all, as if we negativists had much use for the idea of truth. Our ideas of scientific progress and of attempted problem-solving do not seem very closely related to it.

This, I believe, would give quite a mistaken impression of the attitude of our group. Call us negativists, or what you like: but you should realize that we are as much interested in truth as anybody—for example, as the members of a court of justice. When the judge tells a witness that he should speak 'The truth, the *whole truth*, and nothing but the truth', then what he looks for is as much of the *relevant truth* as the witness may be able to offer. A witness who likes to wander off into irrelevancies is unsatisfactory as a witness, even though these irrelevancies may be truisms, and thus part of 'the whole truth'. It is quite obvious that what the judge—or anybody else—wants when he asks for 'the whole truth' is as much *interesting and relevant* true information as can be got; and many perfectly candid witnesses have failed to disclose some important information simply because they were unaware of its relevance to the case.

Thus when we stress, with Busch, that we are not interested in mere truth but in interesting and relevant truth, then, I contend, we only emphasize a point which everybody accepts. And if we are interested in bold conjectures, even if these should soon turn out to be false, then this interest is due to our methodological instinct that tells us that only with the help of such bold conjectures can we hope to discover interesting and relevant truth.

There is a point here which, I suggest, it is the particular task of the logician to analyse. 'Interest', or 'relevance', in the sense here intended, can be *objectively* analysed; it is relative to our problems; and it depends on the explanatory power, and thus on the content, or improbability, of the information. The measures alluded to earlier (and developed in the *Addenda* to this volume) are precisely such measures as take account of some *relative content* of the information—its content relative to a hypothesis or to a problem.

I can therefore gladly admit that falsificationists like myself much prefer an attempt to solve an interesting problem by a bold conjecture, *even (and especially) if it soon turns out to be false*, to any recital of a sequence of irrelevant truisms. We prefer this because we believe that this is the way in which we can learn from our mistakes; and that in finding that our conjecture was false, we shall have learnt much about the truth, and shall have got nearer to the truth.

I therefore hold that both ideas—the idea of truth, in the sense of correspondence with the facts, and the idea of content (which may be measured by the same measure as testability)—play about equally important roles in our considerations, and that both can shed much light on the idea of progress in science.

X

Looking at the progress of scientific knowledge, many people have been moved to say that even though we do not know how near to or how far from the truth we are, we can, and often do, *approach more and more closely to the truth*. I myself have sometimes said such things in the past, but always with a twinge of bad conscience. Not that I believe in being over-fussy about what we say: as long as we speak as clearly as we can, yet do not pretend that what we are saying is clearer than it is, and as long as we do not try to derive apparently exact consequences from dubious or vague premises, there is no harm whatever in occasional vagueness, or in voicing every now and then our feelings and general intuitive impressions about things. Yet whenever I used to write, or to say, something about science as getting nearer to the truth, or as a kind of approach to truth, I felt that I really ought to be writing 'Truth', with a capital 'T', in order to make quite clear that a vague and highly metaphysical notion was involved here, in contradistinction to Tarski's 'truth' which we can with a clear conscience write in the ordinary way with small letters.[17]

It was only quite recently that I set myself to consider whether the idea of truth involved here was really so dangerously vague and metaphysical after all. Almost at once I found that it was not, and that there was no particular difficulty in applying Tarski's fundamental idea to it.

For there is no reason whatever why we should not say that one theory corresponds better to the facts than another. This simple initial step makes everything clear: there really is no barrier here between what at first sight appeared to be Truth with a capital 'T' and truth in a Tarskian sense.

But can we really speak about *better* correspondence? Are there such things as *degrees* of truth? Is it not dangerously misleading to talk as if Tarskian truth were located somewhere in a kind of metrical or at least topological space so that we can sensibly say of two theories—say an earlier theory t_1 and a later theory t_2—that t_2 has superseded t_1, or progressed beyond t_1, by approaching more closely to the truth than t_1?

I do not think that this kind of talk is at all misleading. On the contrary, I believe that we simply cannot do without something like this idea of a better or worse approximation to truth. For there is no doubt whatever that we can say, and often want to say, of a theory t_2 that it corresponds better to the facts, or that as far as we know it seems to correspond better to the facts, than another theory t_1.

I shall give here a somewhat unsystematic list of six types of cases in which we should be inclined to say of a theory t_1 that it is superseded by t_2 in the sense that t_2 seems—as far as we know—to correspond better to the facts than t_1, in some sense or other.

(1) t_2 makes more precise assertions than t_1, and these more precise assertions stand up to more precise tests.

(2) t_2 takes account of, and explains, more facts than t_1 (which will include for example the above case that, other things being equal, t_2's assertions are more precise).

(3) t_2 describes, or explains, the facts in more detail than t_1.

[17] Similar misgivings are expressed by Quine when he criticizes Peirce for operating with the idea of approaching to truth. See W. V. Quine, *Word and Object*, New York, 1960, p. 23.

(4) t_2 has passed tests which t_1 has failed to pass.

(5) t_2 has suggested new experimental tests, not considered before t_2 was designed (and not suggested by t_1, and perhaps not even applicable to t_1); and t_2 has passed these tests.

(6) t_2 has unified or connected various hitherto unrelated problems.

If we reflect upon this list, then we can see that the *contents* of the theories t_1 and t_2 play an important role in it. (It will be remembered that the *logical content* of a statement or a theory *a* is the class of all statements which follow logically from *a*, while I have defined the *empirical content* of *a* as the class of all basic statements which contradict *a*.[18]) For in our list of six cases, the empirical content of theory t_2 exceeds that of theory t_1.

This suggests that we combine here the ideas of truth and of content into one—the idea of a degree of better (or worse) correspondence to truth or of greater (or less) likeness or similarity to truth; or to use a term already mentioned above (in contradistinction to probability) the idea of (degrees of) *verisimilitude*.

It should be noted that the idea that every statement or theory is not only either true or false but has, independently of its truth value, some degree of verisimilitude, does not give rise to any multi-valued logic—that is, to a logical system with more than two truth values, true and false; though some of the things the defenders of multi-valued logic are hankering after seem to be realized by the theory of verisimilitude (and related theories alluded to in section 3 of the *Addenda* to this volume).

XI

Once I had seen the problem it did not take me long to get to this point. But strangely enough, it took me a long time to put two and two together, and to proceed from here to something like a simple

[18] This definition is logically justified by the theorem that, if we confine ourselves to the 'empirical part' of the logical content, comparison of empirical contents and of logical contents always yield the same results; and it is intuitively justified by the consideration that a statement *a* tells the more about our world of experience the more experiences (i.e. possible experiences) it excludes (or forbids). About basic statements see also the *Addendum* 1 to this volume.

definition of verisimilitude in terms of truth and of content. (We can use either logical or empirical content, and thus obtain two closely related ideas of verisimilitude which however merge into one if we consider here only empirical theories, or empirical aspects of theories.)

Let us consider the *content* of a statement *a*; that is, the class of all the logical consequences of *a*. If *a* is true, then this class can consist only of true statements, because truth is always transmitted from a premise to all its conclusions. But if *a* is false, then its content will always consist of both true and false consequences. (Example: 'It always rains on Sundays' is false, but its consequence that it rained last Sunday happens to be true.) Thus whether a statement is true or false, *there may be more truth, or less truth, in what it says*, according to whether its content consists of a greater or a lesser number of true statements.

Let us call the class of the true logical consequences of *a* the 'truth-content' of *a* (a German term '*Wahrheitsgehalt*'—reminiscent of the phrase 'there is truth in what you say'—of which 'truth-content' may be said to be a translation, has been intuitively used for a long time); and let us call the class of the false consequences of *a*—but only these—the 'falsity-content' of *a*. (The 'falsity-content' is not, strictly speaking, a 'content', because it does not contain any of the true conclusions of the false statements which form its elements. Yet it is possible—see the *Addenda*—to define its *measure* with the help of two contents.) These terms are precisely as objective as the terms 'true' or 'false' and 'content' themselves. Now we can say:

Assuming that the truth-content and the falsity-content of two theories t_1 *and* t_2 *are comparable, we can say that* t_2 *is more closely similar to the truth, or corresponds better to the facts, than* t_1, *if and only if either*

(*a*) *the truth-content but not the falsity-content of* t_2 *exceeds that of* t_1, or
(*b*) *the falsity-content of* t_1, *but not its truth-content, exceeds that of* t_2.

If we now work with the (perhaps fictitious) assumption that the content and truth-content of a theory *a* are in principle *measurable*, then we can go slightly beyond this definition and can define $Vs(a)$, that is to say, a measure of the *verisimilitude* or *truthlikeness* of *a*. The simplest definition will be

$$Vs(a) = Ct_T(a) - Ct_F(a)$$

where $Ct_T(a)$ is a measure of the truth-content of a, and $Ct_F(a)$ is a measure of the falsity-content of a. A slightly more complicated but in some respects preferable definition will be found in section 3 of the *Addenda* to the present volume.

It is obvious that $Vs(a)$ satisfies our two demands, according to which $Vs(a)$ should increase

(*a*) if $Ct_T(a)$ increases while $Ct_F(a)$ does not, and
(*b*) if $Ct_F(a)$ decreases while $Ct_T(a)$ does not.

Some further considerations of a slightly technical nature and the definitions of $Ct_T(a)$ and especially $Ct_F(a)$ and $Vs(a)$ will be found in the *Addenda*. Here I want only to discuss three non-technical points.

XII

The first point is this. Our idea of approximation to truth, or of verisimilitude, has the same objective character and the same ideal or regulative character as the idea of objective or absolute truth. It is *not an epistemological or an epistemic idea*—no more than is truth or content. (In Tarski's terminology, it is obviously a 'semantic' idea, like truth, or like logical consequence, and, therefore, content.) Accordingly, we have here again to distinguish between the question 'What do you intend to say if you say that the theory t_2 has a higher degree of verisimilitude than the theory t_1?', and the question 'How do you know that the theory t_2 has a higher degree of verisimilitude than the theory t_1?'

We have so far answered only the first question. The answer to the second question depends on it, and it is exactly analogous to the answer to the analogous (absolute rather than comparative) question about truth: 'I do *not* know—I only guess. But I can examine my guess critically, and if it withstands severe criticism, then this fact may be taken as a good critical reason in favour of it.'

My second point is this. Verisimilitude is so defined that maximum verisimilitude would be achieved only by a theory which is not only true, but completely comprehensively true: if it corresponds to *all* facts, as it were, and, of course, only to *real* facts. This is of course a much more remote and unattainable ideal than a mere correspondence with *some* facts (as in, say, 'Snow is usually white').

But all this holds only for the maximum degree of verisimilitude, and not for the *comparison of theories with respect to their degree of verisimilitude.* This comparative use of the idea is its main point; and the idea of a higher or lower degree of verisimilitude seems less remote and more applicable and therefore perhaps more important for the analysis of scientific methods than the—in itself much more fundamental—idea of absolute truth itself.

This leads me to my third point. Let me first say that I do not suggest that the explicit introduction of the idea of verisimilitude will lead to any changes in the theory of method. On the contrary, I think that my theory of testability or corroboration by empirical tests is the proper methodological counterpart to this new metalogical idea. The only improvement is one of clarification. Thus I have often said that we prefer the theory t_2 which has passed certain severe tests to the theory t_1 which has failed these tests, because a false theory is certainly worse than one which, for all we know, may be true.

To this we can now add that even after t_2 has been refuted in its turn, we can still say that it is better than t_1, for although both have been shown to be false, the fact that t_2 has withstood tests which t_1 did not pass may be a good indication that the falsity-content of t_1 exceeds that of t_2 while its truth-content does not. Thus we may still give preference to t_2, even after its falsification, because we have reason to think that it agrees better with the facts than did t_1.

All cases where we accept t_2 because of experiments which were crucial between t_2 and t_1 seem to be of this kind, and especially all cases where the experiments were found by trying to think out, with the help of t_2, cases where t_2 leads to other results than did t_1. Thus Newton's theory allowed us to predict some deviations from Kepler's laws. Its success in this field established that it did not fail in cases which refuted Kepler's; at least the now known falsity-content of Kepler's theory was not part of Newton's, while it was pretty clear that the truth-content could not have shrunk, since Kepler's theory followed from Newton's as a 'first approximation'.

Similarly, a theory t_2 which is more precise than t_1 can now be shown to have—always provided its falsity content does not exceed that of t_1—a higher degree of verisimilitude than t_1. The same will hold for t_2 whose numerical assertions, though false, come nearer to the true numerical values than those of t_1.

Ultimately, the idea of verisimilitude is most important in cases where we know that we have to work with theories which are *at best* approximations—that is to say, theories of which we actually know that they cannot be true. (This is often the case in the social sciences.) In these cases we can still speak of better or worse approximations to the truth (and we therefore do not need to interpret these cases in an instrumentalist sense).

XIII

It always remains possible, of course, that we shall make mistakes in our relative appraisal of two theories, and the appraisal will often be a controversial matter. This point can hardly be over-emphasized. Yet it is also important that in principle, and as long as there are no revolutionary changes in our background knowledge, the relative appraisal of our two theories, t_1 and t_2, will remain stable. More particularly, our preferences need not change, as we have seen, if we eventually refute the better of the two theories. Newton's dynamics, for example, even though we may regard it as refuted, has of course maintained its superiority over Kepler's and Galileo's theories. The reason is its greater content or explanatory power. Newton's theory continues to explain more facts than did the others; to explain them with greater precision; and to unify the previously unconnected problems of celestial and terrestrial mechanics. The reason for the stability of relative appraisals such as these is quite simple: the logical relation between the theories is of such a character that, first of all, there exist with respect to them those crucial experiments, and these, when carried out, went against Newton's predecessors. And secondly, it is of such a character that the later refutations of Newton's theory could not support the older theories: they either did not affect them, or (as with the perihelion motion of Mercury) they could be claimed to refute the predecessors also.

I hope that I have explained the idea of better agreement with the facts, or of degrees of verisimilitude, sufficiently clearly for the purpose of this brief survey.

XIV

A brief remark on the early history of the confusion between verisimilitude and probability may perhaps be appropriate here.

As we have seen, progress in science means progress towards more interesting, less trivial, and therefore less 'probable' theories (where 'probable' is taken in any sense, such as *lack* of content, or greater statistical frequency, that satisfies the calculus of probability) and this means, as a rule, progress towards less familiar and less comfortable or plausible theories. Yet the idea of greater verisimilitude, of a better approximation to the truth, is usually confused, intuitively, with the totally different idea of a greater probability (in its various senses of 'more likely than not', 'more often than not', 'seems likely to be true', 'sounds plausible', 'sounds convincing'). The confusion is a very old one. We have only to remember some of the other words for 'probable', such as 'likely' which comes originally from 'like the truth' or 'verisimilar' ('*eoikotōs*', '*eikotōs*', '*eikos,*' etc., in Greek; '*verisimilis*' in Latin; '*wahrscheinlich*' in German) in order to see some of the traces, and perhaps some of the sources, of this confusion.

Two at least of the earliest of the Presocratic philosophers used '*eoikota*' in the sense of 'like the truth' or 'similar to the truth'. Thus we read in Xenophanes (DK, B 35): 'These things, let us suppose, are like the truth.'

It is fairly clear that verisimilitude or truthlikeness is meant here, rather than probability or degree of incomplete certainty. (Otherwise the words 'let us suppose' or 'let it be conjectured' or 'let it be imagined' would be redundant, and Xenophanes would have written something like, 'These things, let it be *said*, are probable'.)

Using the same word ('*eoikota*'), Parmenides wrote (DK, B 8, 60):[19]

[19] In this fragment '*eoikota*' has been most frequently translated as 'probable' or 'plausible'. For example W. Kranz, in Diels-Kranz, *Fragmente der Vorsokratiker*, 6th edn., translates it '*wahrscheinlich-einleuchtend*' that is, 'probable and plausible'; he reads the passage thus: 'This world-arrangement (or world-order) I shall expound to you in all its parts as something probable and plausible.' In translating '(wholly) like truth' or '(wholly) like the truth', I

'Now of this world thus arranged to seem wholly like truth I shall tell you . . .'

Yet already in the same generation or the next, Epicharmus, in a criticism of Xenophanes, seems to have used the word '*eikotōs*' in the sense of 'plausible', or something like it (DK, 21 A 15); though the possibility cannot be excluded that he may have used it in the sense of 'like the truth', and that it was Aristotle (our source is *Met.*, 1010a4) who read it in the sense of 'plausible' or 'likely'. Some three generations later, however, '*eikos*' is used quite unambiguously in the sense of 'likely' or 'probable' (or perhaps even of 'more frequently than not') by the sophist Antiphon when he writes (DK, B 60): 'If one begins a thing well it is likely to end well.'

All this suggests that the confusion between verisimilitude and probability goes back almost to the beginning of Western philosophy: and this is understandable if we consider that Xenophanes stressed the fallibility of our knowledge which he described as uncertain guesswork and at best 'like the truth'. This phrase, it seems, lent itself to misinterpretation as 'uncertain and at best of some fair degree of certainty'—that is, 'probable'.

Xenophanes himself seems to have distinguished clearly between degrees of certainty and degrees of truthlikeness. This emerges from another fragment (quoted above towards the end of chapter 5, p. 153) which says that even if by chance we were to hit upon, and pronounce, the final truth (that is, we may add, perfect truthlikeness), we should not know it. Thus great uncertainty is compatible with greatest truthlikeness.

I suggest that we return to Xenophanes and re-introduce a clear distinction between *verisimilitude* and *probability* (using this latter term in a sense laid down by the calculus of probability).

The differentiation between these two ideas is the more important as they have become confused; because both are closely related to the idea of truth, and both introduce the idea of an approach to truth by degrees. Logical probability (we do not discuss here physical

am somewhat influenced by the line (DK, B 35) quoted above from Xenophanes (and also by K. Reinhardt's *Parmenides*, pp. 5 f., where Wilamowitz is referred to). See also section vii of the Introduction to the present volume; the quotation from Osiander in section 1 of ch. 3; section XII of ch. 5, above; and *Addendum* 6, below.

probability) represents the idea of approaching logical certainty, or tautological truth, through a gradual diminution of informative content. Verisimilitude, on the other hand, represents the idea of approaching comprehensive truth. It thus combines truth and content while probability combines truth with lack of content.[20]

The feeling that it is absurd of me to deny that science aims at probability stems, I suggest, from a misguided 'intuition'—from the intuitive confusion between the two notions of verisimilitude and of probability which, as it now turns out, are utterly different.

4. BACKGROUND KNOWLEDGE AND SCIENTIFIC GROWTH

XV

People involved in a fruitful critical discussion of a problem often rely, if only unconsciously, upon two things: the acceptance by all parties of the common aim of getting at the truth, or at least nearer to the truth, and a considerable amount of common background knowledge. This does not mean that either of these two things is an indispensable basis of every discussion, or that these two things are themselves '*a priori*', and cannot be critically discussed in their turn. It only means that criticism never starts from nothing, even though every one of its starting points *may* be challenged, one at a time, in the course of the critical debate.

Yet though every one of our assumptions may be challenged, it is quite impracticable to challenge all of them at the same time. Thus all criticism must be piecemeal (as against the holistic view of Duhem and of Quine); which is only another way of saying that the fundamental maxim of every critical discussion is that we should stick to our problem, and that we should subdivide it, if practicable, and try to solve no more than one problem at a time, although we may, of course, always proceed to a subsidiary problem, or replace our problem by a better one.

[20] This, incidentally, holds for both, absolute probability, $p(a)$, and relative or conditional probability, $p(a, b)$; and we can construct corresponding (but opposed) absolute and relative concepts of verisimilitude.

While discussing a problem we always accept (if only temporarily) all kinds of things as *unproblematic*: they constitute for the time being, and for the discussion of this particular problem, what I call our *background knowledge*. Few parts of this background knowledge will appear to us in all contexts as absolutely unproblematic, and any particular part of it *may* be challenged at any time, especially if we suspect that its uncritical acceptance may be responsible for some of our difficulties. But almost all of the vast amount of background knowledge which we constantly use in any informal discussion will, for practical reasons, necessarily remain unquestioned; and the misguided attempt to question it all—that is to say, *to start from scratch*—can easily lead to the breakdown of a critical debate. (Were we to start the race where Adam started, I know of no reason why we should get any further than Adam did.)

XVI

The fact that, as a rule, we are at any given moment taking a vast amount of traditional knowledge for granted (for almost all our knowledge is traditional) creates no difficulty for the falsificationist or fallibilist. For he does not *accept* this background knowledge; neither as established nor as fairly certain, nor yet as probable. He knows that even its tentative acceptance is risky, and stresses that every bit of it is open to criticism, even though only in a piecemeal way. We can never be certain that we shall challenge the right bit; but since our quest is not for certainty, this does not matter. It will be noticed that this remark contains my answer to Quine's holistic view of empirical tests; a view which Quine formulates (with reference to Duhem) by asserting that our statements about the external world face the tribunal of sense experience not individually but only as a corporate body.[21] Now it has to be admitted that we can often test only a large chunk of a theoretical system, and sometimes perhaps only the whole system, and that, in these cases, it is sheer guess-work which of its ingredients should be held responsible for any falsification; a point which I have tried to emphasize—also with reference to Duhem—for a long time

[21] See W. V. Quine, *From a Logical Point of View*, 1953, p. 41.

past.[22] Though this argument may turn a verificationist into a sceptic, it does not affect those who hold that all our theories are guesses anyway.

This shows that the holistic view of tests, even if it were true, would not create a serious difficulty for the fallibilist and falsificationist. On the other hand, it should be said that the holistic argument goes much too far. It is possible in quite a few cases to find which hypothesis is responsible for the refutation; or in other words, which part, or group of hypotheses, was necessary for the derivation of the refuted prediction. The fact that such logical dependencies may be discovered is established by the practice of *independence proofs* of axiomatized systems; proofs which show that certain axioms of an axiomatic system cannot be derived from the rest. The more simple of these proofs consist in the construction, or rather in the discovery, of a *model*—a set of things, relations, operations, or functions—which satisfies all the axioms except the *one* whose independence is to be shown: for this one axiom—and therefore for the theory as a whole—the model constitutes a counterexample.

Now let us say that we have an axiomatized theoretical system, for example of physics, which allows us to predict that certain things do not happen, and that we discover a counterexample. There is no reason whatever why this counterexample may not be found to satisfy most of our axioms or even all our axioms except one whose independence would be thus established. This shows that the holistic dogma of the 'global' character of all tests or counterexamples is untenable. And it explains why, even without axiomatizing our physical theory, we may well have an inkling of what has gone wrong with our system.

This, incidentally, speaks in favour of operating, in physics, with highly analysed theoretical systems—that is, with systems which, even though they may fuse all the hypotheses into one, allow us to separate various groups of hypotheses, each of which may become an object of refutation by counterexamples. (An excellent recent example is the rejection, in atomic theory, of the law of parity; another is the rejection of the law of commutation for conjugate variables, prior to their interpretation as matrices, and to the statistical interpretation of these matrices.)

[22] See my *L.Sc.D.*, especially sections 19 to 22; and this volume, ch. 3, text to note 28.

XVII

One fact which is characteristic of the situation in which the scientist finds himself is that we constantly add to our background knowledge. If we discard some parts of it, others, closely related to them, will remain. For example, even though we may regard Newton's theory as refuted—that is, his system of ideas, and the formal deductive system which derives from it—we may still assume, as part of our background knowledge, the approximate truth of its quantitative formulae.

The existence of this background knowledge plays an important role in one of the arguments which support (I believe) my thesis that the rational and empirical character of science would vanish if it ceased to progress. I can sketch this argument here only in its barest outline.

A serious empirical test always consists in the attempt to find a refutation, a counterexample. In the search for a counterexample, we have to use our background knowledge; for we always try to refute first the *most risky* predictions, the '*most unlikely* . . . consequences' (as Peirce already saw[23]); which means that we always look in the *most probable kinds* of places for the *most probable* kinds of counterexamples—most probable in the sense that we should expect to find them in the light of our background knowledge. Now if a theory stands up to many such tests, then, owing to the incorporation of the results of our tests into our background knowledge, there may be, after a time, no places left where (in the light of our new background knowledge) counterexamples can with a high probability be expected to occur. But this means that the degree of severity of our test declines. This is also the reason why an often repeated test will no longer be considered as significant or as severe: there is something like a law of diminishing returns from repeated tests (as opposed to tests which, in the light of our background knowledge, are of a *new kind*, and which therefore may still be felt to be significant). These are facts that are inherent in the knowledge-situation; and they have often been described—especially by John Maynard Keynes and by Ernest Nagel—as difficult to explain by an inductivist theory of science. But for us it is all very easy. And we can even explain, by a similar analysis of the knowledge-situation, why

[23] See the *Collected Papers of C. S. Peirce*, vol. VII, 7.182 and 7.206. I owe this reference to W. B. Gallie (cp. *Philosophy*, **35**, 1960, p. 67), and a similar one to David Rynin.

the empirical character of a very successful theory always grows stale, after a time. We may then feel (as Poincaré did with respect to Newton's theory) that the theory is nothing but a set of implicit definitions or conventions—until we progress again and, by refuting it, incidentally re-establish its lost empirical character. (*De mortuis nil nisi bene*: once a theory is refuted, its empirical character is secure and shines without blemish.)

5. THREE REQUIREMENTS FOR THE GROWTH OF KNOWLEDGE

XVIII

But let us return again to the idea of getting nearer to the truth—to the search for theories that agree better with the facts (as indicated by the list of six comparisons in section x above).

What is the general problem situation in which the scientist finds himself? He has before him a scientific problem: he wants to find a new theory capable of explaining certain experimental facts; facts which the earlier theories successfully explained; others which they could not explain; and some by which they were actually falsified. The new theory should also resolve, if possible, some theoretical difficulties (such as how to dispense with certain *ad hoc* hypotheses, or how to unify two theories). Now if he manages to produce a theory which is a solution to all these problems, his achievement will be very great.

Yet it is not enough. I have been asked, 'What more do you want?' My answer is that there are many more things which I want; or rather, which I think are required by the logic of the general problem situation in which the scientist finds himself; by the task of getting nearer to the truth. I shall confine myself here to the discussion of three such requirements.

The first requirement is this. The new theory should proceed from some *simple, new, and powerful, unifying idea* about some connection or relation (such as gravitational attraction) between hitherto unconnected things (such as planets and apples) or facts (such as inertial and gravitational mass) or new 'theoretical entities' (such as fields and particles). This *requirement of simplicity* is a bit vague, and it seems difficult to formulate it very clearly. It seems to be intimately connected with the idea that our theories should describe the structural properties of the world—an idea which it is hard to think out fully without getting involved in an infinite regress. (This is so because any idea of a particular structure of the world—unless, indeed, we think of a purely *mathematical* structure—already presupposes a universal theory; for example, explaining the laws of chemistry by interpreting molecules as structures of atoms, or of subatomic particles, presupposes the idea of universal laws that regulate the properties and the behaviour of the atoms, or of the particles.) Yet one important ingredient in the idea of simplicity can be logically analysed. It is the idea of testability.[24] This leads us immediately to our second requirement.

For, secondly, we require that the new theory should be *independently testable*.[25] That is to say, apart from explaining all the *explicanda* which the new theory was designed to explain, it must have new and testable consequences (preferably consequences of a *new kind*[25a]); it must lead to the prediction of phenomena which have not so far been observed.

This requirement seems to me indispensable since without it our new theory might be *ad hoc*; for it is always possible to produce a theory to fit any given set of explicanda. Thus our two first requirements are needed in order to restrict the range of our choice among the possible solutions (many of them uninteresting) of the problem in hand.

If our second requirement is satisfied then our new theory will represent a potential step forward, whatever the outcome of the new tests may be. For it will be better testable than the previous theory: the fact that it explains all the explicanda of the previous theory, and that, in addition, it gives rise to new tests, suffices to ensure this.

Moreover, the second requirement also ensures that our new theory

[24] See sections 31–46 of my *L.Sc.D.* More recently I have stressed (in lectures) the need to *relativize* comparisons of simplicity to those hypotheses which compete *qua* solutions *of a certain problem, or set of problems*. The idea of simplicity, though intuitively connected with the idea of a unified theory that springs from *one* intuitive picture of the facts, cannot be analysed in terms of numerical paucity of hypotheses. For every (finitely axiomatizable) theory can be formulated in one statement; and it seems that, for every theory and every *n*, there is a set of *n* independent axioms (though not necessarily 'organic' axioms in the Warsaw sense).

[25] For the idea of an *independent test* see my paper 'The Aim of Science', *Ratio*, **1**, 1957.

[25a] See above, ch. 3, section 6, end of p. 157.

will, to some extent, be fruitful as an instrument of exploration. That is to say, it will suggest to us new experiments, and even if these should at once lead to the refutation of the theory, our factual knowledge will have grown through the unexpected results of the new experiments. Moreover, they will confront us with new problems to be solved by new explanatory theories.

Yet I believe that there must be a third requirement for a good theory. It is this. We require that the theory should pass some new, and severe, tests.

XIX

Clearly, this requirement is totally different in character from the previous two. These could be seen to be fulfilled, or not fulfilled, largely by analysing the old and the new theories logically. (They are 'formal requirements'.) The third requirement, on the other hand, can be found to be fulfilled, or not fulfilled, only by testing the new theory empirically. (It is a 'material requirement', a requirement of *empirical success*.)

Moreover, the third requirement clearly cannot be indispensable in the same sense as are the two previous ones. For these two are indispensable for deciding whether the theory in question should be at all accepted as a serious candidate for examination by empirical tests; or in other words, whether it is an interesting and promising theory. Yet on the other hand, some of the most interesting and most admirable theories ever conceived were refuted at the very first test. And why not? The most promising theory may fail if it makes predictions of a new kind. An example is the marvellous theory of Bohr, Kramers and Slater[26] of 1924 which, as an intellectual achievement, might perhaps even rank with Bohr's theory of the hydrogen atom of 1913. Yet unfortunately it was almost at once refuted by the facts—by the coincidence experiments of Bothe and Geiger.[27] This shows that not even the greatest physicist can anticipate the secrets of nature: his inspirations can only be guesses, and it is no fault of his, or of his theory, if it is

[26] *Phil. Mag.*, **47**, 1924, pp. 785 ff.

[27] *Zeitschr. f. Phys.*, **32**, 1925, pp. 63 ff.

refuted. Even Newton's theory was in the end refuted; and indeed, we hope that we shall in this way succeed in refuting, and improving upon, every new theory. And if it is refuted in the end, why not in the beginning? One might well say that it is merely a historical accident if a theory is refuted after six months rather than after six years, or six hundred years.

Refutations have often been regarded as establishing the failure of a scientist, or at least of his theory. It should be stressed that this is an inductivist error. Every refutation should be regarded as a great success; not merely a success of the scientist who refuted the theory, but also of the scientist who created the refuted theory and who thus in the first instance suggested, if only indirectly, the refuting experiment.

Even if a new theory (such as the theory of Bohr, Kramers, and Slater) should meet an early death, it should not be forgotten; rather its beauty should be remembered, and history should record our gratitude to it—for bequeathing to us new and perhaps still unexplained experimental facts and, with them, new problems; and for the services it has thus rendered to the progress of science during its successful but short life.

All this indicates clearly that our third requirement is not indispensable: even a theory that fails to meet it can make an important contribution to science. Yet in a different sense, I hold, it is indispensable none the less. (Bohr, Kramers and Slater rightly aimed at more than making an important contribution to science.)

In the first place, I contend that further progress in science would become impossible if we did not reasonably often manage to meet the third requirement; thus if the progress of science is to continue, and its rationality not to decline, we need not only successful refutations, but also positive successes. We must, that is, manage reasonably often to produce theories that entail new predictions, especially predictions of new effects, new testable consequences, suggested by the new theory and never thought of before.[28] Such a new prediction was that planets would under certain circumstances deviate from Kepler's laws; or that light, in spite of its zero mass, would prove to be subject to

[28] I have drawn attention to 'new' predictions of this kind and to their philosophical significance in ch. 3. See especially pp. 157 f.

gravitational attraction (that is, Einstein's eclipse-effect). Another example is Dirac's prediction that there will be an anti-particle for every elementary particle. New predictions of these kinds must not only be produced, but they must also be reasonably often corroborated by experimental evidence, I contend, if scientific progress is to continue.

We do need this kind of success; it is not for nothing that the great theories of science have all meant a new conquest of the unknown, a new success in predicting what had never been thought of before. We need successes such as that of Dirac (whose anti-particles have survived the abandonment of some other parts of his theories), or that of Yukawa's meson theory. We need the success, the empirical corroboration, of some of our theories, if only in order to appreciate the significance of successful and stirring refutations (like that of parity). It seems to me quite clear that it is only through these temporary successes of our theories that we can be reasonably successful in attributing our refutations to definite portions of the theoretical maze. (For we *are* reasonably successful in this—a fact which must remain inexplicable for one who adopts Duhem's and Quine's views on the matter.) An unbroken sequence of refuted theories would soon leave us bewildered and helpless: we should have no clue about the parts of each of these theories—or of our background knowledge—to which we might, tentatively, attribute the failure of that theory.

XX

Earlier I suggested that science would stagnate, and lose its empirical character, if we should fail to obtain refutations. We can now see that for very similar reasons science would stagnate, and lose its empirical character, if we should fail to obtain verifications of new predictions; that is, if we should only manage to produce theories that satisfy our first two requirements but not the third. For suppose we were to produce an unbroken sequence of explanatory theories each of which would explain all the explicanda in its field, including the experiments which refuted its predecessors; each would also be independently testable by predicted new effects; yet each would be at once refuted when these predictions were put to the test. Thus each would satisfy our first two requirements, but all would fail to satisfy the third.

I assert that, in this case, we should feel that we were producing a sequence of theories which, in spite of their increasing degree of testability, were *ad hoc*, and that we were not getting any nearer to the truth. And indeed, this feeling may well be justified: this whole sequence of theories might easily be *ad hoc*. For if it is admitted that a theory may be *ad hoc* if it is not independently testable by experiments of a new kind but merely explains all the explicanda, including the experiments which refuted its predecessors, then it is clear that the mere fact that the theory is also independently testable cannot as such ensure that it is not *ad hoc*. This becomes clear if we consider that it is always possible, by a trivial stratagem, to make an *ad hoc* theory independently testable, *if we do not also require that it should pass the independent tests in question*: we merely have to connect it (conjunctively) in some way or other with any testable but not yet tested fantastic *ad hoc* prediction which may occur to us (or to some science fiction writer).

Thus our third requirement, like the second, is needed in order to eliminate trivial and other *ad hoc* theories.[29] But it is needed also for what seem to me even more serious reasons.

I think that we are quite right to expect, and perhaps even to hope, that even our best theories will be superseded and replaced by better ones (though we may at the same time feel the need for encouragement in our belief that we are making progress). Yet this should certainly not induce in us the attitude of merely producing theories so that they can be superseded.

For our aim as scientists is to discover the truth about our problem; and we must look at our theories as serious attempts to find the truth. If they are not true, they may be, admittedly, important stepping stones

[29] Dr Jerzy Giedymin (in a paper 'A Generalization of the Refutability Postulate', *Studia Logica*, **10**, 1960, see especially pp. 103 ff.) has formulated a general methodological principle of empiricism which says that our various rules of scientific method must not permit what he calls a 'dictatorial strategy'; that is they must exclude the possibility that we shall always win the game played in accordance with these rules: Nature must be able to defeat us at least sometimes. If we drop our third requirement, then we can always win, and need not consider Nature at all, as far as the construction of 'good' theories is concerned: speculations about answers which Nature may give to our questions will play no role in our problem situation which will always be fully determined by our past failures alone.

towards the truth, instruments for further discoveries. But this does not mean that we can ever be content to look at them as being *nothing but* stepping stones, *nothing but* instruments; for this would involve giving up even the view that they are instruments of theoretical *discoveries*; it would commit us to looking upon them as mere instruments for some observational or pragmatic purpose. And this approach would not, I suspect, be very successful, even from a pragmatic point of view: if we are content to look at our theories as mere stepping stones, then most of them will not even be good stepping stones. Thus we ought not to aim at theories which are mere instruments for the exploration of facts, but we ought to try to find genuine explanatory theories: we should make genuine guesses about the structure of the world. In brief, we should not be satisfied with the first two requirements.

Of course, the fulfilment of our third requirement is not in our own hands. No amount of ingenuity can ensure the construction of a successful theory. We also need luck; and we also need a world whose mathematical structure is not so intricate as to make progress impossible. For indeed, if we should cease to progress in the sense of our third requirement—if we should only succeed in refuting our theories but not in obtaining some verifications of predictions of a new kind—we might well decide that our scientific problems have become too difficult for us because the structure (if any) of the world is beyond our powers of comprehension. Even in this case we might proceed, for a time, with theory construction, criticism, and falsification: the *rational* side of the method of science might, for a time, continue to function. Yet I believe that we should feel that, especially for the functioning of its *empirical* side, both kinds of successes are essential: success in refuting our theories, and success on the part of some of our theories in resisting at least some of our most determined attempts to refute them.

XXI

It may be objected that this is merely good psychological advice about the attitude which scientists ought to adopt—a matter which, after all, is their private affair—and that a theory of scientific method worthy of its name should be able to produce logical or methodological arguments in support of our third requirement. Instead of appealing to the attitude or the psychology of the scientist, our theory of science should even be able to explain his attitude, and his psychology, by an analysis of the logic of the situation in which he finds himself. There is a problem here for our theory of method.

I accept this challenge, and I shall produce three objective reasons: the first from the idea of truth; the second from the idea of getting nearer to the truth (verisimilitude); and the third from our old idea of independent tests and of crucial tests.

(1) The first reason why our third requirement is so important is this. We know that *if we had an independently testable theory which was, moreover, true, then it would provide us with successful predictions* (and *only* with successful ones). Successful predictions—though they are not, of course, *sufficient* conditions for the truth of a theory—are therefore at least necessary conditions for the truth of an independently testable theory. In this sense—and only in this sense—our third requirement may even be said to be 'necessary', if we seriously accept truth as a regulative idea.

(2) The second reason is this. If it is our aim to strengthen the verisimilitude of our theories, or to get nearer to the truth, then we should be anxious not only to reduce the falsity-content of our theories but also to strengthen their truth-content.

Admittedly this may be done in certain cases simply by constructing the new theory in such a way that the refutations of the old theory are explained ('saving the phenomena', in this case the refutations). But there are other cases of scientific progress—cases whose existence shows that this way of increasing the truth content is not the only possible one.

The cases I have in mind are cases in which there was no refutation. Neither Galileo's nor Kepler's theory was refuted before Newton: what Newton tried to do was to explain them from more general assumptions, and thus to unify two hitherto unrelated fields of inquiry. The same may be said of many other theories: Ptolemy's system was not refuted when Copernicus produced his. And though there was, before Einstein, the puzzling experiment of Michelson and Morley, this had been successfully explained by Lorentz and Fitzgerald.

It is in cases like these that *crucial experiments* become decisively important. We have no reason to regard the new theory as better than the old theory—to believe that it is nearer to the truth—until we have derived from the new theory *new predictions* which were unobtainable from the old theory (the phases of Venus, the perturbations, the mass-energy equation) and until we have found that these new predictions were successful. For it is only this success which shows that the new theory had true consequences (that is, a truth-content) where the old theories had false consequences (that is, a falsity-content).

Had the new theory been refuted in any of these crucial experiments then we should have had no reason to abandon the old one in its favour—even if the old theory was not wholly satisfactory. (This was the fate of the Bohr-Kramers-Slater theory.)

In all these important cases we need the new theory in order to find out where the old theory was deficient. Admittedly, the situation is different if the deficiency of the old theory is already known before the new theory is invented; but logically the case has enough similarity with the other cases to regard a new theory which leads to *new* crucial experiments (Einstein's mass-energy equation) as superior to one which can save only the known phenomena (Lorentz-Fitzgerald).

(3) The same point—the importance of crucial tests—can be made without appealing to the aim of increasing the verisimilitude of a theory, by using an old argument of mine—the need to make the tests of our explanations independent.[30] This need is a result of the growth of knowledge—of the incorporation of what was new and problematic knowledge into background knowledge, with a consequent loss of explanatory power to our theories.

These are my main arguments.

XXII

Our third requirement may be divided into two parts: first we require of a good theory that it should be successful in some of its new predictions; secondly we require that it is not refuted too soon—that is, before it has been strikingly successful. Both requirements sound strange. The first because the *logical* relationship between a theory and any corroborating evidence cannot, it seems, be affected by the question whether the theory is temporally prior to the evidence. The second

[30] See especially my paper 'The Aim of Science', *Ratio*, **1**, 1957.

because if the theory is doomed to be refuted, its intrinsic value can hardly depend upon delaying the refutation.

Our explanation of this slightly puzzling difficulty is simple enough: the successful new predictions which we require the new theory to produce are identical with the crucial tests which it must pass in order to become sufficiently interesting to be accepted as an advance upon its predecessor, and to be considered worthy of further experimental examination which may eventually lead to its refutation.

Yet the difficulty can hardly be resolved by an inductivist methodology. It is therefore not surprising that inductivists such as John Maynard Keynes have asserted that the value of predictions (in the sense of facts derived from the theory but previously not known) was imaginary; and indeed if the value of a theory would lie merely in its relation to its evidential basis, then it would be logically irrelevant whether the supporting evidence precedes or follows in time the invention of the theory. Similarly the great founders of the hypothetical method used to stress the 'saving of the phenomena', that is to say, the demand that the theory should explain *known* experience. Successful *new* prediction—of new effects—seems to be a late idea, for obvious reasons; I do not know when and with whom it originated; yet the distinction between the prediction of known effects and the prediction of new effects was hardly ever made explicitly. But it seems to me quite indispensable as a part of an epistemology which views science as progressing to better and better explanatory theories; that is, not merely to instruments of exploration, but to genuine explanations.

Keynes' objection (that it is an historical accident whether this support was known before the theory was proposed, or only afterwards so that it could attain the status of a prediction) overlooks the all-important fact that it is through our theories that we learn to observe, that is to say, *to ask questions* which lead to new observations and to new interpretations. This is the way our observational knowledge grows. And the questions asked are, as a rule, crucial questions which may lead to answers that decide between competing theories. It is my thesis that it is the *growth* of our knowledge, our way of choosing between theories, in a certain problem situation, which makes science rational. Now both the idea of the growth of knowledge and that of a problem situation are, at least partly, historical ideas. This explains why another *partly historical* idea—that of a genuine prediction of evidence (it may be about past facts) not known when the theory was first proposed—may play an important role here, and why the apparently irrelevant time element may become relevant.[31]

I shall now briefly sum up our results with respect to the epistemologies of the two groups of philosophers I have dealt with, the verificationists and the falsificationists.

While the verificationists or inductivists in vain try to show that scientific beliefs can be justified or, at least, established as probable (and so encourage, by their failure, the retreat into irrationalism), we of the other group have found that we do not even want a highly probable theory. Equating rationality with the critical attitude, we look for theories which, however fallible, progress beyond their predecessors; which means that they can be more severely tested, and stand up to some of the new tests. And while the verificationists laboured in vain to discover valid positive arguments in support of their beliefs, we for our part are satisfied that the rationality of a theory lies in the fact that we choose it because it is better than its predecessors; because it can be put to more severe tests; because it may even have passed them, if we are fortunate; and because it may, therefore, approach nearer to the truth.

APPENDIX: A PRESUMABLY FALSE YET FORMALLY HIGHLY PROBABLE NON-EMPIRICAL STATEMENT

In the text of this chapter I have drawn attention to the criterion of progress and of rationality based on the comparison of *degrees of testability*

[31] Verificationists may think that the preceding discussion of what I have called here the third requirement quite unnecessarily elaborates what nobody contests. Falsificationists may think otherwise; and personally I feel greatly indebted to Dr Agassi for drawing my attention to the fact that I have previously never explained clearly the distinction between what are called here the second and third requirements. He thus induced me to state it here in some detail. I should mention, however, that he disagrees with me about the third requirement which, as he explained to me, he cannot accept because he can regard it only as a residue of verificationist modes of thought. (See also his paper in the *Australasian Journal of Philosophy*, **39**, 1961, where he expresses his disagreement on p. 90.) I admit that there may be a whiff of verificationism here; but this seems to me a case where we have to put up with it, if we do not want a whiff of some form of instrumentalism that takes theories to be mere instruments of exploration.

or degrees of the empirical content or explanatory power of theories. I did so because these degrees have been little discussed so far.

I always thought that the comparison of these degrees leads to a criterion which is more important and more realistic than the simpler *criterion of falsifiability* which I proposed at the same time, and which has been widely discussed. But this simpler criterion is also needed. In order to show the need for the falsifiability or testability criterion as a criterion of the empirical character of scientific theories, I will discuss, as an example, a simple, purely existential statement which is formulated in purely empirical terms. I hope this example will also provide a reply to the often repeated criticism that it is perverse to exclude purely existential statements from empirical science and to classify them as metaphysical.

My example consists of the following purely existential theory:

'There exists a finite sequence of Latin elegiac couplets such that, if it is pronounced in an appropriate manner at a certain time and place, this is immediately followed by the appearance of the Devil—that is to say, of a man-like creature with two small horns and one cloven hoof.'

Clearly, this untestable theory is, in principle, verifiable. Though according to my criterion of demarcation it is excluded as non-empirical and non-scientific or, if you like, metaphysical, it is not so excluded by those positivists who consider all well-formed statements and especially all verifiable ones as empirical and scientific.

Some of my positivist friends have indeed assured me that they consider my existential statement about the Devil to be empirical. It is empirical though false, they said. And they indicated that I was mistaking false empirical statements for non-empirical ones.

However, I think that the confusion, if any, is not mine. I too believe that the existential statement is false: but I believe that it is a false *metaphysical* statement. And why, I ask, should anybody who takes it for *empirical* think that it is *false*? Empirically, it is irrefutable. No observation in the world can establish its falsity. There can be no empirical grounds for its falsity.

Moreover, it can be easily shown to be highly probable: like all existential statements, it is in an infinite (or sufficiently large) universe *almost logically true*, to use an expression of Carnap's. Thus, if we take it to be empirical, we have no reason to reject it, and every reason to accept it and to believe in it—especially upon a subjective theory of probable belief.

Probability theory tells us even more: it can be easily proved not only that empirical evidence can *never refute* an almost logically true existential statement, but that it can *never reduce its probability*.[32] (Its probability could be reduced only by some information which is at least 'almost logically false', and therefore not by an observational evidence statement.) So the empirical probability or degree of empirical confirmation (in Carnap's sense) of our statement about the devil-summoning spell must for ever remain equal to unity, whatever the facts may be.

It would of course be easy enough for me to amend my criterion of demarcation so as to include such purely existential statements among the empirical statements. I merely should have to admit not only testable or falsifiable statements among the empirical ones, but also statements which may, in principle, be empirically 'verified'.

But I believe that it is better not to amend my original falsifiability criterion. For our example shows that, if we do not wish to accept my existential statement about the spell that summons the devil, we must deny its empirical character (notwithstanding the fact that it can easily be formalized in any model language sufficient for the expression of even the most primitive scientific assertions). By denying the empirical character of my existential statement, I make it possible to reject it on grounds other than observational evidence. (See chapter 8, section 2, for a discussion of such grounds, and chapter 11, especially pp. 370–3, for a formalization of a similar argument.)

This shows that it is preferable, as I have been trying to make clear for some considerable time, not to assume uncritically that the terms 'empirical' and 'well-formed' (or 'meaningful') must coincide—and that the situation is hardly improved if we assume, uncritically, that probability or probabilistic 'confirmability' may be used as a criterion of the empirical character of statements or theories. For a non-empirical and presumably false statement may have a high degree of probability, as has been shown here.

[32] This is a consequence of the 'principle of stability' of the probability calculus; see theorem (26), section V, of my paper 'Creative and Non-Creative Definitions in the Calculus of Probability', *Synthese*, **15**, 1963, no. 2, pp. 167–86.

Refutations

I think, Socrates, as presumably you do yourself, that in this life it is either altogether beyond our powers, or at least very difficult, to attain certain knowledge about matters such as these. And yet a man would be a coward if he did not try with all his might to refute every argument about them, refusing to give up before he has worn himself out by examining them from all sides. For he must do one of two things: either he must learn, or discover, the truth about these matters, or if that is beyond his powers, he must grasp whatever human theory seems to him to be the best, and to offer the hardest resistance to refutation; and, mounting on it as upon a raft, he must venture into danger and sail upon it through life, unless he can mount on something stronger, less dangerous, and more trustworthy . . .

PLATO

11

THE DEMARCATION BETWEEN SCIENCE AND METAPHYSICS

Summary

Put in a nut-shell, my thesis amounts to this. The repeated attempts made by Rudolf Carnap to show that the demarcation between science and metaphysics coincides with that between sense and nonsense have failed. The reason is that the positivistic concept of 'meaning' or 'sense' (or of verifiability, or of inductive confirmability, etc.) is inappropriate for achieving this demarcation—simply because metaphysics need not be meaningless even though it is not science. In all its variants demarcation by meaninglessness has tended to be *at the same time too narrow and too wide*: as against all intentions and all claims, it has tended to exclude scientific theories as meaningless, while failing to

A paper contributed in January 1955 to the volume, The Philosophy of Rudolf Carnap, *published in 1964 in the Library of Living Philosophers, ed. P. A. Schilpp. It has been, with the permission of Professor Schilpp, distributed in a stencilled version since June 1956. Apart from small stylistic corrections, I have made no changes in the text, although, in the years since it was written, I have further developed a number of points in various publications; see especially my* Logic of Scientific Discovery, *appendix *ix, especially pp. 390 f.; the appendix to chapter 10 of the present volume; an article in* Dialectica, **11**, *1957, pp. 354–374; two Notes in* Mind, **71**, *1962, pp. 69–73, and* **76**, *1967, pp. 103–110; and I. Lakatos (editor),* The Problem of Inductive Logic, *1968. See also Lakatos's and Watkins's contributions to that volume.*

exclude even that part of metaphysics which is known as 'rational theology'.

1. INTRODUCTION

Writing about Carnap—and in criticism of Carnap—brings back to my mind the time when I first met him, at his Seminar, in 1928 or 1929. It brings back even more vividly a later occasion, in 1932, in the beautiful Tyrolese hills, when I had the opportunity of spending part of my holidays in prolonged critical discussions with Carnap and with Herbert Feigl, in the company of our wives. We had a happy time, with plenty of sunshine, and I think we all tremendously enjoyed these long and fascinating talks, interspersed with a little climbing but never interrupted by it. None of us will ever forget, I am sure, how Carnap once led us in a steep climb up a trackless hill, through a beautiful and almost impenetrable thicket of alpine rhododendrons; and how he led us, at the same time, through a beautiful and almost impenetrable thicket of arguments whose topic induced Feigl to christen our hill '*Semantische Schnuppe*' (something like 'Semantical Shooting Star')—though several years had to elapse before Carnap, stimulated by Tarski's criticism, discovered the track which led him from Logical Syntax to Semantics.[1]

I found in Carnap not only one of the most captivating persons I had ever met but also a thinker utterly absorbed in, and devoted to, his problems, and eager to listen to criticism. And indeed, among some other characteristics which Carnap shares with Bertrand Russell—whose influence upon Carnap and upon us all was greater than anyone else's—is his intellectual courage in changing his mind, under the influence of criticism, even on points of fundamental importance to his philosophy.

I had come to the Tyrol with the manuscript of a large book, entitled *Die beiden Grundprobleme der Erkenntnistheorie* ('The Two Fundamental Problems of Epistemology'). It is still unpublished[1a] but an English

[1] In 1932 Carnap used the term 'Semantics' as a synonym for 'logical syntax'; see *Erkenntnis*, **3**, 1932, p. 177.

[1a] Published in 1979 by J.C.B. Mohr (Paul Siebeck), Tübingen.

translation may appear one day; parts of it were later incorporated, in a much abbreviated form, in my *Logik der Forschung*. The 'two problems' were the problems of induction and of demarcation—the *demarcation between science and metaphysics*. The book contained, among much else, a fairly detailed criticism of Wittgenstein's and Carnap's doctrine of the 'elimination' or 'overthrow' (*Ueberwindung*[2]) of metaphysics through meaning-analysis. I criticized this doctrine not from a metaphysical point of view, but from the point of view of one who, interested in science, feared that this doctrine, far from defeating the supposed enemy metaphysics, in effect presented the enemy with the keys of the beleaguered city.

My criticism was directed, largely, against two books of Carnap's, *Der logische Aufbau der Welt* ('*Aufbau*', for short) and *Scheinprobleme in der Philosophie*, and some of his articles in *Erkenntnis*. Carnap accepted part of it,[3] although he felt, as it turned out,[4] that I had exaggerated the differences between my views and those of the Vienna Circle of which he was a leading member.

This silenced me for many years,[5] especially as Carnap paid so much attention to my criticism in his *Testability and Meaning*. But I felt all the time that the differences between our views were far from being imaginary; and my feeling that they were important was much

[2] See Carnap's 'Ueberwindung der Metaphysik durch logische Analyse der Sprache' ('The Overthrow of Metaphysics through Logical Analysis of Language'), *Erkenntnis*, **2**, 1932, pp. 219 ff.

[3] See Carnap's generously appreciative report on certain of my views which were then still unpublished, in *Erkenntnis*, **3**, 1932, pp. 223 to 228, and my discussion of it in my *Logic of Scientific Discovery* (*L.Sc.D.*), 1959, 1960 (originally published in German in 1934 as *Logik der Forschung*, but here always referred to as '*L.Sc.D.*'), note 1 to section 29.

[4] See Carnap's review of my *L.Sc.D.* in *Erkenntnis*, **5**, 1935, pp. 290–4, especially 293: 'By his efforts to characterize his position clearly [Popper] is led to over-emphasize the differences between his views and those . . . which are most closely allied to his . . . [Popper] is very close indeed to the point of view of the Vienna Circle. In his presentation, the differences appear much greater than they are in fact.'

[5] I published nothing even alluding to these differences of opinion during the first ten years after the publication of my *L.Sc.D.* (although I alluded to them in some lectures); and next to nothing during the next ten years, i.e. until I started on the present paper—no more, at any rate, than a few critical remarks on Wittgenstein and Schlick (in my *Open Society*, first published in 1945; see notes 51 f., 46, 26, and 48 to ch. 11; see also chs. 2, 12, and 14 of the present volume).

enhanced by Carnap's most recent papers and books on probability and induction.

The purpose of this paper is to discuss these differences, so far as they concern the problem of demarcation. It is with reluctance that I expose myself again to the charge of exaggerating differences. (But I hope that Professor Carnap won't be prevented from speaking his mind by an apprehension of silencing me for the rest of my days: I promise to be more reasonable this time.) I have, however, accepted the invitation to write this paper; and this leaves me no alternative but to try to characterize our differences as clearly and as sharply as possible. In other words, I must try to defend the thesis that these differences are real—as real as I have felt them to be for the last 25 years.

In Section 2 of this paper I try to give a brief outline of some of my own views which form the basis of my criticism. In the later sections I try to trace the development—as I see it—of Carnap's views on the problem of the demarcation between science and metaphysics. My approach throughout is critical rather than historical; but I have aimed at historical accuracy, if not at historical completeness.

2. MY OWN VIEW OF THE PROBLEM

It was in 1919 that I first faced the problem of *drawing a line of demarcation* between those statements and systems of statements which could be properly described as belonging to empirical science, and others which might, perhaps, be described as 'pseudo-scientific' or (in certain contexts) as 'metaphysical', or which belonged, perhaps, to pure logic or to pure mathematics.

This is a problem which has agitated many philosophers since the time of Bacon, although I have never found an explicit formulation of it. The most widely accepted view was that science was characterized by its *observational basis*, or by its *inductive method*, while pseudo-sciences and metaphysics were characterized by their *speculative method* or, as Bacon said, by the fact that they operated with '*mental anticipations*'—something very similar to hypotheses.

This view I have never been able to accept. The modern theories of physics, especially Einstein's theory (widely discussed in the year 1919), were highly speculative and abstract, and very far removed from what might be called their 'observational basis'. All attempts to show that they were more or less directly 'based on observations' were unconvincing. The same was true even of Newton's theory. Bacon had raised objections against the Copernican system on the ground that it 'needlessly did violence to our senses'; and in general the best physical theories always resembled what Bacon had dismissed as 'mental anticipations'.

On the other hand, many superstitious beliefs, and many rule-of-thumb procedures (for planting, etc.) to be found in popular almanacs and dream books, have had much more to do with observations, and have no doubt often been based on something like induction. Astrologers, more especially, have always claimed that their 'science' was based upon a great wealth of inductive material. This claim is, perhaps, unfounded; but I have never heard of any attempt to discredit astrology by a critical investigation of its alleged inductive material. Nevertheless, astrology was rejected by modern science because it did not fit accepted theories and methods.

Thus there clearly was a need for a different criterion of demarcation; and I proposed (though years elapsed before I published this proposal) that the *refutability or falsifiability* of a theoretical system should be taken as the criterion of demarcation. According to this view, which I still uphold, a system is to be considered as scientific only if it makes assertions which may clash with observations; and a system is, in fact, tested by attempts to produce such clashes; that is to say, by attempts to refute it. Thus testability is the same as refutability, and can therefore likewise be taken as a criterion of demarcation.

This is a view of science which takes its *critical approach* to be its most important characteristic. Thus a scientist should look upon a theory from the point of view of whether it can be critically discussed: whether it exposes itself to criticism of all kinds; and—if it does—whether it is able to stand up to it. Newton's theory, for example, predicted deviations from Kepler's laws (due to the interactions of the planets) which had not been observed at the time. It exposed itself thereby to attempted empirical refutations whose failure meant the success of the theory. Einstein's theory was tested in a similar way. And indeed, all real tests are attempted refutations. Only if a theory successfully withstands the pressure of these attempted refutations can we claim that it is confirmed or corroborated by experience.

There are, moreover (as I found later[6]), *degrees of testability*: some theories expose themselves to possible refutations more boldly than others. For example, a theory from which we can deduce precise numerical predictions about the splitting up of the spectral lines of light emitted by atoms in magnetic fields of varying strength will be more exposed to experimental refutation than one which merely predicts that a magnetic field influences the emission of light. A theory which is more precise and more easily refutable than another will also be the more interesting one. Since it is the more daring one, it will be the one which is *less probable*. But it is better testable, for *we can make our tests more precise and more severe*. And if it stands up to severe tests it will be better confirmed, or better attested, by these tests. *Thus confirmability (or attestability or corroborability) must increase with testability*.

This indicates that the criterion of demarcation cannot be an absolutely sharp one but will itself have degrees. There will be well-testable theories, hardly testable theories, and non-testable theories. Those which are non-testable are of no interest to empirical scientists. They may be described as metaphysical.

Here I must again stress a point which has often been misunderstood. Perhaps I can avoid these misunderstandings if I put my point now in this way. Take a square to represent the class of all statements of a language in which we intend to formulate a science; draw a broad horizontal line, dividing it into an upper and lower half; write 'science' and 'testable' into the upper half, and 'metaphysics' and 'non-testable' into the lower: then, I hope, you will realize that I do *not* propose to draw the line of demarcation in such a way that it coincides with the limits of a language, leaving science inside, and banning metaphysics by excluding it from the class of meaningful statements. On the contrary: beginning with my first publication on this subject,[7] I stressed the fact that it would be *inadequate* to draw the line of demarcation

[6] See *L.Sc.D.*, sections 31 to 46.

[7] See 'Ein Kriterium des empirischen Charakters theoretischer Systeme', *Erkenntnis*, **3**, 1933, pp. 426 ff., now in *L.Sc.D.*, pp. 312–14; see also *L.Sc.D.*, especially sections 4 and 10.

between science and metaphysics so as to exclude metaphysics as nonsensical from a meaningful language.

I have indicated one of the reasons for this by saying that we must not try to draw the line too sharply. This becomes clear if we remember that most of our scientific theories originate in myths. The Copernican system, for example, was inspired by a Neo-Platonic worship of the light of the Sun who had to occupy the 'centre' because of his nobility. This indicates how myths may develop testable components. They may, in the course of discussion, become fruitful and important for science. In my *Logic of Scientific Discovery*[8] I gave several examples of myths which have become most important for science, among them atomism and the corpuscular theory of light. It would hardly contribute to clarity if we were to say that these theories are nonsensical gibberish in one stage of their development, and then suddenly become good sense in another.

Another argument is the following. It may happen—and it turns out to be an important case—that a certain statement belongs to science since it is testable, while *its negation* turns out not to be testable, so that it must be placed below the line of demarcation. This is indeed the case with the most important and most severely testable statements—the *universal laws of science*. I recommended, in my *Logic of Scientific Discovery*, that they should be expressed, for certain purposes, in a form like 'There does not exist any perpetual motion machine' (this is sometimes called 'Planck's formulation of the First Law of Thermodynamics'); that is to say, in the form of a *negation of an existential statement*. The corresponding existential statement—'There exists a perpetual motion machine'—would belong, I suggested, together with 'There exists a sea-serpent' to those below the line of demarcation, as opposed to 'There is a sea-serpent now on view in the British Museum' which is well above the line since it can easily be tested. But we do not know how to test an isolated purely existential assertion.

I cannot in this place argue for the adequacy of the view that isolated purely existential statements should be classed as untestable and as falling outside the scientist's range of interest.[9] I only wish to make

[8] *L.Sc.D.*, section 85, p. 278.

[9] *L.Sc.D.*, section 15. I suppose that some people found it hard to accept the view that a

clear that if this view is accepted, then it would be strange to call metaphysical statements meaningless,[10] or to exclude them from our language. For if we accept the *negation* of an existential statement as meaningful, then we must accept the existential statement itself also as meaningful.

I have been forced to stress this point because my position has repeatedly been described as a proposal to take falsifiability or refutability as the criterion of *meaning* (rather than of demarcation), or as a proposal to exclude existential statements from our language, or perhaps from the language of science. Even Carnap, who discusses my position in considerable detail and reports it correctly, feels himself compelled to interpret it as a proposal to exclude metaphysical statements from some language or other.[11]

But it is a fact that beginning with my first publication on this subject (see note 7 above), I always dismissed the problem of meaninglessness as a pseudo-problem; and I was always opposed to the idea that it may be identified with the problem of demarcation. This is my view still.

pure or isolated existential statement ('There exists a sea-serpent') should be called 'metaphysical', even though it might be deducible from a statement of an empirical character ('There is now a sea-serpent on view in the entrance hall of the British Museum'). But they overlooked the fact that: (*a*) in so far as it was so deducible, it was no longer isolated, but belonged to a testable theory, and (*b*) if a statement is deducible from an empirical or a scientific statement then this fact need not make it empirical or scientific. (Any tautology is so deducible.)

[10] But one may perhaps find in Brouwer's theories a suggestion that a universal statement could be meaningful while its existential negation was meaningless.

[11] See *Testability and Meaning*, section 25, p. 26: 'We may take Popper's principle of falsifiability as an example of the choice of this language' (viz. of a language that excludes existential sentences as meaningless). Carnap continues: 'Popper is however very cautious in the formulation of his . . . principle [of demarcation]; he does not call the [existential] sentences meaningless but only non-empirical or metaphysical.' This second part of the quotation is perfectly correct, and seems quite clear to me; but Carnap continues: 'Perhaps he [Popper] wishes to exclude existential sentences and other metaphysical sentences not from the language altogether, but only from the language of empirical science.' But why does Carnap assume that I should wish to exclude them from *any* language, when I have repeatedly said the opposite?

3. CARNAP'S FIRST THEORY OF MEANINGLESSNESS

One of the theories which I had criticized in my manuscript (and later, more briefly, in my *Logic of Scientific Discovery*) was the assertion that *metaphysics was meaningless, and consisted of nonsensical pseudo-propositions*. This theory[12] was supposed to bring about the 'overthrow' of metaphysics, and to destroy it more radically and more effectively than any earlier anti-metaphysical philosophy. But, as I pointed out in my criticism, the theory was based on a naïve and 'naturalistic'[13] view of the problem of meaningfulness; moreover its propagators, in their anxiety to oust metaphysics, failed to notice that they were throwing all *scientific theories* on the same scrap-heap as the 'meaningless' metaphysical theories. All this, I suggested, was a consequence of trying to destroy metaphysics instead of looking for a criterion of demarcation.

The 'naturalistic' theory (as I called it) of meaningfulness and meaninglessness in Carnap's *Aufbau*, which here followed Wittgenstein's *Tractatus*, was abandoned by Carnap long ago; it has been replaced by the more sophisticated doctrine that a given expression is a meaningful sentence in a certain (artificial) language if, and only if, it complies with the rules of formation for well-formed formulae or sentences in that language.

In my opinion, the development from the naïve or naturalistic theory to the more sophisticated doctrine was a highly important and desirable one. But its full significance has not been appreciated, as far as I can see; apparently it has not been noticed that it simply destroys the doctrine of the meaninglessness of metaphysics.

It is for this reason that I am now going to discuss this development in some detail.

[12] Carnap and the Vienna Circle attributed it to Wittgenstein, but it is much older. The theory goes back to Hobbes, at least; and in the form described below as 'condition (*a*)'—asserting that words purporting to denote unobservable entities cannot have any meaning—it was clearly and forcefully used by Berkeley (and other nominalists). See ch. 6; also my reference to Hume, *L.Sc.D.*, section 4.

[13] Although I called the theory 'naturalistic' (I now also call it 'absolutistic' and 'essentialistic'; cp. note 18 below) for reasons which may perhaps emerge, I do not propose to argue these reasons here; for my criticism of the theory was not, and is not, that it is 'naturalistic' etc., but that it is untenable. See also the passages referred to in note 7 above.

By the naturalistic theory of meaninglessness I mean the doctrine that every linguistic expression purporting to be an assertion is either meaningful or meaningless; not by convention, or as a result of rules which have been laid down by convention, but as a matter of actual fact, or due to its nature, just as a plant is, or is not, green in fact, or by nature, and not by conventional rules.

According to Wittgenstein's famous verifiability criterion of meaning, which Carnap accepted, a sentence-like expression, or a string of words, was a meaningful sentence (or proposition) if, and only if, it satisfied the conditions (*a*) and (*b*)—or a condition (*c*) which will be stated later:

(*a*) all words which occurred in it had meaning, *and*
(*b*) all words which occurred in it fitted together properly.

According to condition (*a*) of the theory (which goes back to Hobbes and Berkeley) a string of words was meaningless if any word in it was meaningless. Wittgenstein formulated it in his *Tractatus* (6.53; italics mine): 'The right method of philosophy is this: when someone . . . wished to say something *metaphysical*, to demonstrate to him that *he had given no meaning to certain signs* in his propositions.' According to Hobbes and Berkeley the only way in which a meaning was given to a word was by linking (associating) the word with certain observable experiences or phenomena. Wittgenstein himself was not explicit on this point, but Carnap was. In his *Aufbau*, he tried to show that *all concepts used in the sciences could be defined on the basis of* ('my own') *observational or perceptual experience*. He called such a definition of a concept its '*constitution*', and the resulting system of concepts a '*constitution system*'. And he asserted that *metaphysical concepts could not be constituted*.

Condition (*b*) of the theory goes back to Bertrand Russell who suggested[14] that certain 'combinations of symbols', which looked like propositions, 'must be absolutely meaningless, not simply false', if certain paradoxes were to be avoided. Russell did not mean to make a *proposal*—that we should *consider* these combinations as contrary to some (partly conventional) rules for forming sentences, in order to avoid the

[14] See, for example, *Principia Mathematica*, 2nd edn., p. 77.

paradoxes. Rather, he *thought that he had discovered the fact* that these apparently meaningful formulae expressed nothing; and that they were, in nature or in essence, meaningless pseudo-propositions. A formula like '*a* is an element of *a*' or '*a* is not an element of *a*' looked like a proposition (because it contained two subjects and a two-termed predicate); but it was not a genuine proposition (or sentence) because a formula of the form '*x* is an element of *y*' could be a proposition only if *x* was one type-level lower than *y*—a condition which obviously could not be satisfied if the same symbol, '*a*', was to be substituted for both, '*x*' and '*y*'.

This showed that a disregard of the type-level of words (or of the entities designated by them) could make sentence-like expressions meaningless; and according to Wittgenstein's *Tractatus* and, more explicitly, Carnap's *Aufbau*, this confusion was a major source of metaphysical nonsense, i.e. of the offering of pseudo-propositions for propositions. It was called 'confusion of spheres' in the *Aufbau*;[15] it is the same kind of confusion which nowadays is often called a 'category mistake'.[16] According to the *Aufbau*, for example, 'my own' experiences ('*das Eigenpsychische*'); physical bodies; and the experiences of others ('*das Fremdpsychische*'), all belong to different spheres or types or categories, and a confusion of these must lead to pseudo-propositions and to pseudo-problems. (Carnap describes the difference between physical and psychological entities as one between '*two types of order*'[17] subsisting within *one* kind or range of *ultimate entities*, which leads him to a solution of the body-mind problem on the lines of 'neutral monism'.)

The outline just given of the 'naïve' or 'naturalistic' theory[18] of

[15] '*Sphaerenvermengung*'; see *Aufbau*, section 30 f.; the '*Sphaere*' is identified with the *logical type* in section 180, p. 254.

[16] See G. Ryle, *The Concept of Mind*, 1949. This use of the expression 'category' may be traced back to Husserl's term 'semantical category' ('*Bedeutungskategorie*'); see his *Logische Untersuchungen*, **2**, Part I (2nd edn.), 1913, pp. 13, 318. Examples of category mistakes given by Husserl are: '*green is or*' (p. 54); '*a round or*'; '*a man and is*' (p. 334). Compare Wittgenstein's example: '*Socrates is identical*'. For a criticism of the theory of category mistakes, see chs. 12 f., below; also J. J. C. Smart's very striking 'A Note on Categories', *B.J.P.S.*, **4**, 1953, pp. 227 f.

[17] '*Ordungsformen*'; see *Aufbau*, section 162, p. 224; see also the bibliography, p. 225.

[18] At present I should be inclined to call it an 'essentialist' theory, in accordance with my book *The Poverty of Historicism*, section 10, and my *Open Society*, especially ch. 11.

meaningful and meaningless linguistic expressions covers only one side of this theory. There is another side to it: the so-called '*verifiability criterion*' which may be formulated as the condition (*c*):

(*c*) an alleged proposition (or sentence) is genuine if, and only if, it is a truth function of, or reducible to, elementary (or atomic) propositions expressing observations or perceptions.

In other words, it is meaningful if, and only if, it is so related to some observation sentences that its truth follows from the truth of these observation sentences. 'It is certain', Carnap writes,[19] 'that a string of words has meaning only if its derivability relations from protocol-sentences [observation sentences] are given . . .'; that is to say, if 'the way to [its] verification . . . is known.'[20]

The conditions (*a*) and (*b*) on the one hand, and (*c*) on the other hand, were asserted by Carnap to be equivalent.[21]

A result of this theory was, in Carnap's words,[22] 'that the alleged sentences of metaphysics stand revealed, by logical analysis, as pseudo-sentences'.

Carnap's theory of the intrinsic meaningfulness or meaninglessness of strings of words was soon to be modified; but in order to prepare a basis for judging these modifications I must say a few words of criticism here.[23]

[19] See his paper on the *Overthrow of Metaphysics*, *Erkenntnis*, **2**, 1932, pp. 222–3. The *Overthrow*-paper belongs, strictly speaking, no longer to the period of the *first* theory of meaninglessness, owing to its recognition of the fact that meaninglessness *depends upon the language in question*; for Carnap writes (p. 220): 'Meaningless in a precise sense is a string of words which, within a certain given language, does not form a sentence.' However the obvious consequences of this remark are not yet drawn, and the theory is still asserted in an absolute sense: our conditions (*a*) and (*b*) are formulated at the bottom of p. 220, and (*c*) on pp. 222–3 (as quoted).

[20] *Ibid.*, p. 224.

[21] *Aufbau*, section 161, p. 222; and section 179 (top of p. 253). See also the important section 2 of Carnap's *Overthrow*-paper, *Erkenntnis*, **2**, 1932, pp. 221–4. (This passage in many ways anticipates, by its general method, the doctrine of reduction in Carnap's *Testability and Meaning*, except that in the latter the demand for verification is weakened.)

[22] *Erkenntnis*, **2**, p. 220. Cp. the foregoing note.

[23] See *L.Sc.D.*, especially sections 4, 10, 14, 20, 25, and 26.

First, a word on (*c*), the verifiability criterion of meaning. This criterion excludes from the realm of meaning all scientific theories (or 'laws of nature'); for these are no more reducible to observation reports than so-called metaphysical pseudo-propositions. *Thus the criterion of meaning leads to the wrong demarcation of science and metaphysics.* This criticism was accepted by Carnap in his *Logical Syntax of Language*[24] and in his *Testability and Meaning*;[25] but even his latest theories are still open to it, as I shall try to show in section 6 below.

Next we consider condition (*a*) of the doctrine, the (nominalistic) view that only empirically definable words or signs have meaning.

Here the situation is even worse, although it is very interesting.

For the sake of simplicity, I begin my criticism with a very simple form of *nominalism*. It is the doctrine that all non-logical (or, as I prefer to say, non-formative) words are names—either such as 'Fido', of a single physical object, or as 'dog', shared by several such objects. Thus 'dog' may be the name shared by the objects Fido, Candy, and Tiffin; and so with all other words.

This view may be said to interpret the various words *extensionally* or *enumeratively*; their 'meaning' is given by *a list or an enumeration of the things they name*: 'this thing here, and that thing over there . . .' We may call such an enumeration an 'enumerative definition' of the meaning of a name; and a language in which all (non-logical or non-formative) words are supposed to be enumeratively defined may be called an 'enumerative language', or a 'purely nominalistic language'.

Now we can easily show that such a purely nominalistic language is completely inadequate for any scientific purpose. This may be expressed by saying that all its sentences are analytic—either analytically true or contradictory—and that no synthetic sentences can be

[24] See the end of the first paragraph and the second paragraph on p. 321, section 82, especially the following remarks of Carnap's on the Vienna Circle: 'It was originally maintained that every sentence, in order to be significant, must be *completely verifiable*. . . . On this view there was no place for the *laws of nature* amongst the sentences of the language. . . . A detailed criticism of the view according to which laws are not sentences is given by Popper.' The continuation of this passage is quoted below, text to note 48. See also note 71, below.

[25] Cp. especially notes 20 and 25 (and the text following note 25) to section 23 of *Testability and Meaning* with note 7 to section 4 (and text), and note I to section 78 of *L.Sc.D.*

expressed in it. Or if we prefer a formulation which avoids the terms 'analytic' and 'synthetic' (which at present are under heavy fire from Professor Quine's guns) then we can put it in this way: in a purely nominalistic language no sentence can be formulated whose truth or falsity could not be decided by merely comparing the defining lists, or enumerations, of the things which are mentioned in the sentence. Thus the truth or falsity of any sentence is decided as soon as the words which occur in it have been given their meaning.

That this is so may be seen from our example. 'Fido is a dog' is true because Fido was one of the things enumerated by us in defining 'dog'. As opposed to this 'Chunky is a dog' must be false, simply because Chunky was not one of the things to which we pointed when drawing up our list defining 'dog'. Similarly, if I give the meaning of 'white' by listing (1) the paper on which I am now writing, (2) my handkerchief, (3) the cloud over there, and (4) our snowman, then the statement, 'I have white hair' will be false, whatever the colour of my hair may be.

It is clear that in such a language hypotheses cannot be formulated. It cannot be a language of science. And conversely, every language adequate for science must contain words whose meaning is not given in an enumerative way. Or, as we may say, every scientific language must make use of *genuine universals*, i.e. of words, whether defined or undefined, with an indeterminate extension, though perhaps with a reasonably definite intensional 'meaning'. (For the intensional analysis of meaning see Carnap's excellent book *Meaning and Necessity*.)

Precisely the same criticism applies to more complicated languages, especially to languages which introduce their concepts by the method of extensional abstraction (used first by Frege and Russell) provided the class of the fundamental elements upon which this method is based, and the fundamental relations between these elements, are supposed to be given extensionally, by lists. Now this was the case in Carnap's *Aufbau*: he operated with one primitive relation, 'Er' ('Experience of remembering'), which was assumed to be given in the form of a list of pairs.[26]

[26] See especially *Aufbau*, section 108. Carnap said there of his *Theorem 1*, which asserts the asymmetry of the primitive relation 'Er', that it is an *empirical theorem*, since its asymmetry can be read off the list of (empirically given) pairs. But we must not forget that this is the same list of pairs which 'constituted', or defined, 'Er'; moreover, a list of pairs which

All concepts belonging to his 'constitution system' were supposed to be extensionally definable on the basis of this primitive relation 'Er', i.e. of the list of pairs which gave a meaning to this relation. Accordingly all statements which could be expressed in his language were true or false simply according to the (extensional) meaning of the words which occurred in them: they were all either analytically true or contradictory,[27] owing to the absence of genuine universal[28] words.

To conclude this section, I turn to the condition (*b*) of the theory, and to the doctrine of meaninglessness due to 'type mistakes' or 'category mistakes'. This doctrine was derived, as we have seen, from Russell's theory that an expression like '*a* is an element of the class *a*' must be meaningless—absolutely, or intrinsically or essentially, as it were.

Now this doctrine has long since turned out to be mistaken. Admittedly, it is true that we can, with Russell, construct a language (embodying a theory of types) in which the expression in question is not a well-formed formula. But we can also, with Zermelo, and his successors (Fraenkel, Behmann, von Neumann, Bernays, Leśniewski, Quine, Ackermann) construct languages in which the expression in question is well-formed and thus meaningful; and in some of them it is even a true statement (for certain values of *a*).

These are, of course, well-known facts. But they completely destroy the idea of an 'inherently' or 'naturally' or 'essentially' meaningless expression. For the expression '*a* is an element of the class *a*' turns out to be meaningless in one language but meaningful in another; and this establishes that a proof that an expression is meaningless in some languages must not be mistaken for a proof of intrinsic meaninglessness.

In order to prove intrinsic meaninglessness we should have to prove a great deal. We should have to prove not merely that an alleged

would lead to the negation of theorem 1, i.e. to the theorem that 'Er' is symmetrical, could not have been interpreted as an adequate list for 'Er', as is particularly clear from sections 153 to 155.

[27] This is the criticism of the *Aufbau* which I put to Feigl when we first met. It was a meeting which for me proved momentous, for it was Feigl who a year or two later arranged the vacation meeting in the Tyrol.

[28] 'The Difference Between Individual Concepts and Universal Concepts' was discussed in the *Aufbau*, section 158; it was criticized briefly in *L.Sc.D.*, sections 14 and 25.

statement, asserted or submitted by some writer or speaker, is meaningless in *all* (consistent) languages, but also that there cannot exist a meaningful sentence (in any consistent language) which would be recognized by the writer or speaker in question as an alternative formulation of what he intended to say. And nobody has ever suggested how such a proof could possibly be given.

It is important to realize that a proof of intrinsic meaninglessness would have to be valid with respect to *every consistent language* and not merely with respect to *every language that suffices for empirical science*. Few metaphysicians assert that metaphysical statements belong to the field of the empirical sciences; and nobody would give up metaphysics because he is told that its statements cannot be formulated within these sciences (or within some language suitable for these sciences). After all, Wittgenstein's and Carnap's original thesis was that metaphysics is absolutely meaningless—that it is sheer gibberish and nothing else; that it is, perhaps, of the character of sighs or groans or tears (or of surrealist poetry), but not of articulate speech. In order to show this, it would be quite insufficient to produce a proof to the effect that it cannot be expressed in languages which suffice for the needs of science.

But even this insufficient proof has never been produced by anybody, in spite of the many attempts to construct metaphysics-free languages for science. Some of these attempts will be discussed in the next two sections.

4. CARNAP AND THE LANGUAGE OF SCIENCE

Carnap's original 'overthrow' of metaphysics was unsuccessful. The naturalistic theory of meaninglessness turned out to be baseless, and the total result was a doctrine which was just as destructive of science as it was of metaphysics. In my opinion this was merely the consequence of an ill-advised attempt to destroy metaphysics wholesale, instead of trying to eliminate, piecemeal as it were, metaphysical elements from the various sciences whenever we can do this without endangering scientific progress by misplaced criticism (such as had been directed by Bacon against Copernicus, or by Duhem and Mach against atomism).

But the naturalistic theory of meaning was abandoned by Carnap a long time ago, as I have said. It has been replaced by the theory that whether a linguistic expression is well-formed or not depends on the rules of the language to which the expression is supposed to belong; together with the theory that the rules of language are often not precise enough to decide the issue, so that we have to introduce more precise rules—and with them, an *artificial language system*.

I wish to repeat that I regard this as a very important development, and as one that provides the key to a considerable number of interesting problems. *But it leaves the problem of demarcation between science and metaphysics exactly where it was*. This is my thesis.

To put it quite differently, the naïve or naturalistic or essentialistic theory of meaningfulness discussed in the previous section is mistaken, and had to be replaced by a theory of well-formed formulae, and with it, of languages which are artificial in being subject to definite rules. This important task has since been carried out by Carnap with great success. *But this reformation of the concept of meaningfulness completely destroys the doctrine of the meaninglessness of metaphysics*. And it leaves us without a hope of ever reconstructing this doctrine on the basis of the reformed concept of meaninglessness.

Unfortunately this seems to have been overlooked. For Carnap and his circle (Neurath was especially influential) tried to solve the problem by constructing a '*language of science*', a language in which every legitimate statement of science would be a well-formed formula, while none of the metaphysical theories would be expressible in it—either because the terminology was not available, or because there was no well-formed formula to express it.

I consider the task of constructing artificial model languages for a language of science an interesting one; but I shall try to show that the attempt to combine this task with that of destroying metaphysics (by rendering it meaningless) has repeatedly led to disaster. The anti-metaphysical bias is a kind of philosophical (or metaphysical) prejudice which prevented the system builders from carrying out their work properly.

I shall try to show this briefly, in this section, for (*a*) the *Physicalistic Language*, (*b*) the *Language of Unified Science*, (*c*) the *languages of the 'Logical Syntax'*, and later, in section 5, more fully for those proposed in '*Testability and Meaning*'.

(*a*) *The Physicalistic Language*. Carnap's *Aufbau* had sponsored what he called a *methodological solipsism*—taking one's own experiences as the basis upon which the concepts of science (and thus the language of science) have to be constructed. By 1931 Carnap had given this up, under Neurath's influence, and had adopted the *thesis of physicalism*, according to which there was *one* unified language which spoke about physical things and their movements in space and time. Everything was to be expressible in this language, or translatable into it, especially psychology in so far as it was scientific. Psychology was to become radically behaviouristic; every meaningful statement of psychology, whether human or animal, was to be translatable into a statement about the spatio-temporal movements of physical bodies.

The tendency underlying this programme is clear: a statement about the human soul was to become as meaningless as a statement about God. Now it may be fair enough to put statements about the soul and about God on the same level. But it seems questionable whether anti-metaphysical and anti-theological tendencies were much furthered by placing all our subjective experiences, or rather all statements about them, on the same level of meaninglessness as the statements of metaphysics. (Theologians or metaphysicians might be very pleased to learn that statements such as 'God exists' or 'The Soul exists' are *on precisely the same level* as 'I have conscious experiences' or 'There exist feelings—such as love or hate—distinguishable from the bodily movements which often, though not always, accompany them'.)

There is no need, therefore, to go into the merits or demerits of the behaviourist philosophy, or the translatability thesis (which, in my opinion, is nothing but materialist metaphysics in linguistic trappings—and I, for one, prefer to meet it without trappings): we see that as an attempt to kill metaphysics this philosophy was not very effective. As usual, the broom of the anti-metaphysicist sweeps away too much, and also too little. As a result we are left with an untidy and altogether untenable demarcation.

For an illustration of 'too much and too little' I may perhaps cite the following passage from Carnap's 'Psychology Within the Physical Language':[29] 'Physics is, altogether, practically free from metaphysics,

[29] See *Erkenntnis*, **3**, 1932, p. 117.

thanks to the efforts of Mach, Poincaré, and Einstein; in psychology, efforts to make it a science free from metaphysics have hardly begun.' Now 'free from metaphysics' means here, for Carnap, reducible to protocol-statements. But not even the simplest physical statements about the functioning of a potentiometer—the example is Carnap's[30]—are so reducible. Nor do I see any reason why we should not introduce mental states in our explanatory psychological theories, if in physics (old or new) we are permitted to explain the properties of a conductor by the hypothesis of an 'electric fluid' or of an 'electronic gas'.

The point is that all physical theories say much more than we can test. Whether this 'more' belongs legitimately to physics, or whether it should be eliminated from the theory as a 'metaphysical element' is not always easy to say. Carnap's reference to Mach, Poincaré, and Einstein was unfortunate, since Mach, more especially, looked forward to the final elimination of atomism which he (with many other positivists) considered to be a metaphysical element of physics. (He eliminated too much.) Poincaré tried to interpret physical theories as implicit definitions, a view which can hardly be more acceptable to Carnap; and Einstein has for a long time been a believer in metaphysics, operating freely with the concept of the 'physically real'; although, no doubt, he dislikes pretentious metaphysical verbiage as much as any of us.[30a] Most of the concepts with which physics works, such as forces, fields, and even electrons and other particles are what Berkeley (for example) called '*qualitates occultae*'. Carnap showed[31] that assuming conscious states in our psychological explanations was exactly analogous to assuming a force—a *qualitas occulta*—in order to explain the 'strength' of a wooden post; and he believed that 'such a view commits the fallacy of hypostasization'[32] of which, he suggested, no physicist is guilty, although it is often committed by psychologists.[33] But the fact is that we cannot explain the strength of the post by its structure alone (as Carnap suggested[34]) but only by its structure together with laws

[30] *Op. cit.*, p. 140.
[30a] (Added in proof). When I wrote this, Albert Einstein was still alive.
[31] *Op. cit.*, p. 115.
[32] *Op. cit.*, p. 116.
[33] *Op. cit.*, p. 115.
[34] *Op. cit.*, p. 114.

which make ample use of 'hidden forces' which Carnap, like Berkeley, condemned as occult.

Before concluding point (*a*) I wish to mention only briefly that this physicalism, although from my point of view too physicalistic in most respects, was not physicalistic enough in others. For I do believe, indeed, that whenever we wish to put a scientific statement to an observational test, *this test must in a sense be physicalistic*; that is to say, that we test our most abstract theories, psychological as well as physical, by deriving from them statements about the behaviour[35] of physical bodies.

I have called simple descriptive statements, describing easily observable states of physical bodies, '*basic statements*', and I asserted that in cases in which tests are needed, it is these basic statements[36] which we try to compare with the 'facts' and that we choose these statements, and these facts, because they are most easily comparable, and intersubjectively most easily testable.

Thus according to my view we do *not*, for the purpose of such basic tests, choose reports (which are difficult to test intersubjectively) about our own observational experiences, but rather reports (which are easy to check) about physical bodies—including potentiometers—which we have observed.

The point is important because this theory of mine concerning the 'physicalist' character of test statements is radically opposed to all those widely accepted theories which hold that we are constructing the

[35] This behaviour, however, is always *interpreted in the light of certain theories* (which creates a danger of circularity). I cannot discuss the problem fully here, but I may mention that the behaviour of men, predicted by psychological theories, nearly always consists, not of purely physical movements, but of physical movements which, if interpreted in the light of theories, are 'meaningful'. (Thus if a psychologist predicts that a patient will have bad dreams, he will feel that he was right, whether the patient reports 'I dreamt badly last night', or whether he reports 'I want to tell you that I have had a shocking dream'; though the two 'behaviours', i.e. the 'movements of the lips' may differ *physically* more widely than the movements corresponding to a negation may differ from those corresponding to an affirmation.)

[36] The terms 'basic statement' ('basic proposition' or 'basic sentence': '*Basissatz*') and 'empirical basis' were introduced in *L.Sc.D.*, sections 7 and 25 to 30; they have often been used since by other authors, in similar and in different senses. (See now also section i of the *Addenda* to the present volume.)

'external world of science' out of 'our own experiences'. I have always believed that this is a prejudice (it is still widely held); and that, quite properly, we never trust 'our own experiences' unless we are confident that they conform with intersubjectively testable views.

Now on this point Carnap's and Neurath's views were much less 'physicalistic' at that time. In fact they still upheld a form of Carnap's original 'methodological solipsism'. For they taught that the sentences which formed the 'empirical basis' (in my terminology) of all tests, and which they called 'protocol-sentences', should be reports about *'our own' observational experiences*, although expressed in a physical language, i.e. as reports about our own bodies. In Otto Neurath's formulation, such a protocol-sentence was to have, accordingly, a very queer form. He wrote:[37] 'A complete protocol-sentence might for example read: "Otto's protocol at 3.17: [Otto's verbalized thinking was at 3.16: (In his room at 3.15 was a table, observed by Otto)]".' One sees that the attempt is made here to incorporate the old starting point—the observer's own subjective experiences, i.e. 'methodological solipsism'.

Carnap later accepted my view; but in the article ('On Protocol-Sentences'[38]) in which he very kindly called this view of mine 'the most adequate of the forms of scientific language at present advocated . . . in the . . . theory of knowledge',[39] he did not yet quite appreciate the fact (appreciated clearly in *Testability and Meaning*, as we shall see) that the difference between my view and Neurath's concerned a fundamental point: whether or not to appeal in our tests to simple, observable, *physical facts* or to *'our own sense-experiences'* (methodological solipsism). He therefore said, in his otherwise admirable report of my views, that the testing subject S will, 'in practice, often stop his tests' when he has arrived at the 'observation statements of the protocolling subject S'; i.e. at statements of *his own sense-experience*; whereas I held that he would stop only when he had arrived at a statement about some easily and intersubjectively observable *behaviour of a physical body* (which, at the moment did not appear to be problematic).[40]

[37] *Erkenntnis*, **3**, 1932, p. 207.

[38] 'Ueber Protokollsätze', *Erkenntnis*, **3**, 1932, pp. 215–28.

[39] *Op. cit.*, p. 228; cp. *Testability and Meaning* (see below, note 60, and the next footnote here).

[40] See also for a brief criticism of Carnap's report, notes 1 and 2 to section 29 of *L.Sc.D.* (The quotation in the text next to note 2 of section 29 is from Carnap's report.)

The point here mentioned is, of course, closely connected with the fact that I never believed in induction (for which it seems natural to start 'from our own experiences') but in a *method of testing predictions* deducible from our theories, while Neurath believed in induction. At that time I thought that, when reporting my views, Carnap had given up his belief in induction. If so, he has returned to it since.

(*b*) *The Language of Unified Science.* Closely connected with physicalism was the view that the physicalist language was a universal language in which everything meaningful could be said. '*The physical language is universal*', Carnap wrote.[41] 'If, because of its character as a universal language, we adopt the language of physics as the . . . language of science, then all science turns into physics. *Metaphysics is excluded as nonsensical.*[42] The various sciences become parts of the unified science.'

It is clear that this *thesis of the one universal language of the one unified science* is closely connected with that of the elimination of metaphysics: if it were possible to express everything that a non-metaphysical scientist may wish to say in one language which, by its rules, makes it impossible to express metaphysical ideas, then something like a *prima facie* case would have been made out in favour of the conjecture that metaphysics cannot be expressed in any 'reasonable' language. (Of course, the conjecture would be still very far from being established.)

Now the queer thing about this thesis of the *one* universal language is that before it was ever published (on the 30th of December 1932) it had been refuted by one of Carnap's colleagues in the Vienna Circle. For Gödel, by his two famous incompleteness theorems, had proved that one unified language would not be sufficiently universal for even the purposes of elementary number theory: although we may construct a language in which all assertions of this theory can be *expressed*, no such language suffices for formalizing all the proofs of those assertions which (in some other language) can be *proved*.

It would have been best, therefore, to scrap forthwith this doctrine of the one universal language of the one universal science (especially in view of Gödel's second theorem which showed that it was pointless to try to discuss the consistency of a language in that language itself). But

[41] *Erkenntnis*, **3**, 1932, p. 108.

[42] *Loc. cit.*, italics mine.

more has happened since to establish the impossibility of the thesis of the universal language. I have in mind, especially, Tarski's proof that every universalistic language is paradoxical (first published in 1933 in Polish, and in 1935 in German). But in spite of all this the doctrine has survived; at least, I have nowhere seen a recantation.[43] And the so-called 'International Encyclopedia of Unified Science', which was founded upon his doctrine (despite my opposition,[44] at the 'First Congress for Scientific Philosophy', in Paris, 1935) is still being continued. It will remain a monument to a metaphysical doctrine, once passionately held by Neurath and brilliantly wielded by him as a major weapon in the anti-metaphysical crusade.

For no doubt the strong philosophical belief which inspired this forceful and lovable person was, by his own standards, purely 'metaphysical'. A unified science in a unified language is really nonsense, I am sorry to say; and demonstrably so, since it has been proved, by Tarski, that no consistent language of this kind can exist. Its logic is outside it. Why should not its metaphysics be outside it too?

[43] The doctrine is still maintained, in all essentials (although in a more cautious mood) in *Testability and Meaning*, and not touched upon in the corrections and additions added to various passages in 1950; see below, note 50, and text. In an excellent and by now famous paragraph of his *Introduction to Semantics* (section 39) Carnap indicated 'how the views exhibited in [his] earlier book, *The Logical Syntax of Language*, have to be modified as a result, chiefly, of the new point of view, of semantics'. But the *Syntax*, although it continued to subscribe to the doctrine of the unified science in a unified language (see especially section 74, the bottom of p. 286, and pp. 280 ff.) did not investigate this doctrine more fully; which may perhaps be the reason why Carnap overlooked the need to modify this doctrine.

[44] In Paris, I opposed the foundation of the *Encyclopedia*. (Neurath used to call me 'the official opposition' of the Circle, although I was never so fortunate as to belong to it.) I pointed out, among other things, that it would have no similarity whatever to an encyclopedia as Neurath conceived it, and that it would turn out to be another series of *Erkenntnis* articles. (For Neurath's ideal of an encyclopedia, see for example his critical article on *L.Sc.D.*, *Erkenntnis*, **5**, pp. 353 to 365, especially section 2.) At the Copenhagen Congress, in 1936, which Carnap did not attend, I tried to show that the doctrine of the unity of science and of the one universal language was incompatible with Tarski's theory of truth. Neurath thereupon suggested in the discussion which followed my lecture that Tarski's theories about the concept of truth must be untenable; and he inspired (if my memory does not deceive me) Arne Naess, who was also present, to undertake an empirical study of the usages of the word 'truth', in the hope of thus refuting Tarski. See also Carnap's appropriate remark on Naess, in the *Introduction to Semantics*, p. 29.

I do not, of course, suggest that Carnap did not know all this; but I suggest that he did not see its devastating effect upon the doctrine of the unified science in the unified language.

It may be objected, perhaps, that I am taking the doctrine of the unified language too seriously, and that a strictly *formalized* science was not intended. (Neurath, for example, used to speak, especially in his later publications, of a 'universal slang', which indicates that he did not think of a *formalized* universal language.) I believe that this is true. But this view, again, destroys *the doctrine of the meaninglessness of metaphysics*. For if there are no strict *rules of formation* for the universal slang, then the assertion that we cannot express metaphysical statements in it is gratuitous; and it can only lead us back to the naïve naturalistic view of meaninglessness, criticized above in section 3.

It may be mentioned in this context that Gödel's (and Church's) discoveries also sealed the fate of another of the pet doctrines of positivism (and of one of my pet aversions[45]). I have in mind Wittgenstein's '*The riddle* does not exist. If a question can be put at all, it *can* also be answered.'[46]

This doctrine of Wittgenstein's, called by Carnap in the '*Aufbau*'[47] 'the proud thesis of the omnipotence of rational science', was hardly tenable even when it first appeared, if we remember Brouwer's ideas, published long before the *Tractatus* was written. With Gödel (especially with his second theorem of undecidability) and Church, its situation became even worse; for from them we learned that we could never complete even our *methods* of solving problems. Thus a well-formed mathematical question may become 'meaningless' if we adopt a

[45] Another is 6.1251 of the *Tractatus* (see also 6.1261): 'Hence there can never be surprises in logic' which is either trivial (viz. if 'logic' is confined to the two-valued propositional calculus) or obviously mistaken, and most misleading in view of 6.234: 'Mathematics is a method of logic'. I think that nearly every mathematical *proof* is surprising. 'By God, this is impossible', Hobbes said when first encountering Euclid's derivation of the Pythagorean theorem.

[46] *Tractatus*, 6.5. We also read there: 'For an answer which cannot be expressed the question too cannot be expressed.' But the question may be 'Is this assertion (for example Goldbach's conjecture) demonstrable?' And the true answer may be, 'We don't know; perhaps we may never know, and perhaps we can never know.'

[47] See *Aufbau*, section 183, p. 261, under '*Literature*'.

criterion of meaning according to which the meaning of a statement lies in the method by which it can be verified (in mathematics: proved or disproved). This shows that we may be able to formulate a question (and, similarly, the possible answers to it) without an inkling as to how we might find out which of the possible answers is true; which demonstrates the superficiality of Wittgenstein's 'proud thesis'.

Carnap was the first philosopher who recognized the immense importance of Gödel's discoveries, and he did his best to make them known to the philosophical world. It is the more surprising that Gödel's result did not produce that change which it should have produced in the Vienna Circle's tenets (in my opinion, undoubtedly and obviously metaphysical tenets, all too tenaciously held) concerning the language and the scope of science.

(c) Carnap's *Logical Syntax* is one of the few philosophical books which can be described as of really first-rate importance. Admittedly, some of its arguments and doctrines are superseded, owing mainly to Tarski's discoveries, as Carnap himself explained frankly in that famous last paragraph of his *Introduction to Semantics*. Admittedly, the book is not easy to read (and even more difficult in English than in German). But it is my firm conviction that, if ever a history of the rational philosophy of the earlier half of this century should be written, this book ought to have a place in it second to none. I cannot even try here (wedged between critical analyses) to do justice to it. But one point at least I must mention. It was through this book that the philosophical world, to the west of Poland, was first introduced to the method of analysing languages in a 'meta-language', and of constructing 'object-languages'—a method whose significance for logic and the foundations of mathematics cannot be overrated; and it was in this book that the claim was first made, and, I believe, completely substantiated, that this method was of the greatest importance for the philosophy of science. If I may speak personally, the book (which came out a few months before my *Logic of Scientific Discovery*, and which I read while my book was in the press) marks the beginning of a revolution in my own philosophical thinking, although I did not understand it fully (because of its real internal difficulties, I believe) before I had read Tarski's great paper on the Concept of Truth (in the German translation, 1935). Then I realized, of course, that a syntactic meta-linguistic analysis was inadequate, and must be replaced by what Tarski called 'semantics'.

Of course I believe that from the point of view of the problem of demarcation, a great step forward was made in the *Syntax*. I say 'of course', since I am alluding to the fact that some of my criticism was accepted in this book. Part of the relevant passage is quoted above (in note 24). But what is most interesting from our present point of view is the passage immediately following the quotation; it shows, I believe, that Carnap did not accept enough of my criticism. 'The view here presented', he writes,[48] 'allows great freedom in the introduction of new primitive concepts and new primitive sentences in the language of physics or of science in general; yet at the same time it retains the *possibility of differentiating pseudo-concepts and pseudo-sentences* from real scientific concepts and sentences, *and thus of eliminating the former*.' Here we find, again, the old thesis of the meaninglessness of metaphysics. But it is mitigated, if only a little, by the immediate continuation of this passage (which Carnap places in square brackets, and which shows the influence of my criticism, mentioned by him on the preceding page). 'This elimination, however, is not so simple as it appeared on the basis of the earlier position of the Vienna Circle, which was in essentials that of Wittgenstein. On that view it was a question of "*the* language" in an absolute sense; it was thought possible to reject both concepts and sentences if they did not fit into *the* language.'

The position indicated by these passages (including the one quoted briefly in note 24 above) may be described as follows:

(1) Some difficulties, especially those of Wittgenstein's verifiability criterion of meaning, are recognized; also the inadequacy of what I have called the 'naturalistic' theory of meaningfulness (which corresponds to the belief in '*the language' in which things simply are, or are not, essentially meaningful by their nature*).

(2) But the belief is still maintained that we can, by some feat of ingenuity, establish *one* language which does the trick of rendering meaningless precisely the 'metaphysical' concepts and sentences and no others.

(3) Even the belief that we can construct one universal language of

[48] *Syntax*, section 82, p. 322 top. (The italics are Carnap's.)

unified science is still upheld, in consequence of (2); but it is not stressed, and not examined in detail. (See point (*b*) in this section, above, and especially the passage from the *Syntax*, section 74, p. 286, mentioned in note 43 above.)

This situation does not call for further criticism on my part: practically all that needs to be said I have said already, especially that this approach renders Tarski's Semantics meaningless, and with it most of the theory of logical inference, i.e. of logic. Only one further—and I believe important—comment has to be made.

One of the difficulties of this great and important book of Carnap's lies in its emphasis upon the fact that the syntax of a language *can* be formulated in that language itself. The difficulty is the greater because the reader has hardly learnt to distinguish between an object-language and a meta-language when he is told that, after all, the distinction is not quite as radical as he supposed it to be, since the meta-language, it is now emphasized, may form part of the object-language.

Carnap's emphasis is, undoubtedly, misplaced. It is a fact that part of the meta-language (viz., its 'syntax') can form part of the object-language. But although this fact is very important, as we know from Gödel's work, its main use is in the construction of self-referring sentences, which is a highly specialized problem. From the point of view of promoting the understanding of the relation between object-language and meta-language, it would no doubt have been wiser to treat the meta-language as distinct from the object-language. It could, of course, have still been shown that at least a part of the meta-language—and enough for Gödel's purposes—may be expressed in the object-language, without stressing the mistaken thesis that the whole of the meta-language can be so expressed.

Now there is little doubt that it was the doctrine of the one universal language in which the one unified science was to be expressed that led Carnap to this emphasis which contributes so much to the difficulties of his book; for he hoped to construct a unified language which would automatically eliminate metaphysics. It is a great pity to find this excellent book spoiled by an anti-metaphysical dogma—and by a wrong demarcation which eliminates, together with metaphysics, the most important parts of logic.

The *Syntax* continues the doctrine of the meaninglessness of metaphysics in the following form: All meaningful sentences either belong to the *language of science*, or (if philosophical) they can be expressed within the *syntax* of that language. This syntax comprises the whole of the philosophy and logic of science so far as these are translatable into the 'formal mode of speech'; moreover, this syntax can, if we wish, be formulated in the same universal ('object-') language in which all the sciences may be formulated.

Here it is not only the doctrine of the one universal language which I cannot accept: I also cannot accept the ruling that what I say must be translatable into the 'formal mode of speech' in order to be meaningful (or to be understood by Carnap). No doubt one should express oneself as clearly as possible; and no doubt what Carnap calls the 'formal mode of speech' is often preferable to what he calls 'the material mode' (and I have often used it, in my *Logic of Scientific Discovery* and before, without having been told to do so). But it is not necessarily preferable. And why should it necessarily be preferable? Perhaps because the *essence* of philosophy is language analysis? But I am not a believer in essences. (Nor in Wittgenstein.) How to make oneself better understood can only be a matter of thought and experience.

And why should *all* philosophy be linguistic analysis? No doubt it may often help to put a question in terms of language-construction. But why should *all* philosophical questions be of this kind? Or is this the one and only non-linguistic thesis of philosophy?

The positivist attack has put, if I may say so, the fear of God into all of us who wish to speak sense. We have become more careful in what we say, and how we say it, and this is all to the good. But let us be clear that *the philosophical thesis that language analysis is everything in philosophy is paradoxical*. (I admit that this criticism of mine no longer applies in this form to *Testability and Meaning* which replaces the *thesis* by a *proposal* that is no longer paradoxical; no reasons, however, are offered in favour of the proposal, except that it is an improved version of the thesis; and this is no reason, it seems to me, for accepting it.)

5. TESTABILITY AND MEANING

Carnap's *Testability and Meaning* is perhaps the most interesting and important of all the papers in the field of the philosophy of the empirical sciences which were written in the period between Wittgenstein's *Tractatus* and the German publication of Tarski's essay on the concept of truth. It was written in a period of crisis, and marks great changes in the author's views. At the same time, its claims are very modest. 'The object of this essay is not to offer . . . solutions . . . It aims rather to stimulate further investigations.' This aim was amply realized: the investigations which sprang from it must number hundreds.

Replacing 'verifiability' by 'testability' (or by 'confirmability'), *Testability and Meaning* is, as its title indicates, very largely a treatise on our central problem. It still attempts to exclude metaphysics from the language of science: '. . . an attempt will be made to formulate the principle of empiricism in a more exact way, by stating a requirement of confirmability or testability as a criterion of meaning', we read in section 1; and in section 27 (p. 33) this hint is elaborated: 'As empiricists, we require the language of science to be restricted in a certain way; we require that descriptive predicates and hence synthetic sentences are not to be admitted unless they have some connection with possible observations . . .' What is 'not to be admitted' is, of course, metaphysics: '. . . even if *L* were to be a language adequate for all science . . . [we] should not wish for example to have [in *L*] . . . sentences [corresponding] to many or perhaps most of the sentences occurring in the books of metaphysicians.'[49]

Thus the main idea—excluding metaphysics from the well-formed formulae of *L*, the language of science—is unchanged. Unchanged, too, is the idea of the *one* language of science: although Carnap now says, very clearly, that we can *choose* our language, and that various scientists can *choose* it in different ways, he proposes that we accept a universal language, and he even defends the *thesis of physicalism* in a modified form. He often speaks (as in the passages quoted) of *the* language of science, or of the possibility of having a language for *all* science, or of the *whole* or the *total* language of science:[50] the impossibility of such a language he still does not realize.

[49] *Testability*, section 18 (p. 5).

[50] See *Testability*, sections 15 (pp. 467 f.) and 27 (p. 33), 18 (p. 5), as quoted, and 16 (pp. 469, 470).

Carnap is however very careful in the formulation of his new ideas. He says that we have a choice between many languages of science, and he says that the 'principle of empiricism'—which turns out to be another name for the principle of the meaninglessness of metaphysics—should preferably be formulated not as an assertion, but as a 'proposal or requirement'[51] for selecting a language of science.

One might think that, with this formulation, the idea of excluding metaphysics as meaningless has in fact been abandoned: for the metaphysician need not, and clearly would not, accept any such proposal; he would simply make another proposal in its place according to which metaphysics would become meaningful (in an appropriate language). But this is not how Carnap sees the situation. He sees it, rather, as the task or duty imposed upon the anti-metaphysician *to justify his view of the meaninglessness of metaphysics by constructing a language of science free from metaphysics*. And this is how the problem is still seen by many, I fear.

It is easy to show, using my old arguments, that no such language can be constructed.

My thesis is that a satisfactory language for science would have to contain, with any well-formed formula, its negation; and since it has to contain universal sentences, it has therefore to contain existential sentences also.

But this means that it must contain sentences which Carnap, Neurath, and all other anti-metaphysicians always considered to be metaphysical. In order to make this quite clear I choose, as an extreme example, what may be called '*the arch-metaphysical assertion*':[52] 'There exists an omnipotent, omnipresent, and omniscient personal spirit.' I shall briefly show how this sentence can be constructed as a well-formed or meaningful sentence in a physicalistic language which is quite similar to those proposed in *Testability and Meaning*.

We can take as primitive the following four physicalistic predicates:

(1) 'The thing *a* occupies a position *b*' or more precisely, '*a* occupies

[51] Section 27 (p. 33).

[52] One need not believe in the 'scientific' character of psycho-analysis (which, I think, is in a metaphysical phase) in order to diagnose the anti-metaphysical fervour of positivism as a form of Father-killing.

a position of which the (point or) region *b* is a part'; in symbols 'Pos(*a*,*b*)'.[53]

(2) 'The thing (machine, or body, or person . . .) *a* can put the thing *b* into position *c*'; in symbols 'Put(*a*,*b*,*c*)'.[54]

(3) '*a* makes the utterance *b*'; in symbols 'Utt(*a*,*b*)'.

(4) '*a* is asked (i.e. adequately stimulated by an utterance combined, say, with a truth drug) whether or not *b*'; in symbols '*Ask*(*a*,*b*)'.

We assume that in our language we have at our disposal *names* of all expressions of the form 'Pos(*a*,*b*)', 'Put(*a*,*b*,*c*)', etc., including some of those introduced below with their help. I shall use for simplicity's sake, *quotation names*. (I am aware, however, of the fact that this procedure is not exact, especially where variables in quotes are bound, as in (14); but this difficulty can be overcome.)

Now we can easily introduce, with the help of explicit definitions using (1) and (2):[55]

[53] 'Pos(*a*,*b*)' is used for the sake of simplicity; we should, really, operate with position *and* momentum, or with the 'state' of *a*. The necessary amendments are trivial. I may remark that I do not presuppose that the variables '*a*', '*b*', etc. all belong to the same type or semantical category.

[54] Or, as Carnap would put it, '*a* is able to make the full sentence "Pos(*b*,*c*)" true'; see Carnap's explanation of his primitive '*realizable*' (a term of the meta-language, however, in contradistinction to my 'Put'), in *Testability*, section 11, p. 455, Explanation 2.

[55] The definitions are: (5) Opos(*a*) ≡ (*b*)Pos(*a*,*b*).—(6) Oput(*a*) ≡ (*b*)(*c*)Put(*a*,*b*,*c*).—Next we have the '*bilateral reduction sentence*': (7) *Ask*(*a*,*b*) ⊃ (Th(*a*,*b*) ≡ Utt(*a*,*b*)).—The remaining definitions are: (8) Thp(*a*) ≡ (E*b*)Th(*a*,*b*).—(9) Sp(*a*) ≡ (Thp(*a*) & ((*b*) ~ Pos(*a*,*b*)) v Opos(*a*)).—An alternative (or an addition to the definiens) might be 'Sp(*a*) ≡ (Thp(*a*) & (*b*) ~ Utt(*a*,*b*))'.—(10) Knpos(*a*,*b*,*c*) ≡ (Pos(*b*,*c*) & Th(*a*,'Pos(*b*,*c*)')).—(11) Knput(*a*,*b*,*c*,*d*) ≡ (Put(*b*,*c*,*d*) & Th(*a*,'Put(*b*,*c*,*d*)')).—(12) Knth(*a*,*b*,*c*) ≡ (Th(*b*,*c*) & Th(*a*,'Th(*b*,*c*)')).—(13) Unkn(*a*) ≡ ((E*b*) (*c*) (Th(*a*,*b*) & (*a* ≠ *c* ⊃ ~Knth(*c*,*a*,*b*))).—(14) Kn(*a*,*b*) ≡ ((*c*)(*d*)(*e*))(*b*) = 'Pos(*c*,*d*)' & Knpos(*a*,*c*,*d*)) v (*b* = 'Put(*c*(*c*,*d*,*e*)' & Knput(*a*,*c*,*d*,*e*)) v (*b* = 'Th(*c*,*d*)' & Knth (*a*,*c*,*d*))).—(15) *Verax*(*a*) ≡ (*b*)(Th(*a*,*b*) ≡ (Kn(*a*,*b*)).—(16) Okn(*a*) ≡ (*b*)(*c*)(*d*)(*e*)(*f*)(*g*)(*h*) (((*a* ≠ *b*) ⊃ (Knput(*a*,*b*,*c*,*d*) ≡ Put(*b*,*c*,*d*))) & ((*a* ≠ *e*) ⊃ (Knpos(*a*,*e*,*f*) ≡ Pos(*e*,*f*))) & ((*a* ≠ *g*) ⊃ (Knth(*a*,*g*,*h*) ≡ Th(*g*,*h*)))) & *Verax* (*a*)).—We can easily prove that 'Unkn(*a*) & Okn(*a*)' implies the uniqueness of *a*; alternatively, we can prove uniqueness, along lines which might have appealed to Spinoza, from 'Opos(*a*)', if we adopt the Cartesian axiom: *a* ≠ *b* ⊃ (E*c*) ((Pos(*a*,*c*) & ~ Pos(*b*,*c*)) v (~ Pos(*a*,*c*) & Pos (*b*,*c*))).

(Added in proofs.) Our definitions can be simplified by employing the Tarskian semantic predicate 'T(*a*)', meaning '*a* is a true statement'. Then (14) may be replaced by Kn(*a*,*b*) ≡ Th(*a*,*b*) & T(*b*); (15) by *Verax*(*a*) ≡ (*b*)Th(*a*,*b*) ⊃ T(*b*); and (16) by Okn(*a*) ≡ (*b*)T(*b*) ⊃ Kn(*a*,*b*).

(5) '*a* is omnipresent' or '*Opos*(*a*)'.

(6) '*a* is omnipotent', or '*Oput*(*a*)'.

Moreover, with the help of (3) and (4), we can introduce, by Carnap's reduction method,

(7) '*a* thinks *b*' or '*Th*(*a*,*b*)'.

Carnap recommends[56] that such a predicate should be admitted. With the help of (7) we can now define explicitly:

(8) '*a* is a thinking person', or '*Thp*(*a*)'.

(9) '*a* is a (personal) spirit', or '*Sp*(*a*)'.

(10) '*a* knows that *b* is in position *c*', or '*Knpos*(*a*,*b*,*c*)'.

(11) '*a* knows that *b* can put *c* into the position *d*', or '*Knput*(*a*,*b*,*c*,*d*)'.

(12) '*a* knows that *b* thinks *c*', or '*Knth*(*a*,*b*,*c*)'.

(13) '*a* is unfathomable', or '*Unkn*(*a*)'.

(14) '*a* knows the fact *b*', or '*Kn*(*a*,*b*)'.

(15) '*a* is truthful', or '*Verax*(*a*)'.

(16) '*a* is omniscient', or '*Okn*(*a*)'.

Nothing is now easier than to give an existential formula expressing *the arch-metaphysical assertion*: that a thinking person *a* exists, positioned everywhere; able to put anything anywhere; thinking all and only what is in fact true; and with nobody else knowing all about *a*'s thinking. (The uniqueness of an *a* of this kind is demonstrable from *a*'s properties. We cannot, however, identify *a* with the God of Christianity. There is a difficulty in defining 'morally good' on a physicalistic basis. But questions of definability are anyway, in my opinion, supremely uninteresting—outside mathematics—except to essentialists: see below.)

It is clear that our purely existential arch-metaphysical formula cannot be submitted to any scientific test: there is no hope whatever of falsifying it—of finding out, if it is false, that it is false. For this reason I describe it as metaphysical—as falling outside the province of science.

But I do not think Carnap is entitled to say that it falls outside science, or outside the language of science, or that it is meaningless. (Its meaning seems to me perfectly clear; it is also clear that some logical analysts must have mistaken its empirical incredibility for meaninglessness. But one could even conceive of experiments which might

[56] *Testability*, section 18, p. 5, S_1.

'confirm' it, in Carnap's sense, that is to say, 'weakly verify' it; see text to note 67.) It helps us very little if we are told, in *Testability*,[57] that 'the meaning of a sentence is in a certain sense identical with the way we determine its truth and falsehood; and a sentence has meaning only if such a determination is possible'. One thing emerges clearly from this passage—that it is not Carnap's intention to allow meaning to a formula like the arch-metaphysical one. But the intention is not realized; it is not realized, I think, because it is not realizable.

I need hardly say that my only interest in constructing our arch-metaphysical existential formula is to show that there is no connection between well-formedness and scientific character. *The problem of how to construct a language of science which includes all we wish to say in science but excludes those sentences which have always been considered as metaphysical is a hopeless one. It is a typical pseudo-problem.* And nobody has ever explained *why it should be interesting to solve it* (*if it is soluble*). Perhaps in order to be able to say, as before, that metaphysics is meaningless? But this would not mean anything like what it meant before.[57a]

[57] *Testability*, section 1, end of first paragraph.

[57a] (Added in Proofs:)

The reaction of my positivist friends to my 'arch-metaphysical formula' (I have not yet seen Carnap's reaction, but I received a report from Bar-Hillel) was this. As this formula is well-formed, it is 'meaningful' and also 'scientific': of course, not scientifically or empirically *true*; but rather scientifically or empirically *false*; or, more precisely, disconfirmed by experience. (Some of my positivist friends also denied that my name 'arch-metaphysical' had any historical justification, and asserted that the anti-metaphysical tendencies of the Vienna Circle never had anything to do with anti-theological tendencies; and this in spite of Neurath's physicalism which was intended as a modern version of either classical or dialectical materialism.)

Now should anyone go so far as to commit himself to the admission that my arch-metaphysical formula is well-formed and therefore empirically true or false then I think he will encounter difficulties in extricating himself from this situation. For how could anybody defend the view that my arch-metaphysical formula is false, or disconfirmed? It is certainly unfalsifiable, and non-disconfirmable. In fact, it is expressible in the form

$$(Ex)\ G(x)$$

—in words: 'there exists something that has the properties of God.' And on the assumption that '$G(x)$' is an empirical predicate, we can *prove* that its probability must equal 1. (See Carnap's *Logical Foundations of Probability*, p. 571.) I can prove, further, that this means that its probability cannot be diminished by any empirical information (that is, by any information whose logical probability differs from zero). But this means, according to

But, it may be said, it may still be possible to realize at least part of the old Wittgensteinian dream, and to make metaphysics meaningless. For perhaps Carnap was simply too generous in allowing us to use *dispositional predicates*, such as '*a* is able to put *b* into *c*' and '*a* thinks *b*' (the latter characterized as a disposition to utter *b*). I cannot hold out any hope to those who pursue this line of thought. As I tried to show when discussing the *Aufbau* in section 3, we need in science *genuine non-extensional universals*. But in my *Logic of Scientific Discovery* I indicated briefly—much too briefly, for I thought that the 'reductionist'[58] ideas of the *Aufbau* had been given up by its author—*that all universals are dispositional*; not only a predicate like 'soluble', but also 'dissolving' or 'dissolved'.

If I may quote from my *Logic of Scientific Discovery* ('*L.Sc.D.*', for short): 'Every descriptive statement uses . . . universals; every statement has the character of a theory, of a hypothesis. The statement, "Here is a glass of water" cannot be verified by any observational experience. The reason is that the *universals* which occur in it cannot be correlated with any particular observational experience . . . By the word "glass", for example, we denote physical bodies which exhibit a certain *law-like behaviour*; and the same applies to the word "water". Universals . . . cannot be "constituted".' (That is, they cannot be defined, in the manner of the *Aufbau*.)[59]

Carnap's *Logical Foundations*, that its degree of confirmation equals 1, and that it *cannot* be disconfirmed—as I asserted here. (See also pp. 249 f., above.)

How then can my positivist friends assert that the empirical statement '$(Ex)\ G(x)$' is false? It is, at any rate, better confirmed than any scientific theory.

My own view is that it is non-testable and therefore non-empirical and non-scientific.

[58] The term '*reductionism*' is, it seems, Quine's. (It corresponds closely to my term 'inductivism'. See, for example, Carnap's report in *Erkenntnis*, 3, 1932, pp. 223–4.) See also my remarks in *L.Sc.D.*, section 4, p. 34, where, in criticism of what Quine calls 'reductionism', I wrote: 'The older positivists accepted as scientific only such concepts (or terms) as . . . could be reduced to elementary experiences (sense-data, impressions, perceptions, experiences of remembrance [Carnap's term in the *Aufbau*], etc.)' See also *L.Sc.D.*, section 14, especially notes 4 and 6, and text.

[59] The passage is from *L.Sc.D.* (end of section 25; see also sections 14 and 20). Although this passage, together with Carnap's related passage about the term 'soluble' (*Testability*, section 7, p. 440) may perhaps have contributed to starting *the so-called 'problem of counterfactual conditionals'*, I have never been able, in spite of strenuous efforts, to understand this problem; or more precisely, what remains of it when one does not subscribe either to essentialism, or to phenomenalism, or to meaning-analysis.

What then is the answer to the problem of defining, or introducing, a dispositional term like 'soluble'? The answer is, simply, that the problem is insoluble. And there is no need whatever to regret this fact.

It is insoluble: for assume we have succeeded in 'reducing' '*x* is soluble in water' by what Carnap calls a 'reduction-sentence', describing an operational test, such as 'if *x* is put into water then *x* is soluble in water if and only if it dissolves'. What have we gained? We have still to reduce '*water*' and 'dissolves'; and it is clear that, among the operational tests which characterize *water*, we should have to include: 'if anything that is soluble in water is put into *x*, then if *x* is water, that thing dissolves'. In other words, not only are we forced, in introducing 'soluble', to fall back upon 'water', which is dispositional in perhaps even a higher degree, but in addition, we are forced into circularity; for we introduce 'soluble' with the help of a term ('water') which in its turn cannot be operationally introduced without 'soluble'; and so on, *ad infinitum*.

The situation with '*x* is dissolving' or '*x* has dissolved' is very similar. We say that *x* has dissolved (rather than that it has disappeared) only if we expect to be able to show (say, by evaporating the water) that certain traces of this process can be found, and that we can, if necessary, even *identify* parts of the dissolved and later reclaimed substance with parts of *x* by tests which will have to establish, among other things, the fact that the reclaimed substance is, again, *soluble*.

There is a very good reason why this circle cannot be broken by establishing a definite order of reduction or introduction. It is this: our actual tests are never conclusive and always tentative. We never should agree to a ruling telling us to stop our tests at any particular point—say, when arriving at primitive predicates. All predicates are for the scientist equally dispositional, i.e. open to doubt, and to tests. This is one of the main ideas of the theory of the *empirical basis* in my *L.Sc.D.*[60]

[60] In *Testability*, Carnap accepts most of my theory of the empirical basis (*L.Sc.D.*, sections 25 to 30) including most of my terminology ('empirical basis', 'basic sentences', etc.; cp. also his introduction and use of the term 'observable' with *L.Sc.D.*, section 28, p. 59). Even the slight but significant discrepancy (which I have here interpreted—see text to notes 38 to 40, above—as a survival from his days of 'methodological solipsism', and which I criticized in *L.Sc.D.*, note 1, and text to note 2 to section 29) is now rectified (*Testability*, section 20; see especially 'Decision 2', p. 12, and text to note 7, p. 13). Some

So much about the fact that 'soluble' cannot be 'reduced' to something that is less dispositional. As to my contention that there is no need to regret this fact, I want only to say (again) that outside mathematics and logic problems of definability are mostly gratuitous. We need many undefined terms[61] whose meaning is only precariously fixed by usage—by the manner in which they are used in the context of theories, and by the procedures and practices of the laboratory. Thus the meaning of these concepts will be changeable. But this is so with all concepts, including defined ones, since a definition can only reduce the meaning of the defined term to that of undefined terms.

What then is behind the demand for definitions? An old tradition, reaching back far beyond Locke to Aristotelian essentialism; and as a result of it, a belief that, if a man was unable to explain what a word meant which he used, then this showed that 'he had given no meaning' to it (Wittgenstein), and had therefore been talking nonsense. But this Wittgensteinian belief is nonsense, since all definitions must ultimately go back to undefined terms. However, since I have discussed all this elsewhere,[62] I shall say nothing further about it here.

In concluding this section, I wish to stress again the point that testability, and confirmability, even if satisfactorily analysed, are in no way better fitted to serve as *criteria of meaning* than the older criterion of verifiability. But I must say that, in addition, I am unable to accept

other points of agreement (apart from those to which Carnap himself refers) are: the thesis that there is a '*conventional component*' in the acceptance or rejection of any (synthetic) sentence (cp. *Testability*, section 3, p. 426, with my *L.Sc.D.*, section 30, p. 108) and the rejection of the doctrine of atomic sentences which state ultimate facts (cp. *Testability*, section 9, p. 448, with my *L.Sc.D.*, section 38, p. 127). Yet in spite of this far-reaching agreement, a decisive difference remains; I stress a *negative view* of testability which, for me, is the same as refutability: and I accept confirmations only if they are the outcome of unsuccessful but genuine attempts at refutation. For Carnap, testability and refutability remain *weakened forms of verification*. The consequences of this difference will become clear in my discussion of probability and induction in section 6 below.

[61] In *Testability*, section 16, p. 470, Carnap hopes that we may introduce all terms on the basis of *one* undefined one-termed predicate (either 'bright', or alternatively 'solid'). But one cannot introduce any other term on this basis with the help of a reduction pair: at least *two* different 'given' predicates are needed even for one bilateral reduction sentence. Moreover, we need at least one *two-termed relation*.

[62] See for example my *Open Society*, ch. 11, section ii.

Carnap's analysis of either 'test', 'testable', etc., or of 'confirmation'. The reason is, again, that his terms are substitutes for 'verification', 'verifiable', etc., slightly weakened so as to escape the objection that laws are not verifiable. But this compromise is inadequate, as we shall see in the next and last section of this paper. *Acceptability in science depends, not upon anything like a truth-surrogate, but upon the severity of tests.*[63]

6. PROBABILITY AND INDUCTION

The full consequences of approaching confirmation as if it was a kind of weakened verification become manifest only in Carnap's two books on probability—the big volume entitled *Logical Foundations of Probability* (referred to here as '*Probability*') and the smaller progress-report, *The Continuum of Inductive Methods* (referred to here as '*Methods*').[64]

[63] As a consequence, the following 'content-condition' or 'entailment condition' is invalid: 'If *x* entails *y* (i.e. if the content of *y* is part of that of *x*), then *y* must be at least as well confirmed as *x*'; the invalidity of the content condition was pointed out in my *L.Sc.D.*, sections of 82 and 83 (cp. sections 33 f.) where content is identified with degree of testability and [*absolute*] logical improbability, and where it was shown that the invalidity of the content condition destroys the identification of degree of confirmation with logical probability. In *Testability*, however, Carnap's whole theory of reduction rests upon this condition. (Cp. paragraph 1 of section 6, p. 434, and Definition I.*a.* on p. 435.) In *Probability*, p. 474 (cp. p. 397), Carnap notes the invalidity of the entailment condition (or 'consequence condition'); but he does not draw from it the (I believe necessary) conclusion that degree of confirmation cannot coincide with probability. (I have re-affirmed this conclusion in appendix *ix to *L.Sc.D.* Cp. notes 74 and 77 f. below, and text.)

[64] There is very little of relevance to the particular problem of demarcation in two of the three books published between *Syntax* and *Probability*—*Introduction to Semantics*, and *Meaning and Necessity* (and nothing, so far as I can see, in *Formalization of Logic*, which comes between them). In the *Introduction* I only find (*a*) what I take to be an allusion to Neurath's opposition to Tarski's concept of truth. (Carnap gives an excellent and tolerant reply to it (pp. vii f.), and (*b*) a just dismissal of the relevance of Arne Naess' questionnaire method (p. 29); see also my note 44 and text, above.) In *Meaning and Necessity* which I for one believe to be the best of Carnap's books (it is also perhaps the one which has been most fiercely attacked), there are a few remarks on ontology and metaphysics (p. 43) which, together with a reference to Wittgenstein (p. 9 f.), appear to indicate that Carnap still believes in the meaninglessness of metaphysics; for this reference reads: '. . . to know the meaning of a sentence is to know in which of the possible cases it would be true and in which not, as Wittgenstein has pointed out.' This passage, however, seems to me to be in conflict with Carnap's main conclusions, which I find convincing. For the cited passage

The topics of these two books are very closely related to our problem. They deal with the theory of induction, and the method of induction has always been one of the most popular criteria of demarcation for science; for the empirical sciences are, as a rule, considered to be characterized by their methods; and these, in turn, are usually characterized as *inductive*.[65]

This is Carnap's view too: his new criterion of demarcation is, as we have seen, *confirmability*. And in these two books, Carnap explains that the methods of confirming a sentence are identical with *the inductive method*. Thus we must conclude that the criterion of demarcation now becomes, more precisely, *confirmability by inductive methods*. In other words, a linguistic expression will belong to the empirical sciences if, and only if, it is logically possible to confirm it by inductive methods, or by inductive evidence.

As I have indicated in section 2, this criterion of demarcation does not satisfy my requirements: all sorts of pseudo-sciences (such as astrology) are clearly not excluded. The answer to this would be, no doubt, that the criterion is not intended to exclude what I call 'pseudo-sciences', and that these consist, simply, of false sentences, or perhaps of *disconfirmed* sentences, rather than of metaphysical *non-confirmable* ones. I am not satisfied by this answer (believing as I do that I have a criterion which excludes for example astrology and which has proved extremely fruitful in connection with a host of problems) but I am prepared to accept it, for argument's sake, and to confine myself to showing, as before, that *the criterion produces the wrong demarcation*.

My criticism of the verifiability criterion has always been this: against the intention of its defenders, *it did not exclude obvious metaphysical*

outlines, it is clear, what Carnap calls an *extensional* approach, as opposed to an *intensional* approach to meaning; on the other hand, 'the main conclusions . . . are' that we must *distinguish* between 'understanding the *meaning* of a given expression and investigating *whether and how it applies*' (p. 202, italics mine), and meaning is explained with the help of *intension*, *application* with the help of *extension*. Relevant to our problem is also Carnap's 'explication' of his concept 'explication', pp. 8 f.; see below.

[65] Our problem of demarcation is not explicitly discussed in these two books except for a remark in *Probability*, p. 31, on the '*principle of empiricism*' (also mentioned on pp. 30 and 71), and a discussion of the *empirical character of the 'principle of uniformity'* of nature, pp. 179 ff. Both passages will be mentioned below.

statements; but it did exclude the most important and interesting of all scientific statements, that is to say, the scientific theories, *the universal laws* of nature. Now let us see how these two groups of statements fare under the new criterion.

As to the first, it turns out that my arch-metaphysical existential formula obtains, in Carnap's system, a high confirmation value; for it belongs to the almost-tautological ('almost L-true') sentences whose confirmation value is 1 or, in a finite world of sufficient size, indistinguishable from 1. Moreover, it is a kind of statement for which even experimental confirmation is conceivable,[66] although *no tests in my sense*: there is no conceivable way of refuting it. Its lack of refutability puts it into the class of metaphysical sentences by my criterion of demarcation. Its high confirmation value in Carnap's sense, on the other hand, should make it vastly superior to, and more scientific than, *any scientific law*.

For *all universal laws have zero confirmation*, according to Carnap's theory, in a world which is in any sense infinite (temporal infinity suffices), as Carnap himself has shown;[67] and even in a finite world their value would be indistinguishable from zero if the number of events or things in this world is sufficiently large. All this is an obvious consequence of the fact that confirmability and confirmation, in Carnap's sense, are just slightly weakened versions of verifiability and verification. The reason why the universal laws are not verifiable is thus identical with the reason why they are not confirmable: they assert a great deal about the world—more than we can ever hope either to 'verify' or to 'confirm'.

[66] There may, conceivably, be seers like Swedenborg who make accurate predictions of future events whenever they tell us (under the influence of truth drugs) that they are now inspired by that *a* for which our existential formula (cf p. 276) is true: and we may, conceivably, be able to build radio receivers to take their place—receivers of *a*-influences—that turn out (under certain circumstances) always to speak, and to predict, the truth.

[67] See *Probability*, section 110 f., p. 571. For a similar result, see my *L.Sc.D.*, section 80, p. 257 f.: 'One might ascribe to a hypothesis [the hypotheses discussed are universal laws] . . . a probability, calculated, say, by estimating the ratio between all the tests passed by it to all those [conceivable] tests which have not [yet] been attempted. But this way too, leads nowhere; for this estimate can be computed with precision, and the result is always that the probability is zero.' (Another passage from this page is quoted in note 70, below.)

In face of the fact that natural laws turn out to be non-confirmable, according to his definition of 'degree of confirmation', Carnap adopts two courses: (*a*) he introduces *ad hoc* a new concept, called the (qualified[68]) 'instance confirmation of the law l', which is so defined that we sometimes obtain, in place of zero, a confirmation value close to 1; (*b*) he explains that natural laws are not really needed in science, and that we can dispense with them. (Verificationism made them meaningless. Confirmationism merely makes them unnecessary: this is the gain which the weakening of the verifiability criterion achieves.)

I shall now discuss (*a*) and (*b*) a little more fully.

(*a*) Carnap realizes, of course, that his zero-confirmation of all laws is counter-intuitive. He therefore suggests measuring the intuitive 'reliability' of a law by the degree of confirmation of an instance of the law (see note 68 above). But he nowhere mentions that this new measure, introduced on p. 572 of *Probability*, satisfies practically none of the criteria of adequacy, and none of the theorems, which have been built up on the preceding 571 pages. This is so, however, and the reason is that the 'instance confirmation' of a law *l* on the evidence *e* is simply *not a probability function* of *l* and *e* (not a 'regular *c*-function' of *l* and *e*).

And it could hardly be otherwise. We are given, up to p. 570, a detailed theory of confirmation (in the sense of probability$_1$). On p. 571, we find that for a law this confirmation is zero. We are now faced with the following alternatives: either (i) we accept the result as correct, and consequently say that the degree of rational belief in a well-supported law cannot differ appreciably from zero—or from that of a refuted law, or even from that of a self-contradictory sentence; or (ii) we take the result as a refutation of the claim that our theory has

[68] I confine my discussion to what Carnap calls (*Probability*, p. 572 f.) the 'qualified' instance confirmation; (*a*) because Carnap prefers it as representing 'still more accurately' our intuitions; and (*b*) because in a sufficiently complex world (with sufficiently many predicates) the non-qualified instance confirmation leads in all interesting cases to extremely low confirmation values. On the other hand, the 'qualified instance confirmation' (this I mention only in passing) is squarely hit by the so-called 'paradox of confirmation' (see *Probability*, p. 469). But this is a defect which (I found) can always be repaired—in this case by making the two arguments of the definiens in (15), p. 573, symmetrical with respect to the two logically equivalent implicative formulations of *l*; they become respectively (after simplification), '$j \supset h'$' and '$e.(h' \supset j)$'. This avoids the paradox.

supplied us with an adequate definition of 'degree of confirmation'. The *ad hoc* introduction of a new measure, in order to escape from an unintended result, is hardly an acceptable third possibility. But what is most unsatisfactory is to take this momentous step—a break with the method of 'explication' (see note 69, below) used so far—without giving any warning to the reader: this may result in the serious misconception that only a minor adjustment has been made.

For if we are to take probability, or confirmation, at all seriously, then the adjustment could not have been more radical; it replaces a confirmation function whose value is 0 by another whose value will be often close to 1. If we permit ourselves the freedom thus to introduce a new measure with no better justification than that the zero probability was counter-intuitive while the probability near to 1 'seems to represent . . . still more accurately what is vaguely meant by the reliability of a law',[69] then we can obtain for any sentence any probability (or degree of confirmation) we like.

Moreover, Carnap nowhere attempts to show that the newly introduced instance confirmation is adequate, or at least consistent (which it is not; see note 68 above). No attempt is made, for example, to show that every *refuted* law obtains a lower instance confirmation than any of those which have stood up to tests.

That this minimum requirement cannot be satisfied (even after repairing the inconsistency) may be shown with the help of Carnap's example, the law 'all swans are white'. This law ought to be considered as *falsified* if our evidence consists of a flock of *one black* and, say, 1000 white swans. But upon this evidence, the instance confirmation, instead of being zero, will be very near to 1. (The precise difference from 1

[69] *Probability*, p. 572, Cp. *Meaning and Necessity*, section 2, pp. 7 f.: 'The task of making more exact a vague or not quite exact concept . . . belongs to the most important tasks of logical analysis . . . We call this the task of . . . giving an *explication* for the earlier concept . . .' (See also *Probability*, section 2, p. 3.) I must say here (again only in passing) that I disagree with Carnap's views on explication. My main point is that I do not believe that one can speak about exactness, except in the relative sense of *exactness sufficient for a particular given purpose* – the purpose of solving a certain given problem. Accordingly, concepts cannot be 'explicated' as such, but only within the framework of a definite problem-situation. Or in other words, adequacy can only be judged if we are given a *genuine problem* (it must not in its turn be a problem of explication) for the solution of which the 'explication' or 'analysis' is undertaken.

will depend upon the choice of the parameter λ discussed below.) More generally, if a theory is again and again falsified, on the average, in every *n*th instance, then its (qualified) 'instance confirmation' approaches $1 - \frac{1}{n}$, instead of 0, as it ought to do, so that the law 'All tossed pennies always show heads' has the instance confirmation $\frac{1}{2}$ instead of 0.

In discussing in my *L.Sc.D.* a theory of Reichenbach's which leads to mathematically equivalent results,[70] I described this unintended consequence of his theory as 'devastating'. After 20 years, I still think it is.

(*b*) With his doctrine that laws may be dispensed with in science, Carnap in effect returns to a position very similar to the one he had held in the heyday of verificationism (viz. that the language of science is 'molecular') and which he had given up in the *Syntax* and in *Testability*. Wittgenstein and Schlick, finding that natural laws are non-verifiable, concluded from this that they are not genuine sentences (overlooking that they were thus committed to calling them 'meaningless pseudo-sentences'). Not unlike Mill they described them as rules for the derivation of genuine (singular) sentences—the *instances* of the law—from other genuine sentences (the initial conditions). I criticized this doctrine in my *L.Sc.D.*; and when Carnap accepted my criticism in the *Syntax* and in *Testability*[71] I thought that the doctrine was dead. But with Carnap's return to verificationism (in a weakened form), it has come to life again (in a weakened form: I do not think that the odds for its survival are good).

In one respect Carnap goes even further than Schlick. Schlick

[70] The confirmation values are identical if Carnap's λ (see below) is zero; and for any finite λ, the value of Carnap's instance confirmation approaches indefinitely, with accumulating evidence, the value criticized by me in my old discussion of Reichenbach's theory. I quote from my *L.Sc.D.*, section 80, p. 257, so far as it fits the present case: 'The probability of this hypothesis [I am speaking quite generally of universal laws] would then be determined by the truth frequency of the [singular] statements which correspond to it [i.e. which are its instances]. A hypothesis would thus have a probability of $\frac{1}{2}$ if, on the average, it is contradicted by every second statement of this sequence [i.e. by every second of its instances]! In order to escape from this devastating conclusion, one might still try two more expedients.' (One of these two leads to the zero probability of all universal laws: the passage is quoted in note 67, above.)

[71] See *L.Sc.D.*, notes 7 and 8 to section 4, and 1 to section 78; and *Testability*, note 20 to section 23, p. 19. See also notes 24 f. above.

believed that without laws we could not make predictions. Carnap however asserts that 'the use of laws is not indispensable for making predictions'.[72] And he continues: 'Nevertheless it is expedient, of course, to state universal laws in books on physics, biology, psychology, etc. Although these laws stated by scientists do not have a high degree of confirmation', he writes (but this is an understatement, since their degree of confirmation could not be lower), 'they have a high qualified instance confirmation . . .'

While reading through this section of my paper, Dr J. Agassi has found a simple (and I believe new) *paradox of inductive confirmation* which he has permitted me to report here.[72a] It makes use of what I propose to call an 'Agassi-predicate'—a factual predicate '$A(x)$' which is so chosen as to hold for all individuals (events, or perhaps things) occurring in the evidence at our disposal; but not for the majority of the others. For example, we may choose (at present) to define '$A(x)$' as 'x has occurred (or has been observed) before 1st January 1965'. (Another choice—'Berkeley's choice', as it were—would be '*x has been perceived*'.) Then it follows from Carnap's theory that, with growing evidence, the degree of confirmation of '$A(a)$' must become indistinguishable from 1 for any individual a in the world (present, past, or future). And the same holds for the (qualified or unqualified) instance confirmation of the universal law, '$(x)A(x)$'—a law stating that all events in the world (present, past, or future) occur before 1965; which makes 1965 an upper bound for the duration of the world. Clearly, the famous cosmological problem of the approximate period of the creation can be equally easily dealt with. Nevertheless, it would hardly be expedient to state universal laws like those of Agassi in books on cosmology—in spite of their high instance confirmation.

In the last pages of *Testability* Carnap discussed the sentence 'If all minds . . . should disappear from the universe, the stars would still go on in their courses'. Lewis and Schlick asserted, correctly, that this sentence was not verifiable; and Carnap replied, equally correctly (in

[72] *Probability*, p. 575.

[72a] (Added in proofs.) Professor Nelson Goodman, to whom I sent a stencilled copy of this paper, has kindly informed me that he has anticipated Dr Agassi in the discovery of this paradox and of what I have here called an 'Agassi predicate'. See Goodman's *Fact, Fiction, & Forecast*, 1955, pp. 74 f.

my opinion) that it was a perfectly legitimate scientific assertion, based as it was on well confirmed *universal laws*. But by now, *universal laws have become dispensable*; and without them the sentence in question cannot possibly be upheld. Moreover, one sees easily from Agassi's argument that a sentence that contradicts it can be maximally confirmed.

But I do not intend to use this one case—the status of natural laws—as my main argument in support of my contention that Carnap's analyses of confirmation, *and with it his criterion of demarcation, are inadequate*. I therefore now proceed to offer in support of this contention arguments which are completely independent of the case of natural laws, although they may allow us to see more clearly why this inadequacy was bound to arise in Carnap's theory.

As motto for my criticism I take the following challenging passage of Carnap's:[73]

> . . . if it could be shown that another method, for instance a new definition for degree of confirmation, leads in certain cases to numerical values more adequate than those furnished by c*, that would constitute an important criticism. Or, if someone . . . were to show that any adequate explicatum must fulfil a certain requirement and that c* does not fulfil it, it might be a helpful first step towards a better solution.

I shall take up both alternatives of this challenge but reverse their order: (1) I shall show that an adequate concept of confirmation cannot satisfy the traditional rules of the calculus of probability. (2) I shall give an alternative definition of degree of confirmation.

Ultimately, I shall show (3) that Carnap's theory of confirmation appears to involve (*a*) an infinite regress, and (*b*) an *a priori* theory of the mutual dependence of all atomic sentences with like predicates.

(1) To begin with, I suggest that we distinguish not only between *logical probability* (probability_1) and *relative frequency* (probability_2), as Carnap does, but between (at least) *three* different concepts—the third being *degree of confirmation*.

Surely, as a first suggestion this is unobjectionable: we could still decide, after due investigation, that *logical probability* can be used as the

[73] *Probability*, section 110, p. 563.

explicandum for *degree of confirmation*. Carnap, unfortunately, prejudges the issue. He assumes, without any further discussion, that his distinction between *two* probability concepts is sufficient, neglecting the warnings of my old book.[74]

It can be shown that confirmation, as Carnap himself understands this concept, cannot be logical probability. I offer three arguments.

(*a*) We can easily agree on the kind of thing we may both call, provisionally, 'probability'; for we both call 'probability' *something that satisfies the laws of the calculus of probability*.[75]

More specifically, Carnap says of the concept of logical probability$_1$ that it satisfies certain axiom systems, and in any case the (special) addition principle and (general) multiplication principle.[76] Now it is an elementary consequence of the latter that *the more a statement asserts, the less probable* it is. This may be expressed by saying that the logical probability of a sentence *x* on a given evidence *y* decreases when the informative content of *x* increases.[77]

[74] *L.Sc.D.*, before section 79; 'Instead of discussing the "probability" of a hypothesis we should try to assess . . . how far it has been corroborated [or confirmed].' Or section 82; 'This shows that it is not so much the number of the corroborating [confirming] instances which determines its degree of corroboration as *the severity of the various tests* to which the hypothesis in question . . . has been subjected. [This] in its turn depends upon the degree of testability . . . of the hypothesis . . .' And section 83: 'A theory can be the better corroborated [confirmed] the better it is testable. Testability, however, is converse to . . . logical probability . . .'

[75] In a note in *Mind*, **47**, 1938, p. 275 f., I said that it was 'desirable to construct a system of axioms' for probability, 'in such a way that it can be . . . interpreted by any of the different interpretations', of which 'the three most discussed are: (1) the classical definition of probability as the ratio of the favourable to the equally possible cases. (2) the frequency theory . . . (3) the logical theory, defining probability as the degree of a logical relation between sentences. . . .' (I took this classification from *L.Sc.D.*, section 48, reversing the order of (2) and (3). A similar classification can be found in *Probability*, p. 24. Contrast also the discussion of the *arguments* of the probability function in my *Mind* note with *Probability*, section 10, A & B, and section 52. In this note I gave an independent formal axiom system which, however, I have much simplified since. It was published in the *B.J.P.S.*, **6**, 1955, p. 53. (My *Mind* note has now been reprinted in *L.Sc.D.*, pp. 320–2.)

[76] *Probability*, section 53, p. 285; see also section 62, pp. 337 ff.

[77] This is equivalent to the 'content condition' (see note 63 above). Since Carnap considers this condition to be invalid (*Probability*, section 87, p. 474 'consequence condition'), he is, I believe, committed to agreeing that 'degree of confirmation' cannot be a 'regular confirmation function', i.e. a probability$_1$.

But this is sufficient to show that a high probability cannot be one of the aims of science. For the scientist is most interested in theories with a high content. He does not care for highly probable trivialities but for bold and severely testable (and severely tested) hypotheses. If (as Carnap tells us) a high degree of confirmation is one of the things we aim at in science, then degree of confirmation cannot be identified with probability.

This may sound paradoxical to some people. But if high probability were an aim of science, then scientists should say as little as possible, and preferably utter tautologies only. But their aim is to 'advance' science, that is to *add* to its content. Yet this means lowering its probability. And in view of the high content of universal laws, it is neither surprising to find that their probability is zero, nor that those philosophers who believe that science must aim at high probabilities cannot do justice to facts such as these: that the formulation (and testing) of *universal laws* is considered their most important aim by most scientists: or that the intersubjective testability of science depends upon these laws (as I pointed out in section 8 of my *L.Sc.D.*).

From what has been said it should be clear that an adequately defined 'degree of confirmation' cannot satisfy the general multiplication principle for probabilities.[78]

To sum up point (*a*). *Since we aim in science at a high content, we do not aim at a high probability*.

(*b*) The severity of possible tests of a statement or a theory depends (among other factors) on the precision of its assertions and upon its predictive power; in other words, upon its informative content (which increases with these two factors). This may be expressed by saying that *the degree of testability of a statement increases with its content*. But the better a statement can be tested, the better it can be confirmed, i.e. attested by its tests. Thus we find that the opportunities of confirming a statement,

[78] See sections 4–5 of my note 'Degree of Confirmation', *L.Sc.D.*, pp. 396–8. Dr Y. Bar-Hillel has drawn my attention to the fact that some of my examples were anticipated by Carnap in *Probability*, section 71, p. 394 f., case 3*b*. Carnap infers from them that the content condition (see notes 63 and 77 above) is 'invalid', but fails to infer that all 'regular confirmation functions' are inadequate.

and accordingly the degree of its confirmability or corroborability or attestability, increase with its testability, and with its content.[79]

To sum up point (*b*). *Since we want a high degree of confirmation (or corroboration), we need a high content (and thus a low absolute probability).*

(*c*) Those who identify confirmation with probability must believe that a high degree of probability is desirable. They implicitly accept the rule: 'Always choose the most probable hypothesis!'

Now it can be easily shown that this rule is equivalent to the following rule: 'Always choose the hypothesis which goes as little beyond the evidence as possible!' And this, in turn, can be shown to be equivalent, not only to 'Always accept the hypothesis with the lowest content (within the limits of your task, for example, your task of predicting)!', but also to 'Always choose the hypothesis which has the highest degree of *ad hoc* character (within the limits of your task)!' This is an unintended consequence of the fact that a highly probable hypothesis is one which fits the known facts, going as little as possible beyond them.

But it is well known that *ad hoc* hypotheses are disliked by scientists: they are, at best, stop-gaps, not real aims. (Scientists prefer a bold hypothesis because it can be more severely tested, and *independently* tested.)

To sum up point (*c*). *Aiming at high probability entails a counter-intuitive rule favouring ad hoc hypotheses.*

These three arguments exemplify my own point of view, for I see in a *confirming instance* the result of a severe test, or of an attempted (but unsuccessful) refutation of the theory. Those, on the other hand, who do not look for severe tests, but rather for 'confirmation' in the sense of the old idea of 'verification' (or a weakened version of it), come to a different idea of confirmability: a sentence will be the better confirmable the more nearly verifiable it is, or the more nearly deducible from observation sentences. It is clear, in this case, that universal laws are not (as in our analysis) highly confirmable, but that owing to their high content their confirmability will be zero.

(2) In taking up the challenge to construct a better definition of confirmation, I wish to say first that I do not believe that it is possible to

[79] For a fuller argument see *L.Sc.D.*, sections 82 f.

give a completely satisfactory definition. My reason is that a theory which has been tested with great ingenuity and with the sincere attempt to refute it will have a higher degree of confirmation than one which has been tested with laxity; and I do not think that we can completely formalize what we mean by an ingenious and sincere test.[80] Nor do I think that it is an important task to give an adequate definition of degree of confirmation. (In my view the importance, if any, of giving the best possible definition lies in the fact that such a definition shows clearly the inadequacy of all probability theories posing as theories of induction.) I have given what I consider a reasonably adequate definition elsewhere.[81] I may give here a slightly simpler definition (which satisfies the same *desiderata* or conditions of adequacy):

$$C(x,y) = \frac{p(y,x) - p(y)}{p(y,x) - p(x,y) + p(y)}$$

Here '$C(x,y)$' means 'the degree of confirmation of x by y', while '$p(x,y)$' and '$p(x)$' are relative and absolute probabilities, respectively. The definition can be relativized:

$$C(x,y,z) = \frac{p(y,x,z) - p(y,z)}{p(y,x,z) - p(x,y,z) + p(y,z)}$$

Here z should be taken as the general 'background knowledge' (the old evidence, and old and new initial conditions) including, if we wish, accepted theories, while y should be taken as representing those (new) observational results (excluded from z) which may be claimed to confirm the (new) explanatory hypothesis, x.[82]

[80] See the end of my note 'Degree of Confirmation' referred to in note 78 (*L.Sc.D.*, p. 402).

[81] 'Degree of Confirmation', *L.Sc.D.*, p. 395 f. Cp. my remark, p. 402: 'The particular way in which $C(x,y)$ is here defined I consider unimportant. What may be important are the *desiderata, and the fact that they can be satisfied together*.'

[82] That is to say, the total evidence e is to be partitioned into y and z; and y and z should be so chosen as to give $C(x,y,z)$ the highest value possible for x, on the available total evidence.

My definition satisfies, among other conditions of adequacy,[83] the condition that the *confirmability* of a statement—its highest possible degree of confirmation—equals its content (i.e. the degree of its testability).

Another important property of this concept is that it satisfies the condition that the severity of a test (measured by the improbability of the test-instance) has an almost-additive influence upon the resultant degree of confirmation of the theory. This shows that some at least of the intuitive demands are satisfied.

My definition does not automatically exclude *ad hoc* hypotheses, but it can be shown to give most reasonable results if combined with a rule excluding *ad hoc* hypotheses.[84]

So much about my own present positive theory (which goes very considerably beyond my *L.Sc.D.*). But I must return to my critical task: I believe that my positive theory strongly suggests that the fault lies with the verificationist and inductivist approach which—in spite of the attention paid to my criticism—has never been completely abandoned by Carnap. *But inductive logic is impossible.* I shall try to show this (following my old *L.Sc.D.*) as my last critical point.

(3) I asserted, in my *L.Sc.D.*, that an inductive logic must involve either (*a*) an infinite regress (discovered by Hume), or (*b*) the acceptance (with Kant) of some synthetic principle as valid *a priori*. I have a strong suspicion that Carnap's theory of induction can be criticized as involving both (*a*) and (*b*).

(*a*) If we need, in order to justify induction as probable, a (probable) *principle of induction*, such as a *principle of the uniformity of nature*, then we

[83] Called '*desiderata*' in the note in question. Kemeny has rightly stressed that the conditions of adequacy should not be introduced to fit the explicatum. That this is not the case here is perhaps best proved by the fact that I have now improved my definition (by simplifying it) without changing my *desiderata*.

[84] The rule for the exclusion of *ad hoc* hypotheses may take the following form: the hypothesis *must not repeat* (except in a completely generalized form) the evidence, or any conjunctive component of it. That is to say x = 'This swan is white', is not acceptable as a hypothesis to explain the evidence y = 'This swan is white' although 'All swans are white' would be acceptable; and no explanation x of y must be circular in this sense with respect to any (non-redundant) conjunctive component of y. This leads to an emphasis upon *universal laws as indispensable*, while Carnap believes, as we have seen (see above, and *Probability*, section 110, H. esp. p. 575) that universal laws can be dispensed with.

also need a second such principle in order to justify the induction of the first. Carnap, in his section on the 'Presuppositions of Induction'[85] introduces a principle of uniformity. He does not mention the objection of a regress, but a remark in his exposition seems to indicate that he has it in mind: 'The opponents', he writes (p. 181), 'would perhaps say that the statement of the probability of uniformity must be taken as a factual statement. . . . Our reply is: . . . this statement is itself analytic.' I was far from convinced by Carnap's arguments; but since he indicates that 'the whole problem of the justification and the presupposition of inductive method' will be treated in a later volume 'in more exact, technical terms', it is perhaps better to suppress, at this stage, my inclination to offer a proof that no such principle of uniformity can be analytic (except in a Pickwickian sense of 'analytic'); especially since my discussion of point (*b*) will perhaps indicate the lines on which a proof of this kind might proceed.

(*b*) Natural laws, or more generally, scientific theories, whether of a causal or a statistical character, are hypotheses about some *dependence*. They assert, roughly speaking, that certain events (or statements describing them) are *in fact not independent* of others, although so far as their purely logical relations go they are independent. Let us take two possible facts which are, we first assume, completely unconnected (say 'Chunky is clever' and 'Sandy is clever'), described by the two statements x and y. Then somebody may conjecture—perhaps mistakenly—that they are connected (that Chunky is a relation of Sandy's); and that the information or evidence y increases the probability of x. If he is wrong, that is, if x and y are independent, then we have

(1) $$p(x,y) = p(x)$$

which is equivalent to

(2) $$p(x,y) = p(x)p(y)$$

This is the usual definition of independence.

[85] *Probability*, section 41, F., pp. 177 ff., especially pp. 179, 181. For the passages from *L.Sc.D.*, see section 1, pp. 28 f., and 81, pp. 263 f.

If the conjecture that the events are connected or inter-dependent is correct, then we have

$$(3) \qquad p(x,y) > p(x)$$

that is, the information y raises the probability of x above its 'absolute' or 'initial' value $p(x)$.

I believe—as I think most empiricists do—that any such conjecture about the inter-dependence or correlation of events should be formulated as a separate hypothesis, or as a natural law ('Cleverness runs in families') to be submitted first to a process of careful formulation, with the aim of making it as highly testable as possible, and after that to severe empirical tests.

Carnap is of a different opinion. He proposes that we accept (as probable) a principle to the effect that the evidence 'Sandy is clever' increases the probability of '*A* is clever' for any individual *A*—whether '*A*' is the name of a cat, a dog, an apple, a tennis ball, or a cathedral. This is a consequence of the definition of 'degree of confirmation' which he proposes. According to this definition, any two sentences with the same predicate ('clever' or 'sick') and different subjects are inter-dependent or positively correlated, whatever the subjects may be, and wherever they may be situated in the world; this is the actual content of his principle of uniformity.

I am far from certain whether he has realized these consequences of his theory, for he nowhere mentions them explicitly. But he introduces a universal parameter which he calls λ; and $\lambda + 1$ turns out, on a simple mathematical calculation, to be the reciprocal of the 'logical correlation coefficient'[86] for any two sentences with the same predicate and different subjects.[87] (The assumption that λ is infinite corresponds to the assumption of independence.)

[86] The 'logical correlation coefficient' of x and y can be defined as $(p(xy) - p(x)p(y))/(p(x)p(y)p(\bar{x})p(\bar{y}))^{\frac{1}{2}}$. Admitting this formula for all ('regular') probability functions means a slight generalization of a suggestion which is made in Kemeny and Oppenheim, 'Degree of Factual Support', *Philos. of Sci.*, **19**, p. 314, formula (7), for a special probability function in which all atomic sentences are (absolutely) independent. (It so happens that I think that this special function is the only one which is adequate.)

[87] We can prove this for example, by taking *Methods*, p. 30, formula (9–8), putting $s = S_M =$

According to Carnap, we are bound to choose a finite value of λ when we wish to choose our *definition* of the probability$_1$ function. The choice of λ, and with it of the degree of correlation between any two sentences with the same predicate, thus appears to be part of a 'decision' or 'convention': the choice of a definition of probability. It looks, therefore, as if no statement about the world was involved in the choice of λ. But it is a fact that our choice of λ is equivalent to the most sweeping assertion of dependence that one can imagine. It is equivalent to the acceptance of as many natural laws as there are predicates, each asserting the same degree of dependence of any two events with like predicates in the world. And since such an assumption about the world is made in the form of a non-testable act—the introduction of a definition—there seems to me an element of *apriorism* involved.

One might still say, perhaps, that there is no *apriorism* here since the dependencies mentioned are a consequence of a definition (that of probability or degree of confirmation), which rests on a convention or a 'decision', and is therefore *analytic*. But Carnap gives two reasons for his choice of his confirmation function which do not seem to fit this view. The first of the two reasons I have in mind is that his confirmation function, as he remarks, is the only one (among those which suggest themselves) 'which is not entirely inadequate';[88] inadequate, that is, for explaining (or 'explicating') the undoubted *fact that we can learn from experience*. Now this *fact* is empirical; and a theory whose adequacy is judged by its ability to explain or cohere with this fact does not quite look like being analytic. It is interesting to see that Carnap's argument in favour of his choice of λ (which I am suspecting of *apriorism*) is the same as Kant's or Russell's, or Jeffreys's; it is what Kant called a 'transcendental' argument ('How is knowledge possible?'), the appeal to the fact that we possess empirical knowledge, i.e. that we can learn

1; $w/\kappa = c(x) = c(\bar{x}) = c(y)$; and replacing '$c(h_M,e_M)$' by '$c(x,y)$'. We obtain $\lambda = c(\bar{x}y)/(c(xy) - c(x)c(y))$, which shows that λ is the reciprocal of a dependence-measure, and from this $1/(\lambda + 1) = (c(xy) - c(x)c(y))/c(\bar{x})c(y)$, which, as $c(x) = c(\bar{x}) = c(y)$, is the logical correlation coefficient.—I may perhaps say here that I prefer the term 'dependence' to Keynes' and Carnap's term 'relevance': looking (like Carnap) at probability as a generalized deductive logic, I take probabilistic dependence as a generalization of logical dependence.

[88] *Probability*, section 110, p. 565; cp. *Methods*, section 18, p. 53.

from experience. The second of the two reasons is Carnap's argument that the adoption of an appropriate λ (one which is neither infinite, for an infinite λ is equivalent to independence, nor zero) would be more successful in nearly all universes (except in the two extreme cases in which all individuals are independent or have like properties). All these reasons seem to me to suggest that the choice of λ, i.e. of a confirmation function, is to depend upon its success, or upon the probability of its success, in the world. But then it would not be analytic—in spite of the fact that it is also a 'decision' concerning the adoption of a definition. I think that it can be explained how this may be so. We can, if we like, define the word 'truth' so that it comprises some of those statements we usually call 'false'. Similarly we can define 'probable' or 'confirmed' so that absurd statements get a 'high probability'. All this is purely conventional or verbal, as long as we do not take these definitions as 'adequate explications'. But if we do, then the question is no longer conventional, or analytic. For to say of a contingent or factual statement *x* that it is true, *in an adequate sense of the word 'true'*, is to make a factual statement; and so it is with '*x* is (now) highly probable'. It is the same with '*x* is strongly dependent upon *y*' and '*x* is independent of *y*'—the statements whose fate is decided upon when we choose λ. The choice of λ is therefore indeed equivalent to that of adopting a sweeping though unformulated statement about the general interdependence or uniformity of the world.

But this statement is adopted without any empirical evidence. Indeed, Carnap shows[89] that without adopting it we can never learn from empirical evidence (according to his theory of knowledge). Thus empirical evidence does not and cannot count *before* the adoption of a finite λ. This is why it has to be adopted *a priori*.

'The principle of empiricism', Carnap writes in another context,[90] 'can be violated only by the assertion of a factual (synthetic) sentence without a sufficient empirical foundation, or by the thesis of *apriorism* when it contends that for knowledge with respect to certain factual sentences no empirical foundation is required.' I believe that what we have observed here shows that there is a third way of violating the

[89] *Probability*, section 110, p. 556.

[90] *Probability*, section 10, p. 31.

principle of empiricism. We have seen how it can be violated by constructing a theory of knowledge which cannot do without a principle of induction—a principle that tells us in effect that the world is (or very probably is) a place in which men can learn from experience; and that it will remain (or very probably remain) so in future. I do not believe that a cosmological principle of this kind can be a principle of pure logic. But it is introduced in such a way that it cannot be based upon experience either. It therefore seems to me that it cannot be anything else but a principle of *a priori* metaphysics.

Nothing but the synthetic, the factual, character of λ seems to be able to explain Carnap's suggestion that we may try out which value of λ is most successful in a given world. But since empirical evidence does not count without the prior adoption of a finite λ, there can be no clear procedure for testing the λ chosen by the method of trial and error. My own feeling is that I prefer in any case to apply the method of trial and error to the *universal laws* which are indispensable for intersubjective science; which are clearly, and admittedly, factual; and which we may succeed in making severely testable, with the aim of eliminating all those theories that can be discovered to be erroneous.

I am glad to have been given an opportunity to get these matters off my mind—or off my chest, as physicalists might say. I do not doubt that, with another vacation in the Tyrol, and another climb up the *Semantische Schnuppe*, Carnap and I could reach agreement on most of these points; for we both, I trust, belong to the fraternity of rationalists—the fraternity of those who are eager to argue, and to learn from one another. But since the physical gap between us seems unbridgeable I now send to him across the ocean—knowing that soon I shall be at the receiving end—these my best barbed arrows, with my best brotherly regards.

12

LANGUAGE AND THE BODY-MIND PROBLEM

A Restatement of Interactionism

1. INTRODUCTION

This is a paper on the impossibility of a physicalistic causal theory of the human language.[1]

1.1 It is *not* a paper on linguistic analysis (the analysis of word-usages). For I completely reject the claim of certain language analysts that the source of philosophical difficulties is to be found in the misuse of language. No doubt some people talk nonsense, but I claim (*a*) that there does not exist a logical or language-analytical method of detecting philosophical nonsense (which, by the way, does not stop short of the ranks of logicians, language analysts and semanticists); (*b*) that the belief that such a method exists—the belief more especially that philosophical nonsense can be unmasked as due to what Russell might have called 'type-mistakes' and what nowadays are sometimes called

First published in the Proceedings of the 11th International Congress of Philosophy, **7**, 1953.

[1] This issue was first discussed by Karl Bühler in his *Sprachtheorie*, 1934, pp. 25–28.

'category-mistakes'—is the aftermath of a philosophy of language which has since turned out to be baseless.

1.2 It is the result of Russell's early belief that a formula like '*x* is an element of *x*' is (essentially or intrinsically) meaningless. We now know that this is not so. Although we can, indeed, construct a formalism F_1 ('theory of types') in which the formula in question is 'not well-formed' or 'meaningless', we can construct another formalism (a type-free formalism) F_2, in which the formula is 'well-formed' or 'meaningful'. The fact that a doubtful expression cannot be translated into a meaningful expression of a given F_1 does not therefore establish that there exists no F_2 such that the doubtful formula in question can be translated into a meaningful statement of F_2. In other words, we are never able to say, in doubtful cases, that a certain formula, as used by some speaker, is 'meaningless' in any precise sense of this term; for somebody may invent a formalism such that the formula in question can be rendered by a well-formed formula of that formalism, to the satisfaction of the original speaker. The most one can say is, 'I do not see how such a formalism can be constructed'.

1.3 As for the body-mind problem, I wish to reject the following two different theses of the language analyst. (1) The problem can be solved by pointing out that there are two languages, a physical and a psychological language, but not two kinds of entities, bodies and minds. (2) The problem is due to a faulty way of talking about minds, i.e. it is due to talking as if mental states exist *in addition to* behaviour, while all that exists is behaviour of varying character, for example, intelligent and unintelligent behaviour.

1.31 I assert that (1), the two-language solution, is no longer tenable. It arose out of 'neutral monism', the view that physics and psychology are two ways of constructing theories, or languages, out of some neutral 'given' material, and that the statements of physics and of psychology are (abbreviated) statements about that given material, and therefore *translatable* into one another; that they are two ways of talking about the same facts. But the idea of a mutual translatability had to be given up long ago. With it, the two-language solution disappears. For if the two languages are not inter-translatable, then they deal with different kinds of facts. The relation between these kinds of facts constitutes our problem, which can therefore only be formulated by constructing *one* language in which we can speak about *both* kinds of facts.

1.32 Since (2) is so vague, we must ask: Is there, or is there not, the station-master's belief that the train is leaving, *in addition to* his belief-like behaviour? Is there his intention to communicate a fact about the train to the signalman, *in addition to* his making the appropriate movements? Is there the signalman's understanding of the message *in addition to* his understanding-like behaviour? Is it possible that the signalman understood the message perfectly well but behaved (for some reason or other) as if he had misunderstood it?

1.321 If (as I think) the answer to these questions is 'yes', then the body-mind problem arises in an approximately Cartesian form. If the answer 'no' is upheld, we are faced with a philosophical theory that may be called 'physicalism' or 'behaviourism'. If the questions are not answered but dismissed as 'meaningless'; if, more especially, we are told that to ask whether Peter has a toothache *in addition to* his toothache-like behaviour is meaningless because all that can be known about his toothache is known through observing his behaviour, then we are faced with the positivist's mistaken belief that a fact is (or is reducible to) the sum total of the evidence in its favour—i.e. with the verifiability dogma of meaning. (Cf. 4.3, below, and my *Logic of Scientific Discovery*, 1959.)

1.4 An important assumption of what follows here is that the *deterministic interpretation of physics, even of classical physics, is a misinterpretation*, and that there are no 'scientific' reasons in favour of determinism. (Cf. my paper 'Indeterminism in Quantum Physics and in Classical Physics', *Brit. Journ. Philos. of Science*, 7, 1950.)

2. FOUR MAJOR FUNCTIONS OF LANGUAGE

2. Karl Bühler appears to have been the first to propose, in 1918,[2] the doctrine of the three functions of language: (1) the expressive or symptomatic function; (2) the stimulative or signal function; (3) the descriptive function. To these I have added (4) the argumentative function, which can be distinguished[3] from function (3). It is not asserted

[2] Referred to in his *Sprachtheorie*, loc. cit.

[3] Cf. ch. 4, above, esp. p. 181.

that there are no other functions (such as prescriptive, advisory, etc.) but it is asserted that these four functions mentioned constitute a hierarchy, in the sense that each of the higher ones *cannot* be present without all those which are lower, while the lower ones *may* be present without the higher ones.

2.1 An argument, for example, serves as an expression in so far as it is an outward symptom of some internal state (whether physical or psychological is here irrelevant) of the organism. It is also a signal, since it may provoke a reply, or agreement. In so far as it is *about* something, and supports a view of some *situation or state of affairs*, it is descriptive. And lastly, there is its argumentative function, its giving *reasons* for holding this view, e.g. by pointing out difficulties or even inconsistencies in an alternative view.

3. A GROUP OF THESES

3.1 The primary interest of science and philosophy lies in their descriptive and argumentative functions; our interest in behaviourism or physicalism, for example, depends on the cogency of their critical arguments.

3.2 Whether a person does in fact describe or argue, or whether he merely expresses or signals, depends on whether he speaks intentionally *about* something, or intentionally supports (or attacks) some view.

3.3 The linguistic *behaviour* of two persons (or of the same person at two different dates) may be indistinguishable; yet the one may, in fact, describe or argue, while the other may only express (and stimulate).

3.4 *Any causal physicalistic theory of linguistic behaviour can only be a theory of the two lower functions of language.*

3.5 *Any such theory is therefore bound either to ignore the difference between the higher and lower functions, or to assert that the two higher functions are 'nothing but' special cases of the two lower functions.*

3.6 This holds, more especially, for such philosophies as behaviourism, and the philosophies which try to rescue the causal completeness or self-sufficiency of the physical world, such as epiphenomenalism, psycho-physical parallelism, the two-language solutions, physicalism, and materialism. (All these are self-defeating in so far as their arguments establish—unintentionally, of course—the non-existence of arguments.)

4. THE MACHINE ARGUMENT

4.1 A wall-thermometer may be said not only to express its internal state, but also to signal, and even to describe. (A self-registering one does so even in writing.) Yet we do not attribute the responsibility for the description to it; we attribute it to its maker. Once we understand this situation, we see that it does not describe, any more than my pen does: like my pen it is only an instrument for describing. But it expresses its own state; and it signals.

4.2 The situation outlined in 4.1 is fundamentally the same for all physical machines, however complicated.

4.21 It may be objected that example 4.1 is too simple, and that by complicating the machine and the situation we may obtain true descriptive behaviour. Let us therefore consider more complex machines. By way of concession to my opponents, I shall even assume that machines can be constructed to *any behaviouristic specification*.

4.22 Consider a machine (invested with a lens, an analyser, and a speaking apparatus) which pronounces, whenever a physical body of medium size appears before its lens, the name of this body ('cat'; 'dog', etc.) or says, in some cases, 'I don't know'. Its behaviour can be made even more human-like (1) by making it do this not always, but only in response to a stimulus question, 'Can you tell me what this thing is?', etc.; (2) by making it in a percentage of cases reply, 'I am getting tired, let me alone for a while', etc. Other responses can be introduced, and varied—perhaps according to inbuilt probabilities.

4.23 If the behaviour of such a machine becomes very much like that of a man, then we may mistakenly believe that the machine describes and argues; just as a man who does not know the working of a radio receiver may mistakenly think that the receiver describes and argues. Yet an analysis of its mechanism teaches us that nothing of this kind happens. The radio does not argue, although it expresses its physical states, and signals.

4.24 There is, in principle, no difference between a wall-thermometer and the 'observing' and 'describing' machine discussed.

Even a man who is conditioned to react to appropriate stimuli with the sounds 'cat' and 'dog', *without intention* to describe or to name, does not describe, although he expresses and signals.

4.25 But let us assume that we find a physical machine whose mechanism we do not understand and whose behaviour is very human. We may then wonder whether it does not, perhaps, act intentionally, rather than mechanically (causally, or probabilistically), i.e. whether it does not have a mind after all; whether we should not be very careful to avoid causing it pain, etc. But once we realize completely how it is constructed, how it can be copied, who is responsible for its design, etc., no degree of complexity will make it different in kind from an automatic pilot, or a watch, or a wall-thermometer.

4.3 Objections to this view, and to the view 3.3, are usually based on the positivistic doctrine of the identity of empirically indistinguishable objects.

Two clocks, the argument goes, may look alike, although the one works mechanically and the other electrically, but their difference *can* be discovered by observation. If no difference can be so discovered, then there simply is none. Reply: if we find two pound notes which are physically indistinguishable (even as to the number) we may have good reason to believe that *one* of them at least is forged; and a forged note does not become genuine because the forgery is perfect or because all historical traces of the act of forgery have disappeared.

4.4 Once we understand the causal behaviour of the machine, we realize that its behaviour is purely expressive or symptomatic. For amusement we may continue to ask the machine questions, but we shall not seriously argue with it—unless we believe that it transmits the arguments, both from a person and back to a person.

4.5 This, I think, solves the so-called problem of 'other minds'. If we talk to other people, and especially if we argue with them, then we *assume* (sometimes mistakenly) that they also argue: that they speak *intentionally* about things, seriously wishing to solve a problem, and not merely behaving as if they were doing so. It has often been seen that language is a social affair and that solipsism, and doubts about the existence of other minds, become self-contradictory if formulated in a language. We can put this now more clearly. In arguing with other people (a thing which we have learnt from other people), for example about other minds, we cannot but attribute to them intentions, and this means, mental states. We do not argue with a thermometer.

5. THE CAUSAL THEORY OF NAMING

5.1 But there are stronger reasons. Consider a machine which, every time it sees a ginger cat, says 'Mike'. It represents, we may be tempted to say, a *causal model* of naming, or of the name-relation.

5.2 But this causal model is deficient. We shall express this by saying that it is not (and cannot be) a *causal realization* of the name-relation. Our thesis is that a causal realization of the name-relation cannot exist.

5.21 We admit that the machine may be described as realizing what we may loosely call a 'causal chain'[4] of events joining Mike (the cat) with 'Mike' (its name). But there are reasons why we cannot accept this causal chain as a representation or realization of the relation between a thing and its name.

5.3 It is naïve to look at this chain of events as beginning with the appearance of Mike and ending with the enunciation 'Mike'.

It 'begins' (if at all) with a state of the machine prior to the appearance of Mike, a state in which the machine is, as it were, ready to respond to the appearance of Mike. It 'ends' (if at all) not with the enunciation of a word, since there is a state following this. (All this is true of the corresponding human response, if causally considered.) It is *our interpretation* which makes Mike and 'Mike' the extremes (or terms) of the causal chain, and not the 'objective' physical situation. (Moreover, we might consider *the whole process of reaction* as name, or only the last letters of 'Mike', say, 'Ike'.) Thus, although those who know or understand the name-relation may choose to interpret a causal chain as a model of it, it is clear that the name-relation is not a causal relation, and cannot be realized by any causal model. (The same holds for all 'abstract', e.g. logical relations, even for the simplest one-one relation.)

5.4 The name-relation is therefore clearly not to be realized by, say, an association model, or a conditioned reflex model, of whatever complexity. It involves some kind of *knowledge that* 'Mike' is (by some

[4] It does not matter for our present purposes whether or not the expression 'causal chain' is adequate for a more thorough analysis of causal relations.

convention) the name of the cat Mike, and some kind of *intention* to use it as a name.

5.5 Naming is by far the simplest case of a descriptive use of words. Since no causal realization of the name-relation is possible, *no causal physical theory of the descriptive and argumentative functions of language is possible.*

6. INTERACTION

6.1 It is true that the presence of Mike in my environment may be one of the physical 'causes' of my saying, 'Here is Mike'. But if I say, 'Should this be your argument, then it is contradictory', because I have grasped or realized that it is so, then there was no physical 'cause' analogous to Mike; I do not need to hear or see your words in order to realize that a certain theory (it does not matter whose) is contradictory. The analogy is not to Mike, but rather to *my realization that* Mike is here. (This realization of mine may be causally, but not purely physically, connected with the physical presence of Mike.)

6.2 Logical relationships, such as consistency, do not belong to the physical world. They are abstractions (perhaps 'products of the mind'). But my realization of an inconsistency may lead me to act, in the physical world, precisely as may my realization of the presence of Mike. Our mind may be said to be as capable of being swayed by logical (or mathematical, or, say, musical) relationships as by a physical presence.

6.3 There is no reason (except a mistaken physical determinism) why mental states and physical states should not interact. (The old argument that things so different could not interact was based on a theory of causation which has long been superseded.)

6.4 If we act through being influenced by the grasp of an abstract relationship, we initiate physical causal chains which have no sufficient *physical* causal antecedents. We are then 'first movers', or creators of a physical 'causal chain'.

7. CONCLUSION

The fear of obscurantism (or of being judged an obscurantist) has prevented most anti-obscurantists from saying such things as these. But this fear has produced, in the end, only obscurantism of another kind.

13

A NOTE ON THE BODY-MIND PROBLEM

I am very grateful to Professor Wilfrid Sellars for bringing[1] my paper 'Language and the Body-Mind Problem',[2] to the attention of philosophers, and even more for his kindness in describing it as 'challenging', and as 'telling, if uneven'. Of its unevenness nobody can be more aware than I. I think I am more sensitive to it than Andersen's princess was to the pea. And although I am inclined to count its three leaves among my scanty laurels, I could not rest on them even if I wished to. But the small hard peas which bother me and keep me awake at night seem to have been well hidden, and in a spot far removed from Professor Sellars' two largish lumps of stuffing which I believe are not at all hard to smooth out.

I

As to the first lump, Professor Sellars, after quoting me correctly at some length, proceeds to 'focus attention', as he puts it, 'On the

First published in Analysis, N.S., **15**, 1955, *as a reply to Professor Wilfrid Sellars.*

[1] By way of his 'A Note on Popper's argument for Dualism', *Analysis*, **15**, pp. 23 f.

[2] Not 'Mind-body problem' as Professor Sellars writes. My paper is included in this volume as ch. 12.

statement [Popper's statement] quoted above, that ". . . if the two languages are not translatable, they deal with different sets of facts".' And Professor Sellars then goes on to say that a 'fact' may be either a '*descriptive fact*' or else something like 'the "fact" that we ought to fulfil our undertakings', which I may be permitted to call a '*quasi fact*'. And he says that my argument would be valid if only it would contain 'the premise that both languages in question *have the business of describing*', i.e. of stating '*descriptive facts*'.

Now I agree with every word of this but I completely fail to see its relevance: in focusing attention upon *one* statement, Professor Sellars, understandably enough, got its context out of focus.

For (*a*), the premise which, according to Professor Sellars, makes my argument valid, was clearly enough indicated in my own argument which therefore is itself valid, according to Professor Sellars. Moreover, my argument has the form of a *reductio ad absurdum* of the 'two language theory', and the premise correctly demanded by Professor Sellars is not mine but part of that theory. It is, indeed, referred to in my argument as part of the 'two language solution'—of 'the view that . . . the statements of physics and of psychology are . . . two ways of talking about the same facts' (which clearly indicates that these 'facts' are 'descriptive facts' in Professor Sellars' terminology). (*b*) My own contribution consisted, simply, in pointing out that, once the two languages (of physics and of psychology) are admitted not to be translatable into each other, they cannot any longer be said to talk about the same facts, and must be admitted to talk about different facts—where 'facts' means whatever the two-language theorists meant when they said that physics and psychology talked about the same facts.

Thus the problem of 'quasi facts' simply does not arise.

All this can be verified by reading more closely the passage from my paper which Professor Sellars himself quotes at the beginning of his paper: it is the passage which gets out of focus once he focuses attention on part of it. (There is a not very important misquotation—'set' instead of 'kind'—in the focused passage.)

So no hard core, no difference of opinion as far as I can see, underlies Professor Sellars' first lump—although we seem to differ greatly about the relevance of his comments.

II

Now to smooth out the second lump. 'In the later sections of his paper', Professor Sellars writes, 'Professor Popper makes a telling, if uneven, defence of the thesis that *aboutness* or *reference* cannot be defined in Behaviourese.' (Professor Sellars himself believes in the truth of this alleged thesis of mine.) I must confess that I was surprised when I read this. I was not aware of having ever tried to defend anything of the kind. It happens to be one of my oldest convictions that a thesis of the kind here attributed to me—that such and such *cannot be defined* in somebody's language—is *nearly always irrelevant*. (It is not irrelevant, of course, if the opponent's thesis was one about definability. Definability may be interesting in certain contexts, but to say a term is not definable never implies that it cannot be legitimately used; for it may be legitimately used as an undefined term.) There was no need for me to read through my paper in order to be sure that I never maintained anything like the 'thesis' attributed to me by Professor Sellars. But to make doubly sure I did read through my paper and I found no trace of such a thesis on definability. And to make trebly sure, I herewith publicly recant any theory I may ever have advanced based upon the thesis attributed to me by Professor Sellars: not because the thesis is false (I agree with Professor Sellars that it is true, and I even agree that my arguments might be used to support its truth—which may perhaps explain the misunderstanding) but because I should hate the idea of philosophizing with the help of arguments about non-definability.

Professor Sellars goes on to say 'And he [Popper] is surely right [in holding the thesis I have just repudiated]. However, at this stage he [Popper] tacitly adds the premise "'E is about *x*' is a descriptive assertion".'

It is hard for me to check whether or not I have added this premise tacitly at this stage, since 'this stage' is not indicated by Professor Sellars—or only indicated with the help of a reference to that alleged thesis of mine which I fail to find anywhere in my paper. (I may here warn readers that seven of the passages in quotation marks in this second part of Professor Sellars' paper are not quotations from my paper, as some might think. Two others, 'Name relation' and

'Causal-physicalistic', did occur in my paper, but the former hyphened, the latter unhyphened.)

If, however, I have somewhere 'tacitly' and unconsciously added the premise which Professor Sellars says I have added (I cannot detect any trace of it) then I wish, again, to recant. For I am in complete agreement with Professor Sellars' thesis that if a statement *A* says that another statement E is about something, then *A* usually does not, to use Professor Sellars' words play 'the same sort of a role as "The Moon is round"'. *A* need not be, and usually is not, 'descriptive' in the same sense as the statement about the moon (although it may be: 'What was your last lecture about?'—'It was a lecture about probability', is an instance of descriptive usage).

I also agree entirely with Professor Sellars' concluding remark that 'from the fact, and it is a fact, that what Professor Popper calls the "name relation" (paragraph 5 ff.) is not definable in "causal-physicalistic" terms, we cannot conclude to the truth of Dualism'. Exactly. This is why I never said anything about definability. Indeed, had I no stronger arguments in favour of my dualistic faith than this completely irrelevant fact (for I agree that it is a fact, though completely irrelevant), then I should be ready—nay, most anxious—to give up dualism. As it happens, my arguments were quite different. They were about[3] the possible scope of deductive physical theories rather than about definability; and *my thesis* was that '*no causal physical theory of the descriptive and argumentative functions of language is possible*'.

I wish to make it perfectly clear that I have no objection whatever to Professor Sellars' thesis—that a statement such as 'E is about *x*' is (ordinarily, or frequently) 'a device whereby we *convey* to the hearer how a *mentioned* expression is used, by *using* an equivalent expression'. Nor do I deny that this thesis of Professor Sellars' is relevant to my own thesis. All I wish to say here is that my thesis is not based on the argument about definability which Professor Sellars ascribes to me. If it were, I should retract it.

[3] This is another instance of an about-statement *A* which describes an argument E.

III

There is a remark on Professor Ryle's views in Professor Sellars's paper which seems wrong to me. Professor Sellars writes: 'I also agree that "the idea of a mutual translatability" of . . . mind talk and behaviour talk "had to be given up long since", in spite of Ryle's valiant efforts to the contrary.'

To this I should like to say that I am not aware of the fact that Professor Ryle has ever held what I call 'the two languages theory'. How could he, believing as he does that the problem arises out of category-mistakes within the one natural language? It is not to him I was alluding in that place.

At the same time, it is perfectly true that I had Professor Ryle in mind when, in another paragraph of my paper, I tried to show briefly that the theory of 'category-mistakes' is also untenable.

If I might here add to my arguments another, then I should say this. Assuming that, by the usages of our language, expressions naming physical states are put in a category different from that in which expressions naming mental states are put, I should be inclined to see in this fact an indication, or a suggestion (not more than this, to be sure), that these two categories of expression name entities which are *ontologically* different—or in other words, that they are *different kinds of entities*. Thus I should be inclined (not more than this) to entertain the opposite conclusion to the one drawn by Professor Ryle although, admittedly, the premises would be insufficient for a formal derivation of the conclusion.

However, I am not prepared to grant the truth of this assumption, quite apart from my (and from Professor Smart's[4]) objections to arguments based upon the idea of category-mistakes. I find very many of Professor Ryle's analyses most illuminating, but I can only say that ordinary English very often treats mental states and physical states on a par with each other; not only where it speaks of a 'mental disease', of a 'hospital for the mentally sick' or of a man who is 'both physically and mentally well balanced', etc. (these cases might be dismissed as deriving from a philosophical dualism) but especially where we say:

[4] See his excellent brief 'A Note on Categories' in the *British Journal for the Philosophy of Science*, **4**, 1953, pp. 227 f.

'Thinking of sheep always helps me to fall asleep' or 'Reading Mr Smith's novels always helps me to fall asleep' (which does not mean 'training my eyes on one of Mr Smith's novels always helps me to fall asleep' and yet is completely analogous to 'taking bromide always helps me to fall asleep'). There are countless similar examples. They certainly do not establish that ordinary English words describing mental states and physical states *always* belong to the same 'category' (Professor Ryle has succeeded in showing that they don't). But my examples establish, I think, that the words are often used in ways which are strikingly alike. The uncertainty of the language-situation may be illustrated by an example of Professor Ryle's.[5] He says, rightly, that a child who has just watched the parade of all the battalions, batteries, and squadrons, which constitute a division, makes a mistake (in the sense that he has not quite got the meaning of the words) when he then asks 'And when will the division come?'—'He would', Professor Ryle says, 'be shown his mistake by being told that in watching the battalions, batteries and squadrons marching past he had been watching the division marching past. The march-past was not a parade of battalions . . . *and* a division; it was a parade of the battalions . . . *of* a division.' This is absolutely true. But are there no contexts, of perfectly good English usage, in which battalions are treated on a par with divisions? Could there not be a parade of, say, one division *and* three battalions *and* two batteries? I can imagine that this might be an outrage to military usage (although a battle in which a division attacks a battalion is, I suppose, perfectly good military usage). But is it an outrage to ordinary English usage? And if not, can the mistake which the child undoubtedly committed be a category mistake? And if not, do we not commit a category mistake (assuming such a thing exists) if we wrongly diagnose that the child's mistake was a category mistake?

[5] *The Concept of Mind*, pp. 16 f. The example of the Colleges and the University is precisely analogous: the foreigner who wants to see the University asks, of course, for a University *building* (perhaps one like the Senate House in London); and this *building* would be of the same category as the college *buildings*. Is it not therefore a category mistake to suggest that he has made a category mistake?

14

SELF-REFERENCE AND MEANING IN ORDINARY LANGUAGE

Theaetetus. Now listen to me attentively, Socrates, for what I shall put before you is not a little tricky.

Socrates. I promise to do my best, Theaetetus, so long as you spare me the details of your achievements in the theory of numbers, and speak in a language which I, an ordinary man, can understand.

Th. The very next question which I am going to ask you is an extraordinary one, although expressed in perfectly ordinary language.

S. There is no need to warn me: I am all ears.

Th. What did I say between your last two interruptions, Socrates?

S. You said: 'The very next question which I am going to ask you is an extraordinary one, although expressed in perfectly ordinary language.'

Th. And did you understand what I was saying?

S. I did, of course. Your warning referred to a question which you intended to ask me.

First published in Mind, N.S. **63**, 1954. (*See also my* Open Society, *vol.* ii, *note 7 to ch*. 24.)

Th. And what was this question of mine to which my warning referred? Can you repeat it?

S. Your question? Let me see . . . Oh, yes, your question was: 'What did I say between your last two interruptions, Socrates?'

Th. I see you have kept your promise, Socrates: you did attend to what I was saying. But did you understand this question of mine which you have just quoted?

S. I think I can prove that I understood your question at once. For did I not reply correctly when you first put it to me?

Th. That is so. But do you agree that it was an extraordinary question?

S. No. Admittedly, it was not very politely put, Theaetetus, but this, I am afraid, is nothing out of the ordinary. No, I can't see anything extraordinary in it.

Th. I am sorry if I was rude, Socrates; believe me, I only wanted to be brief, which was of some importance at that stage of our discussion. But I find it interesting that you think my question an ordinary one (apart from its rudeness); for some philosophers might say that it is an impossible question—at any rate one which it is impossible to understand properly, since it can have no meaning.

S. Why should your question have no meaning?

Th. Because indirectly it referred to itself.

S. I do not see this. As far as I can see, your question only referred to the warning you gave me, just before you asked it.

Th. And what did my warning refer to?

S. Now I see what you mean. Your warning referred to your question, and your question to your warning.

Th. But you say that you understood both, my warning and my question?

S. I had to trouble at all in understanding what you said.

Th. This seems to prove that two things a person says may be perfectly meaningful in spite of the fact that they are indirectly self-referring—that the first refers to the second and the second to the first.

S. It does seem to prove it.

Th. And don't you think that this is extraordinary?

S. To me it does not appear extraordinary. It seems obvious. I do not see why you should bother to draw my attention to such a truism.

Th. Because it has been denied, at least implicitly, by many philosophers.

S. Has it? You surprise me.

Th. I mean the philosophers who say that a paradox such as the *Liar* (the Megaric version of the *Epimenides*) cannot arise because a meaningful and properly constructed statement cannot refer to itself.

S. I know the *Epimenides* and the *Liar* who says, 'What I am now saying is untrue' (and nothing else); and I find the solution you just mentioned attractive.

Th. But it cannot solve the paradox if you admit that indirect self-reference is admissible. For, as Russell and Jourdain and Langford have shown (and Buridan before them), the *Liar* or the *Epimenides* can be formulated by using indirect self-reference instead of direct self-reference.

S. Please give me this formulation at once.

Th. The next assertion I am going to make is a true one.

S. Don't you always speak the truth?

Th. The last assertion I made was untrue.

S. So you wish to withdraw it? All right, you may begin again.

Th. *You* don't seem to realize what my two assertions taken together amounted to.

S. Oh, now I see the implications of what you were saying. You are quite right. It is old *Epimenides* all over again.

Th. I have used indirect self-reference instead of direct self-reference; that is the only difference. And this example establishes, I believe, that such paradoxes as the *Epimenides* cannot be solved by dwelling on the impossibility of self-referring assertions. For even if direct self-reference were impossible, or meaningless, indirect self-reference is certainly quite a common thing. I may, for example, make the following comment: I am confidently looking forward to a clever and appropriate remark from you, Socrates.

S. This expression of your confidence, Theaetetus, is highly flattering.

Th. This shows how easily it may occur that a comment is a comment upon another one, which in its turn is a comment upon the first. But once we see that we cannot solve the paradoxes in this way, we shall also see that even direct self-reference may be perfectly in order. In fact, many examples of non-paradoxical although directly self-referring assertions have been known for a long time; both of self-referring statements of a more or less empirical character and of self-referring statements whose truth or falsity can be established by logical reasoning.

S. Could you produce an example of a self-referring assertion which is empirically true?

Th. . . . .

S. I could not hear what you were saying, Theaetetus. Please repeat it a little louder. My hearing is no longer what it used to be.

Th. I said: 'I am now speaking so softly that dear old Socrates cannot make out what I am saying.'

S. I like this example; and I cannot deny that, when you were speaking so softly, you were speaking truthfully. Nor can I deny the empirical character of this truth; for had my ears been younger, it would have turned out an untruth.

Th. The truth of my next assertion will be even logically demonstrable, for example by a *reductio ad absurdum*, a method most beloved of Euclid the Geometrician.

S. I do not know him; you don't mean the man from Megara, I presume. But I think I know what you mean by a *reductio*. Will you now state your theorem?

Th. What I am now saying is meaningful.

S. If you don't mind I shall try to prove your theorem myself. For the purpose of a *reductio* I begin with the assumption that your last utterance was meaningless. This, however, turns out to contradict your utterance, and thus to entail the falsity of your utterance. But if an utterance is false, then it must clearly be meaningful. Thus my assumption is absurd; which proves your theorem.

Th. You have got it, Socrates. You have proved my theorem, as you insist on calling it. But some philosophers may not believe you. They will say that my theorem (or the demonstrably false antitheorem 'What I am now saying is meaningless') was paradoxical, and that, since it is paradoxical, you can 'prove' whatever you like about it—its truth as well as its falsity.

S. As I showed, the assumption of the truth of your antitheorem 'What I am now saying is meaningless' leads to an absurdity. Let them show, by a similar argument, that the assumption of its falsity (or of the truth of your theorem) leads to an absurdity also. When they succeed in this, then they may claim its paradoxical character or, if you like, its meaninglessness, and the meaninglessness of your theorem also.

Th. I agree, Socrates; moreover, I am perfectly satisfied that they will not succeed—at least as long as by 'a meaningless utterance' they mean something like an expression which is formulated in a manner which violates the rules of grammar or, in other words, a badly constructed expression.

S. I am glad that you feel so sure, Theaetetus; but are you not just a little too sure of our case?

Th. If you don't mind, I'll postpone the answer to that question for a minute or two. My reason is that I should like first to draw your attention to the fact that even if somebody did show that my theorem, or else my antitheorem, was paradoxical, he would not thereby have succeeded in showing that it is to be described as 'meaningless', in the best and most appropriate sense of the word. For in order to succeed he would have to show that, if we assume the truth of my theorem (or the falsity of my antitheorem 'What I am now saying is meaningless'), an absurdity follows. But I should be inclined to argue that such a derivation cannot be attempted by anybody who does not understand the meaning of my theorem and my antitheorem. And I should also be inclined to argue that, if the meaning of an utterance can be understood, then the utterance *has* a meaning; and again, that, if it has any implications (that is to say, if anything follows from it), it must also have a meaning. This view, at least, seems to be in accordance with ordinary usage, don't you think so?

S. I do.

Th. Of course, I do not wish to say that there may not be other ways of using the word 'meaningful'; for example, one of my fellow-mathematicians has suggested that we call an assertion 'meaningful' only if we possess a valid proof of it. But this would have the consequence that we could not know of a conjecture such as Goldbach's—'Every even number (except 2) is the sum of two primes'—whether it is at all meaningful, before we have found a valid proof of it; moreover, even the discovery of a counter example would not disprove the conjecture but only confirm its lack of meaning.

S. I think this would be both a strange way and an awkward way of using the word 'meaningful'.

Th. Other people have been a little more liberal. They suggested that we should call an assertion 'meaningful' if, and only if, there is a method which can either prove it or disprove it. This would make a conjecture such as Goldbach's meaningful the moment we have found a counter example (or a method of constructing one). But as long as we have not found a method of proving or disproving it, we cannot know whether or not it is meaningful.

S. It does not seem right to me to denounce all conjectures or hypotheses as 'meaningless' or 'nonsensical' simply because we don't know yet how to prove them or disprove them.

Th. Others again have suggested calling an assertion 'meaningful' only if we know how to find out whether it is true or false; a suggestion which amounts more or less to the same.

S. It does look to me very similar to your previous suggestion.

Th. If, however, we mean by 'a meaningful assertion or question' something like an expression which is understandable by anybody knowing the language, because it is formed in accordance with the grammatical rules for the formation of statements or questions in that language, then, I believe, we can give a correct answer to my next question which again will be a self-referring one.

S. Let me see whether I can answer it.

Th. Is the question I am now asking you meaningful or meaningless?

S. It is meaningful, and demonstrably so. For assume my answer to be false and the answer, 'It is meaningless', to be true. Then a true answer to your question can be given. But a question to which an answer can be given (and a true answer at that) must be meaningful. Therefore your question was meaningful, *quod erat demonstrandum*.

Th. I wonder where you picked up all this Latin, Socrates. Still, I can find no flaw in your demonstration; it is, after all, only a version of your proof of what you call my 'theorem'.

S. I think you have disposed of the suggestion that self-referring assertions are always meaningless. But I am sad at this admission, for it seemed such a straightforward way of getting rid of the paradoxes.

Th. You need not be sad: there simply was no way out in this direction.

S. Why not?

Th. Some people seem to think that there is a way of solving the paradoxes by dividing our utterances or expressions into meaningful statements which, in turn, can be either true or false, and utterances which are meaningless or nonsensical or not properly constructed (or 'pseudostatements', or 'indefinite propositions' as some philosophers preferred to call them), and which can be neither true nor false. If they could only show that a paradoxical utterance falls into the third of these three exclusive and exhaustive classes—true, false, and meaningless—then, they believe, the paradox in question would have found its solution.

S. Precisely. This was the way I had in mind, though I was not so clear about it; and I found it attractive.

Th. But these people don't ask themselves whether it is at all possible to solve a paradox such as that of the liar on the basis of a classification into these three classes, even if we could prove that it belongs to this third class of meaningless utterances.

S. I don't follow you. Assume they have succeeded in finding a proof which establishes that an utterance of the form 'U is false' is meaningless, whenever 'U' is a name of this very utterance 'U is false'. Why should this not solve the paradox?

Th. It would not. It would only shift it. For under the assumption that U is itself the utterance 'U is false', I can disprove the hypothesis that U is meaningless with the help of precisely this threefold classification of utterances.

S. If you are right, then a proof of the hypothesis that U is meaningless would indeed only establish a new statement which can be proved as well as disproved, and therefore a new paradox. But how can you disprove the hypothesis that U is meaningless?

Th. Again by a *reductio*. Quite generally, we can read off from our classification two rules. (i) From the truth of 'X is meaningless' we can derive the falsity of 'X is true' and also (what interests us here), the falsity of 'X is false'. (ii) From the falsity of any utterance Y, we can conclude that Y is meaningful. According to these rules, we find that from the truth of our hypothesis, '*U* is meaningless', we can derive by (i) the falsity of '*U* is false'; concluding by (ii) that '*U* is false' is meaningful. But since '*U* is false' is nothing but *U* itself, we have shown (by (ii) again) that *U* is meaningful; which concludes the *reductio*. (Incidentally, since the truth of our hypothesis entails the falsity of '*U* is false', it also entails our original paradox.)

S. This is a surprising result: a *Liar* who comes back by the window, just when you think you have got rid of him by the door. Is there no way whatever of eliminating these paradoxes?

Th. There is a very simple way, Socrates.

S. What is it?

Th. Just avoid them, as nearly everybody does, and don't worry about them.

S. But is this sufficient? Is this safe?

Th. For ordinary language and for ordinary purposes it seems sufficient and quite safe. At any rate, you can do nothing else in ordinary language, since paradoxes can be constructed in it, and are understandable, as we have seen.

S. But could we not legislate, say, that any kind of self-reference, whether direct or indirect, should be avoided, and thereby purify our language of paradoxes?

Th. We might try to do this (although it might lead to new difficulties). But a language for which we legislate in this way is no longer our ordinary language; artificial rules make an artificial language. Has not our discussion shown that at least indirect self-reference is quite an ordinary thing?

S. But for mathematics, say, a somewhat artificial language would be appropriate, would it not?

Th. It would; and for the construction of a language with artificial rules which, if it is properly done, might be called a 'formalized language', we shall take hints from the fact that paradoxes (which we wish to avoid) can occur in ordinary language.

S. And you would legislate for your formalized language, I suppose, that all self-reference must be strictly excluded, would you not?

Th. No. We can avoid paradoxes without using such drastic measures.

S. Do you call them drastic?

Th. They are drastic because they would exclude some very interesting uses of self-reference, especially Gödel's method of constructing self-referring statements, a method which has most important applications in my own field of interest, the theory of numbers. They are drastic, moreover, because we have learned from Tarski that in any consistent language—let us call it 'L'—the predicates 'true in L' and 'false in L' cannot occur (as opposed to 'meaningful in L', and 'meaningless in L' which may occur), and that without predicates such as these, paradoxes of the type of the *Epimenides*, or of Grelling's paradox of the heterological adjectives, cannot be formulated. This hint turns out to be sufficient for the construction of formalized languages in which these paradoxes are avoided.

S. Who are all these mathematicians? Theodorus never mentioned their names.

Th. Barbarians, Socrates. But they are very able. Gödel's 'method of arithmetization', as it is called, is especially interesting in the context of our present discussion.

S. Another self-reference, and quite an ordinary one. I am getting a bit hyper-sensitive to these things.

Th. Gödel's method, one might say, is to translate certain non-arithmetical assertions into arithmetical ones; they are turned into an arithmetical code, as it were; and among the assertions which can be so coded there happens to be also the one which you have jokingly described as my theorem. To be a little more exact, the assertion which can be turned into Gödel's arithmetical code is the self-referring statement, 'This expression is a well-formed formula'; here 'well-formed formula' replaces, of course, the word 'meaningful'. I felt, you will remember, a little too sure for your liking that my theorem cannot be disproved. My reason was, simply, that when turned into the Gödelian code, my theorem becomes a theorem of arithmetic. It is demonstrable, and its negation is refutable. Now if anybody were to succeed, by a valid argument (perhaps by one similar to your own proof) in disproving my theorem—for example, by deriving an absurdity from the assumption that the negation of my theorem is false—then this argument could be used to show the same of the corresponding arithmetical theorem; and since this would at once provide us with a method of proving '0 = 1', I feel that I have good reasons for believing that my theorem cannot be disproved.

S. Could you explain Gödel's method of coding without getting involved in technicalities?

Th. There is no need to do this since it has been done before—I do not mean before now, the supposed dramatic date of this little dialogue of ours (which is about 400 B.C.), but I mean before our dialogue will ever be concocted by its author, which won't take place before another 2,350 years have elapsed.

S. I am shocked, Theaetetus, by these latest self-references of yours. You talk as if we were actors reciting the lines of a play. This is a trick which, I am afraid, some playwrights think witty, but hardly their victims; anyway, I don't. But even worse than any such self-referring joke is this preposterous, nay, this nonsensical chronology of yours. Seriously, I must draw a line somewhere, Theaetetus, and I am drawing it here.

Th. Come, Socrates, who cares about chronology? Ideas are timeless.

S. Beware of metaphysics, Theaetetus!

15

WHAT IS DIALECTIC?

1. DIALECTIC EXPLAINED

> There is nothing so absurd or incredible that it has not been asserted by one philosopher or another.
>
> DESCARTES

The above motto can be generalized. It applies not only to philosophers and philosophy, but throughout the realm of human thought and enterprise, to science, technology, engineering and politics. Indeed, the tendency to try anything once, suggested by the motto, can be discerned in a still wider realm—in the stupendous variety of forms and appearances which are produced by life on our planet.

Thus if we want to explain why human thought tends to try out every conceivable solution for any problem with which it is faced, then we can appeal to a highly general sort of regularity. The method by which a solution is approached is usually the same; it is the *method of trial and error*. This, fundamentally, is also the method used by living organisms in the process of adaptation. It is clear that the success of this method depends very largely on the number and variety of the trials:

A paper read to a philosophy seminar at Canterbury University College, Christchurch, New Zealand, in 1937. First published in Mind, *N.S.*, **49**, 1940.

the more we try, the more likely it is that one of our attempts will be successful.

We may describe the method employed in the development of human thought, and especially of philosophy, as a particular variant of the trial and error method. Men seem inclined to react to a problem either by putting forward some theory and clinging to it as long as they can (if it is erroneous they may even perish with it rather than give it up[1]), or by fighting against such a theory, once they have seen its weaknesses. This struggle of ideologies, which is obviously explicable in terms of the method of trial and error, seems to be characteristic of anything that may be called a development in human thought. The cases in which it does not occur are, in the main, those in which a certain theory or system is dogmatically maintained throughout some long period; but there will be few if any examples of a development of thought which is slow, steady, and continuous, and proceeds by successive degrees of improvement rather than by trial and error and the struggle of ideologies.

If the method of trial and error is developed more and more consciously, then it begins to take on the characteristic features of 'scientific method'. This 'method'[2] can briefly be described as follows. Faced with a certain problem, the scientist offers, tentatively, some sort of solution—a theory. This theory science accepts only provisionally, if at all; and it is most characteristic of the scientific method that scientists will spare no pains to criticize and test the theory in question. Criticizing and testing go hand in hand; the theory is criticized from very many different sides in order to bring out those points which may be vulnerable. And the testing of the theory proceeds by exposing these vulnerable points to as severe an examination as possible. This, of

[1] The dogmatic attitude of sticking to a theory as long as possible is of considerable significance. Without it we could never find out what is in a theory—we should give the theory up before we had a real opportunity of finding out its strength; and in consequence no theory would ever be able to play its role of bringing order into the world, of preparing us for future events, of drawing our attention to events we should otherwise never observe.

[2] It is not a method in the sense that, if you practise it, you will succeed; or if you don't succeed, you can't have practised it; that is to say, it is not a definite way to results: a method in this sense does not exist.

course, is again a variant of the method of trial and error. Theories are put forward tentatively and tried out. If the outcome of a test shows that the theory is erroneous, then it is eliminated; the method of trial and error is essentially a method of elimination. Its success depends mainly on three conditions, namely, that sufficiently numerous (and ingenious) theories should be offered, that the theories offered should be sufficiently varied, and that sufficiently severe tests should be made. In this way we may, if we are lucky, secure the survival of the fittest theory by the elimination of those which are less fit.

If this description[3] of the development of human thought in general and of scientific thought in particular is accepted as more or less correct, then it may help us to understand what is meant by those who say that the development of thought proceeds on '*dialectic*' lines.

Dialectic (in the modern[4] sense, i.e. especially in the sense in which Hegel used the term) is a theory which maintains that something—more especially, human thought—develops in a way characterized by what is called the dialectic triad: *thesis, antithesis*, and *synthesis*. First there is some idea or theory or movement which may be called a 'thesis'. Such a thesis will often produce opposition, because, like most things in this world, it will probably be of limited value and will have its weak spots. The opposing idea or movement is called the '*antithesis*', because it is directed against the first, the thesis. The struggle between the thesis and the antithesis goes on until some solution is reached which, in a certain sense, goes beyond both thesis and antithesis by recognizing their respective values and by trying to preserve the merits and to avoid the limitations of both. This solution, which is the third step, is called the *synthesis*. Once attained, the synthesis in its turn may become the first step of a new dialectic triad, and it will do so if the particular synthesis reached turns out to be one-sided or otherwise unsatisfactory. For in

[3] A more detailed discussion can be found in *L. Sc.D.*

[4] The Greek expression '*Hē dialektikē (technē)*' may be translated '(the art of) the argumentative usage of language'. This meaning of the term goes back to Plato; but even in Plato it occurs in a variety of different meanings. One at least of its ancient meanings is very close to what I have described above as 'scientific method'. For it is used to describe the method of constructing explanatory theories and of the critical discussion of these theories, which includes the question whether they are able to account for empirical observations, or, using the old terminology, whether they are able to '*save the appearances*'.

this case opposition will be aroused again, which means that the synthesis can then be described as a new thesis which has produced a new antithesis. The dialectic triad will thus proceed on a higher level, and it may reach a third level when a second synthesis has been attained.[5]

So much for what is called the 'dialectic triad'. It can hardly be doubted that the dialectic triad describes fairly well certain steps in the history of thought, especially certain developments of ideas and theories, and of social movements which are based on ideas or theories. Such a dialectic development may be 'explained' by showing that it proceeds in conformity with the method of trial and error which we have discussed above. But it has to be admitted that it is not exactly the same as the development (described above) of a theory by trial and error. Our earlier description of the trial and error method dealt only with an idea and its criticism, or, using the terminology of dialecticians, with the struggle between a thesis and its antithesis; originally we made no suggestions about a further development, we did not imply that the struggle between a thesis and an antithesis would lead to a synthesis. Rather we suggested that the struggle between an idea and its criticism or between a thesis and its antithesis would lead to the elimination of the thesis (or, perhaps, of the antithesis) if it is not satisfactory; and that the competition of theories would lead to the adoption of new theories only if enough theories are at hand and are offered for trial.

Thus the interpretation in terms of the trial and error method may be said to be slightly wider than that in terms of dialectic. It is not confined to a situation where only one thesis is offered to start with, and so it can easily be applied to situations where from the very beginning a number of different theses are offered, independently of one another, and not only in such a way that the one is opposed to the other. But admittedly it happens very frequently, perhaps usually, that the development of a certain branch of human thought starts with one

[5] In Hegel's terminology, both the thesis and the antithesis are, by the synthesis, (1) *reduced to components (of the synthesis)* and they are thereby (2) *cancelled* (or *negated*, or *annulled*, or *set aside*, or *put away*) and, at the same time, (3) *preserved* (or *stored*, or *saved*, or *put away*) and (4) *elevated* (or *lifted to a higher level*). The italicized expressions are renderings of the four main meanings of the one German word '*aufgehoben*' (literally 'lifted up') of whose ambiguity Hegel makes much use.

single idea only. If so, then the dialectic scheme may often be applicable because this thesis will be open to criticism and in this way 'produce', as dialecticians usually say, its antithesis.

The dialectician's emphasis involves still another point where dialectic may differ slightly from the general trial-and-error theory. For the trial-and-error theory as suggested above will be content to say that an unsatisfactory view will be refuted or eliminated. The dialectician insists that there is more to be said than this. He emphasizes that although the view or theory under consideration may have been refuted, there will most probably be an element in it which is worthy of preservation, for otherwise it is not very likely that it would have been offered at all and taken seriously. This valuable element of the thesis is likely to be brought out more clearly by those who defend the thesis against the attacks of their opponents, the adherents of the antithesis. Thus the only satisfactory solution of the struggle will be a synthesis, i.e. a theory in which the best points of both thesis and antithesis are preserved.

It must be admitted that such a dialectical interpretation of the history of thought may sometimes be quite satisfactory, and that it may add some valuable details to an interpretation in terms of trial and error.

Let us take the development of physics as an example. We can find very many instances which fit the dialectic scheme, such as the corpuscular theory of light which, after first having been replaced by the wave theory, remains 'preserved' in the new theory which replaces them both. To put it more precisely, the old formulae can usually be described, from the standpoint of the new ones, as approximations; that is to say, they appear to be very nearly correct, so that they can be applied, either if we do not demand a very high degree of exactitude, or even, within certain limited fields of application, as perfectly exact formulae.

All this can be said in favour of the dialectic point of view. But we have to be careful not to admit too much.

We must be careful, for instance, about a number of metaphors used by dialecticians and unfortunately often taken much too seriously. An example is the dialectical saying that the thesis 'produces' its antithesis. Actually it is only our critical attitude which produces the antithesis, and where such an attitude is lacking—which often enough is the case—no antithesis will be produced. Similarly, we have to be careful not to think that it is the 'struggle' between a thesis and its antithesis which 'produces' a synthesis. The struggle is one of minds; and these minds must be productive of new ideas: there are many instances of futile struggles in the history of human thought, struggles which ended in nothing. And even when a synthesis has been reached, it will usually be a rather crude description of the synthesis to say that it 'preserves' the better parts of both the thesis and the antithesis. This description will be misleading even where it is true, because in addition to older ideas which it 'preserves', the synthesis will, in every case, embody some new idea which cannot be reduced to earlier stages of the development. In other words, the synthesis will usually be much more than a construction out of material supplied by thesis and antithesis. Considering all this, the dialectic interpretation, even where it may be applicable, will hardly ever help to develop thought by its suggestion that a synthesis should be constructed out of the ideas contained in a thesis and an antithesis. This is a point which some dialecticians have stressed themselves; nevertheless, they nearly always assume that dialectic can be used as a technique that will help them to promote, or at least to predict, the future development of thought.

But the most important misunderstandings and muddles arise out of the loose way in which dialecticians speak about contradictions.

They observe, correctly, that contradictions are of the greatest importance in the history of thought—precisely as important as is criticism. For criticism invariably consists in pointing out some contradiction; either a contradiction within the theory criticized, or a contradiction between the theory and another theory which we have some reason to accept, or a contradiction between the theory and certain facts—or more precisely, between the theory and certain statements of fact. Criticism can never do anything except either point out some such contradiction, or, perhaps, simply contradict the theory (i.e. the criticism may be simply the statement of an antithesis). But criticism is, in a very important sense, the main motive force of any intellectual development. Without contradictions, without criticism, there would be no rational motive for changing our theories: there would be no intellectual progress.

Having thus correctly observed that contradictions—especially, of course, the contradiction between a thesis and an antithesis, which 'produces' progress in the form of a synthesis—are extremely fertile, and indeed the moving forces of any progress of thought, dialecticians conclude—wrongly as we shall see—that there is no need to avoid these fertile contradictions. And they even assert that contradictions cannot be avoided, since they occur everywhere in the world.

Such an assertion amounts to an attack upon the so-called 'law of contradiction' (or, more fully, upon the 'law of the exclusion of contradictions') of traditional logic, a law which asserts that two contradictory statements can never be true together, or that a statement consisting of the conjunction of two contradictory statements must always be rejected as false on purely logical grounds. Appealing to the fruitfulness of contradictions, dialecticians claim that this law of traditional logic must be discarded. They claim that dialectic leads in this way to a new logic—a dialectical logic. Dialectic, which I have so far presented as a merely historical doctrine—a theory of the historical development of thought—would turn out in this way to be a very different doctrine: it would be at the same time a logical theory and (as we shall see) a general theory of the world.

These are tremendous claims, but they are without the slightest foundation. Indeed, they are based on nothing better than a loose and woolly way of speaking.

Dialecticians say that contradictions are fruitful, or fertile, or productive of progress, and we have admitted that this is, in a sense, true. It is true, however, only so long as we are determined not to put up with contradictions, and to change any theory which involves contradictions; in other words never to accept a contradiction: it is solely due to this determination of ours that criticism, i.e. the pointing out of contradictions, induces us to change our theories, and thereby to progress.

It cannot be emphasized too strongly that if we change this attitude, and decide to put up with contradictions, then contradictions must at once lose any kind of fertility. They would no longer be productive of intellectual progress. For if we were prepared to put up with contradictions, pointing out contradictions in our theories could no longer induce us to change them. In other words, all criticism (which consists in pointing out contradictions) would lose its force. Criticism would be answered by 'And why not?' or perhaps even by an enthusiastic 'There you are!'; that is, by welcoming the contradictions which have been pointed out to us.

But this means that if we are prepared to put up with contradictions, criticism, and with it all intellectual progress, must come to an end.

Thus we must tell the dialectician that he cannot have it both ways. Either he is interested in contradictions because of their fertility: then he must not accept them. Or he is prepared to accept them: then they will be barren, and rational criticism, discussion, and intellectual progress will be impossible.

The only 'force' which propels the dialectic development is, therefore, our determination not to accept, or to put up with, the contradiction between the thesis and the antithesis. It is not a mysterious force inside these two ideas, not a mysterious tension between them which promotes development—it is purely our decision, our resolution, not to admit contradictions, which induces us to look out for a new point of view which may enable us to avoid them. And this resolution is entirely justified. For it can easily be shown that if one were to accept contradictions then one would have to give up any kind of scientific activity: it would mean a complete breakdown of science. This can be shown by proving that *if two contradictory statements are admitted, any statement whatever must be admitted*; for from a couple of contradictory statements any statement whatever can be validly inferred.

This is not always realized,[6] and will therefore be fully explained here. It is one of the few facts of elementary logic which are not quite trivial, and deserves to be known and understood by every thinking man. It can easily be explained to those readers who do not dislike the use of symbols which look like mathematics; but even those who dislike such symbols should understand the matter easily if they are not too impatient, and prepared to devote a few minutes to this point.

Logical inference proceeds according to certain *rules of inference*. It is

[6] See for example H. Jeffreys, 'The Nature of Mathematics', *Philosophy of Science*, **5**, 1938, 449, who writes: 'Whether a contradiction entails any proposition is doubtful.' See also Jeffreys' reply to me in *Mind*, **51**, 1942, p. 90, my rejoinder in *Mind*, **52**, 1943, pp. 47 ff., and *L.Sc.D.*, note *2 to section 23. All this was known, in effect, to Duns Scotus (*ob.* 1308), as has been shown by Jan Łukasiewicz in *Erkenntnis*, **5**, p. 124.

valid if the rule of inference to which it appeals is valid; and *a rule of inference is valid if, and only if, it can never lead from true premises to a false conclusion*; or, in other words, if it unfailingly transmits the truth of the premises (provided they are all true) to the conclusion.

We shall need two such rules of inference. In order to explain the first and more difficult one, we introduce the idea of a *compound statement*, that is to say, of a statement such as 'Socrates is wise *and* Peter is a King', or perhaps '*Either* Socrates is wise *or* Peter is a King (but not both)' or perhaps 'Socrates is wise *and/or* Peter is a King'. The two statements ('Socrates is wise'; and 'Peter is a king') of which such a compound statement is composed are called component statements.

Now there is one kind of compound statement which interests us here—the one which is so constructed that it is *true if and only if at least one of its two components is true*. The ugly expression '*and/or*' has precisely the effect of producing such a compound: the assertion 'Socrates is wise *and/or* Peter is a King' is one which will be true if and only if one or both of its component statements are true; and it will be false if and only if both of its component statements are false.

It is customary in logic to replace the expression '*and/or*' by the symbol '**v**' (to be pronounced 'vel') and to use such letters as '*p*' and '*q*' to represent any statement we like. We can then say that a statement of the form '*p* **v** *q*' will be true if one at least of its two components, *p* and *q*, is true.

We are now in a position to formulate our first rule of inference. It may be formulated in this way:

(1) From a premise *p* (for example, 'Socrates is wise') any conclusion of the form '*p* **v** *q*' (for example, 'Socrates is wise **v** Peter is a King') may be validly deduced.

That this rule must be valid can be seen at once if we remember the meaning of '**v**'. This symbol makes a compound which is true whenever at least one of the components is true. Accordingly, if *p* is true, *p* **v** *q* must also be true. Thus our rule can never lead from a true premise to a false conclusion, which means that it is valid.

In spite of its validity, our first rule of inference often strikes those who are not used to such things as strange. And it is indeed a rule which is rarely used in everyday life, since the conclusion contains much less information than the premise. But it is sometimes used, for example, in betting. I may, say, toss a penny twice, betting that heads will turn up *at least once*. This, obviously, is tantamount to my betting on the truth of the compound statement 'Heads turn up at the first toss **v** heads turn up at the second toss'. The probability of this statement equals 3/4 (according to usual calculations); it is thus different, for example, from the statement 'Heads turn up at the first toss *or* heads turn up at the second toss (but not both)', whose probability is 1/2. Now everybody will say that I have won my bet if heads turned up at the first toss—in other words, that the compound statement on whose truth I was betting must be true if its first component was true; which shows that we argued in accordance with our first rule of inference.

We can also state our first rule in this way

$$\frac{p}{p \mathbf{v} q}$$

which may be read: 'from the premise p we obtain the conclusion p **v** q.'

The second rule of inference which I am going to use is more familiar than the first. If we denote the negation of p by '*non-p*', then it can be stated in this way

$$\frac{\begin{array}{c}\text{non-}p\\ p \mathbf{v} q\end{array}}{q}$$

which may be put in words:

(2) From the two premises *non-p*, and p **v** q, we obtain the conclusion q.

The validity of this rule can be established if we consider that *non-p* is a statement which is true if and only if p is false. Accordingly, if the first premise *non-p*, is true, then the first component of the second premise is false; thus if both premises are true, the second component of the second premise must be true; that is to say, q must be true whenever the two premises are true.

In reasoning that, if *non-p* is true, *p* must be false, we have made implicit use, it may be said, of the 'law of contradiction' which asserts that *non-p* and *p* cannot be true together. Thus if it were my task at this moment to argue in favour of contradiction, we should have to be more cautious. But at this moment, I am only trying to show that *using valid rules of inference, we can infer from a couple of contradictory premises any conclusion we like.*

Using our two rules we can indeed show this. For assume we have the two contradictory premises—say

(*a*) The sun is shining now,

(*b*) The sun is not shining now.

From these two premises any statement—for example, 'Caesar was a traitor' can be inferred, as follows.

From the first premise (*a*) we can infer, in accordance with rule (1), the following conclusion:

(*c*) The sun is shining now **v** Caesar was a traitor.

Taking now (*b*) and (*c*) as premises, we can ultimately deduce, in accordance with rule (2)

(*d*) Caesar was a traitor.

It is clear that by the same method we might have inferred any other statement we wanted to infer; for example, 'Caesar was not a traitor'. We may thus infer '$2 + 2 = 5$' and '$2 + 2 \neq 5$'—not only every statement we like, but also its negation, which we may not like.

We see from this that if a theory contains a contradiction, then it entails everything, and therefore, indeed, nothing. A theory which adds to every information which it asserts also the negation of this information can give us no information at all. A theory which involves a contradiction is therefore entirely useless *as a theory*.

In view of the importance of the logical situation analysed, I shall now present some other rules of inference which lead to the same result. In contradistinction to rule (1), the rules now to be examined, and to be used, form part of the classical theory of the syllogism, with the exception of the following rule (3) which we shall discuss first.

(3) From any two premises, *p* and *q*, we may derive a conclusion which is identical with one of them—say *p*; or schematically,

$$\frac{\begin{array}{c}p\\q\end{array}}{p}$$

In spite of its unfamiliarity, and of the fact that some philosophers[7] have not accepted it, this rule is undoubtedly valid; for it must infallibly lead to a true conclusion whenever the premises are true. This is obvious, and indeed trivial; and it is this very triviality which makes the rule, in ordinary discourse, redundant, and therefore unfamiliar. But redundancy does not mean invalidity.

In addition to this rule (3) we shall need another rule which I have called 'the rule of indirect reduction' (because in the classical theory of the syllogism it is implicitly used for the indirect reduction of the 'imperfect' figures to the first or 'perfect' figure).

Assume we have a valid syllogism such as

$$\frac{\begin{array}{l}(a)\ \text{All men are mortal}\\(b)\ \text{All Athenians are men}\end{array}}{(c)\ \text{All Athenians are mortal.}}$$

Now the rule of indirect reduction says:

(4) If $\frac{\begin{array}{c}a\\b\end{array}}{c}$ is a valid inference, then $\frac{\begin{array}{c}a\\\text{non-}c\end{array}}{\text{non-}b}$ is a valid inference too.

For example, owing to the validity of the inference of (c) from the premises (a) and (b), we find that

$$\frac{\begin{array}{l}(a)\ \text{All men are mortal}\\(\text{non-}c)\ \text{Some Athenians are non-mortal}\end{array}}{(\text{non-}b)\ \text{Some Athenians are non-men}}$$

must also be valid.

The rule we are going to use is a slight variant of the one just stated; it is this:

[7] Notably G. E. Moore.

(5) If $\dfrac{\begin{matrix} a \\ \text{non-}b \end{matrix}}{c}$ is a valid inference, then $\dfrac{\begin{matrix} a \\ \text{non-}c \end{matrix}}{b}$ is a valid inference too.

Rule (5) may be obtained, for example, from the rule (4) together with the law of double negation which tells us that from *non-non-b* we may deduce *b*. Now if rule (5) is valid for any statements *a*, *b*, *c*, which we choose (and only then is it valid) then it must also be valid in case *c* happens to be identical with *a*; that is to say, the following must be valid

(6) If $\dfrac{\begin{matrix} a \\ \text{non-}b \end{matrix}}{a}$ is a valid inference, then $\dfrac{\begin{matrix} a \\ \text{non-}a \end{matrix}}{b}$ is a valid inference too.

But we know from (3) that $\dfrac{\begin{matrix} a \\ \text{non-}b \end{matrix}}{a}$ is indeed a valid inference. Thus (6) and (3) together yield

(7) $\dfrac{\begin{matrix} a \\ \text{non-}a \end{matrix}}{b}$ is a valid inference, whatever the statements *a* and *b* may assert.

But (7) states exactly what we wanted to show—that from a couple of contradictory premises, *any* conclusion may be deduced.

The question may be raised whether this situation holds good in any system of logic, or whether we can construct a system of logic in which contradictory statements do not entail every statement. I have gone into this question, and the answer is that such a system can be constructed. The system turns out, however, to be an extremely weak system. Very few of the ordinary rules of inference are left, not even the *modus ponens* which says that from a statement of the form 'If *p* then *q*' together with *p*, we can infer *q*. In my opinion, such a system[8] is of no

[8] The system alluded to is the 'dual-intuitionist calculus'; see my paper 'On the Theory of Deduction I and II', *Proc. of the Royal Dutch Academy*, **51**, Nos. 2 and 3, 1948, 3.82 on p. 182, and 4.2 on p. 322, and 5.32, 5.42, and note 15. Dr Joseph Kalman Cohen has developed the system in some detail. I have a simple interpretation of this calculus. All the statements may be taken to be modal statements asserting possibility. From '*p* is possible' and '"if *p* then *q*" is possible', we cannot indeed derive '*q* is possible' (for if *p* is false, *q* may be

use for drawing inferences although it may perhaps have some appeal for those who are specially interested in the construction of formal systems as such.

It has sometimes been said that the fact that from a couple of contradictory statements anything we wish follows does not establish the uselessness of a contradictory theory: first, this theory may be interesting in itself even though contradictory; secondly, it may give rise to corrections which make it consistent; and ultimately, we may develop a method, even if it is an *ad hoc* method (such as, in Quantum Theory, the methods of avoiding the divergencies), which prevents us from obtaining the false conclusions which admittedly are logically entailed by the theory. All this is quite true; but such a make-shift theory gives rise to the grave dangers previously discussed: if we seriously intend to put up with it then nothing will make us search for a better theory; and also the other way round: if we look for a better theory, then we do so because we think the theory we have described is a bad one, *owing to the contradictions involved*. The acceptance of contradictions must lead here as everywhere to the end of criticism, and thus to the collapse of science.

One sees here the danger of loose and metaphorical ways of speaking. The looseness of the dialectician's assertion that contradictions are not avoidable and that it is not even desirable to avoid them because they are so fertile is dangerously misleading. It is misleading because what may be called the fertility of the contradictions is, as we have seen, merely the result of our decision not to put up with them (an attitude which accords with the law of contradiction). And it is dangerous, because to say that the contradictions need not be avoided, or perhaps even that they cannot be avoided, must lead to the breakdown of science, and of criticism, i.e. of rationality. This should make it clear to anybody wishing to further truth and enlightenment that it is a necessity and even a duty to train himself in the art of expressing things clearly and unambiguously—even if this means giving up certain niceties of metaphor and clever double meanings.

It is therefore better to avoid certain formulations. For instance,

an impossible statement). Similarly, from '*p* is possible' and '*non-p* is possible' we clearly cannot deduce the possibility of all statements.

instead of the terminology we have used in speaking of thesis, antithesis, and synthesis, dialecticians often describe the dialectic triad by using the term 'negation (of the thesis)' instead of 'antithesis' and 'negation of the negation' instead of 'synthesis'. And they like to use the term 'contradiction' where terms like 'conflict' or perhaps 'opposing tendency' or 'opposing interest', etc., would be less misleading. Their terminology would do no harm if the terms 'negation' and 'negation of the negation' (and similarly, the term 'contradiction') had not clear and fairly definite logical meanings, different from the dialectical usage. In fact the misuse of these terms has contributed considerably to the confusion of logic and dialectic which so often occurs in the discussions of the dialecticians. Frequently they consider dialectic to be a part—the better part—of logic, or something like a reformed, modernized logic. The deeper reason for such an attitude will be discussed later. At present I shall only say that our analysis does not lead to the conclusion that dialectic has any sort of similarity to logic. For logic can be described—roughly, perhaps, but well enough for our present purposes—as a theory of deduction. We have no reason to believe that dialectic has anything to do with deduction.

To sum up: What dialectic is—dialectic in the sense in which we can attach a clear meaning to the dialectic triad—can be described thus. Dialectic, or more precisely, the theory of the dialectic triad, maintains that certain developments, or certain historical processes, occur in a certain typical way. It is, therefore, an empirical descriptive theory, comparable, for instance, with the theory which maintains that most living organisms increase their size during some stage of their development, then remain constant, and finally decrease until they die; or with the theory which maintains that opinions are held first dogmatically, then sceptically, and only afterwards, in a third stage, in a scientific, i.e. critical, spirit. Like such theories, dialectic is not applicable without exceptions—unless we force the dialectic interpretations—and like such theories, dialectic has no special affinity to logic.

The vagueness of dialectic is another of its dangers. It makes it only too easy to force a dialectic interpretation on all sorts of developments and even on quite different things. We find, for instance, a dialectic interpretation which identifies a seed of corn with a thesis, the plant which develops from this seed with the antithesis, and all the seeds which develop from this plant with the synthesis. That such an application expands the already too vague meaning of the dialectic triad in a way which dangerously increases its vagueness is obvious; it leads to a point where by describing a development as dialectic we convey no more than by saying that it is a development in stages—which is not saying very much. But to interpret this development by saying that germination of the plant is the negation of the seed because the seed ceases to exist when the plant begins to grow, and that the production of a lot of new seeds by the plant is the negation of the negation—a new start on a higher level—is obviously a mere playing with words. (Is this the reason why Engels said of this example that any child can understand it?)

The standard examples presented by dialecticians from the field of mathematics are even worse. To quote a famous example used by Engels in the brief form given to it by Hecker,[9] 'The law of the higher synthesis . . . is commonly used in mathematics. The negative $(-a)$ multiplied by itself becomes a^2, i.e. the negation of the negation has accomplished a new synthesis.' But even assuming a to be a thesis and $-a$ its antithesis or negation, one might expect that the negation of the negation is $-(-a)$, i.e. a, which would not be a 'higher' synthesis, but identical with the original thesis itself. In other words, why should the synthesis be obtained just by multiplying the antithesis with itself? Why not, for example, by adding thesis and antithesis (which would yield 0)? Or by multiplying thesis and antithesis (which would yield $-a^2$ rather than a^2)? And in what sense is a^2 'higher' than a or $-a$? (Certainly not in the sense of being numerically greater, since if $a = \frac{1}{2}$ then $a^2 = \frac{1}{4}$.) The example shows the extreme arbitrariness with which the vague ideas of dialectic are applied.

A theory like logic may be called 'fundamental', thereby indicating that, since it is the theory of all sorts of inferences, it is used all the time by all sciences. We can say that dialectic in the sense in which we found that we could make a sensible application of it is not a fundamental but merely a descriptive theory. It is therefore about as inappropriate to regard dialectic as part and parcel of logic, or else as opposed to logic, as it would be so to regard, say, the theory of evolution. Only the loose

[9] Hecker, *Moscow Dialogues*, London, 1936, p. 99. The example is from the *Anti-Dühring*.

metaphorical and ambiguous way of speaking which we have criticized above could make it appear that dialectic can be both a theory describing certain typical developments and a fundamental theory such as logic.

From all this I think it is clear that one should be very careful in using the term 'dialectic'. It would be best, perhaps, not to use it at all—we can always use the clearer terminology of the method of trial and error. Exceptions should be made only where no misunderstanding is possible, and where we are faced with a development of theories which does in fact proceed along the lines of a triad.

2. HEGELIAN DIALECTIC

So far I have tried to outline the idea of dialectic in a way which I hope makes it intelligible, and it was my aim not to be unjust about its merits. In this outline dialectic was presented as a way of describing developments; as one way among others, not fundamentally important, but sometimes quite suitable. As opposed to this, a theory of dialectic has been put forward, for example by Hegel and his school, which exaggerates its significance, and which is dangerously misleading.

In order to make Hegel's dialectic intelligible it may be useful to refer briefly to a chapter in the history of philosophy—in my opinion not a very creditable one.

A major issue in the history of modern philosophy is the struggle between Cartesian rationalism (mainly continental) on the one hand, and empiricism (mainly British) on the other. The sentence from Descartes which I have used as a motto for this paper was not intended by its author, the founder of the rationalist school, in the way in which I have made use of it. It was not intended as a hint that the human mind has to try everything in order to arrive at something—i.e. at some useful solution—but rather as a hostile criticism of those who dare to try out such absurdities. What Descartes had in mind, the main idea behind his sentence, is that the real philosopher should carefully avoid absurd and foolish ideas. In order to find truth he has only to accept those rare ideas which appeal to reason by their lucidity, by their clarity and distinctness, which are, in short, 'self-evident'. The

Cartesian view is that we can construct the explanatory theories of science without any reference to experience, just by making use of our reason; for every reasonable proposition (i.e. one recommending itself by its lucidity) must be a true description of the facts. This, in brief outline, is the theory which the history of philosophy has called '*rationalism*'. (A better name would be '*intellectualism*'.) It can be summed up (using a formulation of a much later period, namely that of Hegel) in the words: 'That which is reasonable must be real.'

Opposed to this theory, empiricism maintains that only experience enables us to decide upon the truth or falsity of a scientific theory. Pure reasoning alone, according to empiricism, can never establish factual truth; we have to make use of observation and experiment. It can safely be said that empiricism, in some form or other, although perhaps in a modest and modified form, is the only interpretation of scientific method which can be taken seriously in our day. The struggle between the earlier rationalists and empiricists was thoroughly discussed by Kant, who tried to offer what a dialectician (but not Kant) might describe as a *synthesis* of the two opposing views, but what was, more precisely, a modified form of empiricism. His main interest was to reject pure rationalism. In his *Critique of Pure Reason* he asserted that the scope of our knowledge is limited to the field of possible experience, and that speculative reasoning beyond this field—the attempt to build up a metaphysical system out of pure reason—has no justification whatever. This criticism of pure reason was felt as a terrible blow to the hopes of nearly all continental philosophers; yet German philosophers recovered and, far from being convinced by Kant's rejection of metaphysics, hastened to build up new metaphysical systems based on '*intellectual intuition*'. They tried to use certain features of Kant's system, hoping thereby to evade the main force of his criticism. The school which developed, usually called the school of the German idealists, culminated in Hegel.

There are two aspects of Hegel's philosophy which we have to discuss—his idealism and his dialectic. In both cases Hegel was influenced by some of Kant's ideas, but tried to go further. In order to understand Hegel we must therefore show how his theory made use of Kant's.

Kant started from the fact that science exists. He wanted to explain this fact; that is, he wanted to answer the question, 'How is science possible?' or, 'How are human minds able to gain knowledge of the world', or, 'How can our minds grasp the world?' (We might call this question the epistemological problem.)

His reasoning was somewhat as follows. The mind can grasp the world, or rather the world as it appears to us, because this world is not utterly different from the mind—because it is mind-like. And it is so, because in the process of obtaining knowledge, of grasping the world, the mind is, so to speak, actively digesting all that material which enters it by the senses. It is forming, moulding this material; it impresses on it its own intrinsic forms or laws—the forms or laws of our thought. What we call 'nature'—the world in which we live, the world as it appears to us—is already a world digested, a world formed, by our minds. And being thus assimilated by the mind, it is mind-like.

The answer, 'The mind can grasp the world because the world *as it appears to us* is mind-like' is an idealistic argument; for what idealism asserts is just that the world has something of the character of mind.

I do not intend to argue here for or against this Kantian epistemology and I do not intend to discuss it in detail. But I want to point out that it certainly is not entirely idealistic. It is, as Kant himself points out, a mixture or a *synthesis*, of some sort of realism and some sort of idealism—its realist element being the assertion that the world, as it appears to us, is some sort of *material* formed by our mind, whilst its idealist element is the assertion that it is some sort of material *formed by our mind*.

So much for Kant's rather abstract but certainly ingenious epistemology. Before I proceed to Hegel, I must beg those readers (I like them best) who are not philosophers and who are used to relying on their common sense to bear in mind the sentence which I chose as a motto for this paper; for what they will hear now will probably appear to them—in my opinion quite rightly—absurd.

As I have said, Hegel in his idealism went further than Kant. Hegel, too, was concerned with the epistemological question, 'How can our minds grasp the world?' With the other idealists, he answered: 'Because the world is mind-like.' But his theory was more radical than Kant's. He did not say, like Kant, 'Because the mind *digests* or *forms* the world'. He said, 'Because the mind *is* the world'; or in another formulation, 'Because the reasonable *is* the real; because reality and reason are identical'.

This is Hegel's so-called 'philosophy of the identity of reason and reality', or, for short, his 'philosophy of identity'. It may be noted in passing that between Kant's epistemological answers, 'Because the mind forms the world', and Hegel's philosophy of identity, 'Because the mind is the world', there was, historically, a bridge—namely Fichte's answer, 'Because the mind creates the world'.[10]

Hegel's philosophy of identity, 'That which is reasonable is real, and that which is real is reasonable; thus, reason and reality are identical', was undoubtedly an attempt to re-establish rationalism on a new basis. It permitted the philosopher to construct a theory of the world out of pure reasoning and to maintain that this must be a true theory of the real world. Thus it allowed exactly what Kant had said to be impossible. Hegel, therefore, was bound to try to refute Kant's arguments against metaphysics. He did this with the help of his dialectic.

To understand his dialectic, we have to go back to Kant again. To avoid too much detail, I shall not discuss the triadic construction of Kant's table of categories, although no doubt it inspired Hegel.[11] But I have to refer to Kant's method of rejecting rationalism. I mentioned above that Kant maintained that the scope of our knowledge is limited to the field of possible experience and that pure reasoning beyond this field is not justified. In a section of the *Critique* which he headed 'Transcendental Dialectic' he showed this as follows. If we try to construct a theoretical system out of pure reason—for instance, if we try to argue that the world in which we live is infinite (an idea which obviously goes beyond possible experience)—then we can do so; but we shall find to our dismay that we can always argue, with the help of analogous arguments, to the opposite effect as well. In other words, given such a metaphysical thesis, we could always construct and defend an exact antithesis; and for any argument which supports the thesis, we can easily construct its opposite argument in favour of the antithesis. And both arguments will carry with them a similar force and conviction—

[10] This answer was not even original, because Kant had considered it previously; but he of course rejected it.

[11] MacTaggart has made this point the centre of his interesting *Studies in Hegelian Dialectic*.

both arguments will appear to be equally, or almost equally, reasonable. Thus, Kant said, reason is bound to argue against itself and to contradict itself, if used to go beyond possible experience.

If I were to give some sort of modernized reconstruction, or reinterpretation, of Kant, deviating from Kant's own view of what he had done, I should say that Kant showed that the metaphysical principle of reasonableness or self-evidence does not lead unambiguously to one and only one result or theory. It is always possible to argue, with similar apparent reasonableness, in favour of a number of different theories, and even of opposite theories. Thus if we get no help from experience, if we cannot make experiments or observations which at least tell us to eliminate certain theories—namely those which although they may seem quite reasonable, are contrary to the observed facts—then we have no hope of ever settling the claims of competing theories.

How did Hegel overcome Kant's refutation of rationalism? Very easily, by holding that contradictions do not matter. They just have to occur in the development of thought and reason. They only show the insufficiency of a theory which does not take account of the fact that thought, that is reason, and with it (according to the philosophy of identity) reality, is not something fixed once and for all, but is developing—that we live in a world of evolution. Kant, so says Hegel, refuted metaphysics, but not rationalism. For what Hegel calls 'metaphysics', as opposed to 'dialectic', is only such a rationalistic system as does not take account of evolution, motion, development, and thus tries to conceive of reality as something stable, unmoved and free of contradictions. Hegel, with his philosophy of identity, infers that since reason develops, the world must develop, and since the development of thought or reason is a dialectic one, the world must also develop in dialectic triads.

Thus we find the following three elements in Hegel's dialectic.

(*a*) An attempt to evade Kant's refutation of what Kant called 'dogmatism' in metaphysics. This refutation is considered by Hegel to hold only for systems which are metaphysical in his more narrow sense, but not for dialectical rationalism, which takes account of the development of reason and is therefore not afraid of contradictions. In evading Kant's criticism in this way, Hegel embarks on an extremely dangerous venture which must lead to disaster, for he argues something like this: 'Kant refuted rationalism by saying that it must lead to contradictions. I admit that. But it is clear that this argument draws its force from the law of contradiction: it refutes only such systems as accept this law, i.e. such as try to be free from contradictions. It is not dangerous for a system like mine which is prepared to put up with contradictions—that is, for a dialectic system.' It is clear that this argument establishes a dogmatism of an extremely dangerous kind—a dogmatism which need no longer be afraid of any sort of attack. For any attack, any criticism of any theory whatsoever, must be based on the method of pointing out some sort of contradiction, either within the theory itself or between the theory and some facts, as I said above. Hegel's method of superseding Kant, therefore, is effective, but unfortunately too effective. It makes his system secure against any sort of criticism or attack and thus it is dogmatic in a very peculiar sense, so that I should like to call it a 'reinforced dogmatism'. (It may be remarked that similar reinforced dogmatisms help to support the structures of other dogmatic systems as well.)

(*b*) The description of the development of reason in terms of dialectic is an element in Hegel's philosophy which had a good deal of plausibility. This becomes clear if we remember that Hegel uses the word 'reason' not only in the subjective sense, to denote a certain mental capacity, but also in the objective sense, to denote all sorts of theories, thoughts, ideas and so on. Hegel, who holds that philosophy is the highest expression of reasoning, has in mind mainly the development of philosophical thought when he speaks of the development of reasoning. And indeed hardly anywhere can the dialectic triad be more successfully applied than in the study of the development of philosophical theories, and it is therefore not surprising that Hegel's most successful attempt at applying his dialectic method was his History of Philosophy.

In order to understand the danger connected with such a success, we have to remember that in Hegel's time—and even much later—logic was usually described and defined as the theory of reasoning or the theory of thinking, and accordingly the fundamental laws of logic were usually called the 'laws of thought'. It is therefore quite understandable that Hegel, believing that dialectic is the true description of our actual procedure when reasoning and thinking, held that he must alter logic so as to make dialectic an important, if not the most important, part of logical theory. This made it necessary to discard the 'law of contradiction', which clearly was a grave obstacle to the acceptance of dialectic. Here we have the origin of the view that dialectic is 'fundamental' in the sense that it can compete with logic, that it is an improvement upon logic. I have already criticized this view of dialectic, and I only want to repeat that any sort of logical reasoning, whether before or after Hegel, and whether in science or in mathematics or in any truly rational philosophy, is always based on the law of contradiction. But Hegel writes (*Logic*, Section 81, (1)): 'It is of the highest importance to ascertain and understand rightly the nature of Dialectic. Wherever there is movement, wherever there is life, wherever anything is carried into effect in the actual world, there Dialectic is at work. It is also the soul of all knowledge which is truly scientific.'

But if by dialectic reasoning Hegel means a reasoning which discards the law of contradiction, then he certainly would not be able to give any instance of such reasoning in science. (The many instances quoted by dialecticians are without exception on the level of Engels's examples referred to above—the grain and $(-a)^2 = a^2$—or even worse.) It is not scientific reasoning itself which is based on dialectic; it is only the history and development of scientific theories which can with some success be described in terms of the dialectic method. As we have seen, this fact cannot justify the acceptance of dialectic as something fundamental, because it can be explained without leaving the realm of ordinary logic if we remember the working of the trial and error method.

The main danger of such a confusion of dialectic and logic is, as I said, that it helps people to argue dogmatically. For we find only too often that dialecticians, when in logical difficulties, as a last resort tell their opponents that their criticism is mistaken because it is based on logic of the ordinary type instead of on dialectic; if they would only use dialectic, they would see that the contradictions which they have found in some arguments of the dialecticians are quite legitimate (namely from the dialectic point of view).

(c) A third element in Hegelian dialectic is based on his philosophy of identity. If reason and reality are identical and reason develops dialectically (as is so well exemplified by the development of philosophical thought) then reality must develop dialectically too. The world must be ruled by the laws of dialectical logic. (This standpoint has been called 'panlogism'.) Thus, we must find in the world the same contradictions as are permitted by dialectic logic. It is this very fact that the world is full of contradictions which shows us from another angle that the law of contradiction has to be discarded. For this law says that no self-contradictory proposition, or no pair of contradictory propositions, can be true, that is, can correspond to the facts. In other words, the law implies that a contradiction can never occur in nature, i.e. in the world of facts, and that facts can never contradict each other. But on the basis of the philosophy of the identity of reason and reality, it is asserted that facts can contradict each other since ideas can contradict each other and that facts develop through contradictions, just as ideas do; so that the law of contradiction has to be abandoned.

But apart from what appears to me to be the utter absurdity of the philosophy of identity (about which I shall say something later), if we look a little closer into these so-called contradictory facts, then we find that all the examples proffered by dialecticians just state that the world in which we live shows, sometimes, a certain structure which could perhaps be described with the help of the word 'polarity'. An instance of that structure would be the existence of positive and negative electricity. It is only a metaphorical and loose way of speaking to say, for instance, that positive and negative electricity are contradictory to each other. An example of a true contradiction would be two sentences: 'This body here was, on the 1st of November, 1938, between 9 and 10 a.m., positively charged', and an analogous sentence about the same body, saying that it was at the same time *not* positively charged.

This would be a contradiction between two sentences and the corresponding contradictory fact would be the fact that a body is, as a whole, at the same time both positively and not positively charged, and thus at the same time both attracts and does not attract certain negatively charged bodies. But we need not say that such contradictory facts do not exist. (A deeper analysis might show that the non-existence of such facts is not a law which is akin to laws of physics, but is based on logic, that is, on the rules governing the use of scientific language.)

So there are three points: (*a*) the dialectic opposition to Kant's anti-rationalism, and consequently the re-establishment of rationalism supported by a reinforced dogmatism; (*b*) the incorporation of dialectic in logic, grounded on the ambiguity of expressions like 'reason', 'laws of thought', and so on; (*c*) the application of dialectic to 'the whole world', based on Hegel's panlogism and his philosophy of identity. These three points seem to me to be the main elements within Hegelian dialectic. Before I proceed to outline the fate of dialectic after Hegel, I should like to express my personal opinion about Hegel's philosophy, and especially about his philosophy of identity. I think it represents the worst of all those absurd and incredible philosophic theories to which Descartes refers in the sentence which I have chosen as the motto for this paper. It is not only that philosophy of identity is offered without any sort of serious argument; even the problem which it has been invented to answer—the question, 'How can our minds grasp the world?'—seems to me not to be at all clearly formulated. And the idealist answer, which has been varied by different idealist philosophers but remains fundamentally the same, namely, 'Because the world is mind-like', has only the appearance of an answer. We shall see clearly that it is not a real answer if we only consider some analogous argument, like: 'How can this mirror reflect my face?'—'Because it is face-like.' Although this sort of argument is obviously utterly unsound, it has been formulated again and again. We find it formulated by Jeans, for instance, in our own time, along lines like these: 'How can mathematics grasp the world?'—'Because the world is mathematics-like.' He argues thus that reality is of the very nature of mathematics—that the world is a mathematical thought (and therefore ideal). This argument is obviously no sounder than the following: 'How can language describe the world?'—'Because the world is language-like—it is linguistic', and no sounder than: 'How can the English language describe the world?'—'Because the world is intrinsically British.' That this latter argument really is analogous to the one advanced by Jeans is easily seen if we recognize that the mathematical description of the world is just a certain way of describing the world and nothing else, and that mathematics supplies us with the means of description—with a particularly rich language.

Perhaps one can show this most easily with the help of a trivial example. There are primitive languages which do not employ numbers but try to express numerical ideas with the help of expressions for one, two, and many. It is clear that such a language is unable to describe some of the more complicated relationships between certain groups of objects, which can easily be described with the help of the numerical expressions 'three', 'four', 'five', and so on. It can say that A has many sheep, and more than B, but it cannot say that A has 9 sheep and 5 more than B. In other words, mathematical symbols are introduced into a language in order to describe certain more complicated relationships which could not be described otherwise; a language which contains the arithmetic of natural numbers is simply richer than a language which lacks the appropriate symbols. All that we can infer about the nature of the world from the fact that we have to use mathematical language if we want to describe it is that the world has a certain degree of complexity, so that there are certain relationships in it which cannot be described with the help of too primitive instruments of description.

Jeans was uneasy about the fact that our world happens to suit mathematical formulae originally invented by pure mathematicians who did not intend at all to apply their formulae to the world. Apparently he originally started off as what I should call an 'inductivist'; that is, he thought that theories are obtained from experience by some more or less simple procedure of inference. If one starts from such a position it obviously is astonishing to find that a theory which has been formulated by pure mathematicians, in a purely speculative manner, afterwards proves to be applicable to the physical world. But for those who are not inductivists, this is not astonishing at all. They know that it happens quite often that a theory put forward originally as a pure speculation, as a mere possibility, later proves to have its empirical applications. They know that often it is this speculative anticipation which prepares the way for the empirical theories. (In this way the problem of induction, as it is called, has a bearing on the problem of idealism with which we are concerned here.)

3. DIALECTIC AFTER HEGEL

> The thought that facts or events might mutually contradict each other appears to me as the very paradigm of thoughtlessness.
>
> DAVID HILBERT

Hegel's philosophy of the identity of reason and reality is sometimes characterized as (absolute) idealism, because it states that reality is mind-like or of the character of reason. But clearly such a dialectical philosophy of identity can easily be turned round so as to become a kind of materialism. Its adherents would then argue that reality is in fact of material or physical character, as the ordinary man thinks it is; and by saying that it is identical with reason, or mind, one would imply that the mind is also a material or physical phenomenon—or if not, that the difference between the mental and the physical cannot be of great importance.

This materialism can be regarded as a revival of certain aspects of Cartesianism, modified by links with dialectic. But in discarding its original idealistic basis, dialectic loses everything which made it plausible and understandable; we have to remember that the best arguments in favour of dialectic lay in its applicability to the development of thought, especially of philosophical thought. Now we are faced blankly with the statement that physical reality develops dialectically—an extremely dogmatic assertion with so little scientific support that materialistic dialecticians are forced to make a very extensive use of the dangerous method we have already described whereby criticism is rejected as non-dialectical. Dialectical materialism is thus in agreement with points (*a*) and (*b*) discussed above, but it alters point (*c*) considerably, although I think with no advantage to its dialectic features. In expressing this opinion, I want to stress the point that although I should not describe myself as a materialist, my criticism is not directed against materialism, which I personally should probably prefer to idealism if I were forced to choose (which happily I am not). It is only the combination of dialectic and materialism that appears to me to be even worse than dialectic idealism.

These remarks apply particularly to the 'Dialectical Materialism' developed by Marx. The materialistic element in this theory could be comparatively easily reformulated in such a way that no serious objections to it could be made. As far as I can see the main point is this: there is no reason to assume that whilst the natural sciences can proceed on the basis of the common man's realistic outlook the social sciences need an idealist background like the one offered by Hegelianism. Such an assumption was often made in Marx's time, owing to the fact that Hegel with his idealist theory of the State appeared strongly to influence, and even to further, the social sciences, while the futility of views which he held within the field of the natural sciences was—at least for natural scientists—only too obvious.[12] I think it is a fair interpretation of the ideas of Marx and Engels to say that one of their chief interests in emphasizing materialism was to dismiss any theory which, referring to the rational or spiritual nature of man, maintains that sociology has to be based on an idealist or spiritualist basis, or on the analysis of reason. In opposition they stressed the material side of human nature—such as our need for food and other material goods—and its importance for sociology.

This view was undoubtedly sound; and I hold Marx's contributions on this point to be of real significance and lasting influence. Everyone learned from Marx that the development even of ideas cannot be fully understood if the history of ideas is treated (although such a treatment may often have its great merits) without mentioning the conditions of their origin and the situation of their originators, among which conditions the economic aspect is highly significant. Nevertheless I person-

[12] At least it should be obvious to everybody who considers, as an instance, the following surprising analysis of the essence of *electricity* which I have translated as well as I could, even to the extent of trying to render it more understandable than Hegel's original:

'Electricity . . . is the purpose of the form from which it emancipates itself, it is the form that is just about to overcome its own indifference; for, electricity is the immediate emergence, or the actually just emerging, from the proximity of the form, and still determined by it—not yet the dissolution, however, of the form itself, but rather the more superficial process by which the differences desert the form which, however, they still retain, as their condition, having not yet grown into independence of and through them.' (No doubt it ought to have been 'of and through it'; but I do not wish to suggest that this would have made much difference to the differences.) The passage is from Hegel's *Philosophy of Nature*. See also the passages on Sound and on Heat, quoted in my *Open Society*, note 4 to ch. 12.

ally think that Marx's economism—his emphasis on the economic background as the ultimate basis of any sort of development—is mistaken and in fact untenable. I think that social experience clearly shows that under certain circumstances the influence of ideas (perhaps supported by propaganda) can outweigh and supersede economic forces. Besides, granted that it is impossible fully to understand mental developments without understanding their economic background, it is at least as impossible to understand economic developments without understanding the development of, for instance, scientific or religious ideas.

For our present purpose it is not so important to analyse Marx's materialism and economism as to see what has become of the dialectic within his system. Two points seem to me important. One is Marx's emphasis on historical method in sociology, a tendency which I have called 'historicism'. The other is the anti-dogmatic tendency of Marx's dialectic.

As for the first point, we have to remember that Hegel was one of the inventors of the historical method, a founder of the school of thinkers who believed that in describing a development historically one has causally explained it. This school believed that one could, for example, explain certain social institutions by showing how mankind has slowly developed them. Nowadays it is often recognized that the significance of the historical method for social theory has been much over-rated; but the belief in this method has by no means disappeared. I have tried to criticize this method elsewhere (especially in my book *The Poverty of Historicism*). Here I merely want to stress that Marx's sociology adopted from Hegel not only the view that its method has to be historical, and that sociology as well as history have to become theories of social development, but also the view that this development has to be explained in dialectical terms. To Hegel history was the history of ideas. Marx dropped idealism but retained Hegel's doctrine that the dynamic forces of historical development are the dialectical 'contradictions', 'negations', and 'negations of negations'. In this respect Marx and Engels followed Hegel very closely indeed, as may be shown by the following quotations. Hegel in his *Encyclopaedia* (Part I ch. vi, p. 81) described Dialectic as 'the universal and irresistible power before which nothing can stay, however secure and stable it may deem itself'.

Similarly, Engels writes (*Anti-Dühring*, Part I, 'Dialectics: Negation of the Negation'): 'What therefore is the negation of the negation? An extremely general . . . law of development of Nature, history and thought; a law which . . . holds good in the animal and plant kingdom, in geology, in mathematics, in history, and in philosophy.'

In Marx's view it is the main task of sociological science to show how these dialectic forces are working in history, and thus to prophesy the course of history; or, as he says in the preface to *Capital*, 'It is the ultimate aim of this work to lay bare the economic law of motion of modern society'. And this dialectic law of motion, the negation of the negation, furnishes the basis of Marx's prophecy of the impending end of capitalism (*Capital*, I, ch. XXIV, § 7): 'The capitalist mode of production . . . is the first negation . . . But capitalism begets, with the inexorability of a law of Nature, its own negation. It is the negation of the negation.'

Prophecy certainly need not be unscientific, as predictions of eclipses and other astronomical events show. But Hegelian dialectic, or its materialistic version, cannot be accepted as a sound basis for scientific forecasts. ('But all Marx's predictions have come true', Marxists usually answer. They have not. To quote one example out of many: In *Capital*, immediately after the last passage quoted, Marx said that the transition from capitalism to socialism would naturally be a process incomparably less 'protracted, violent, and difficult' than the industrial revolution, and in a footnote he amplified this forecast by referring to the 'irresolute and non-resisting bourgeoisie'. Few Marxists will say nowadays that these predictions were successful.) Thus if forecasts based on dialectic are made, some will come true and some will not. In the latter case, obviously, a situation will arise which has not been foreseen. But dialectic is vague and elastic enough to interpret and to explain this unforeseen situation just as well as it interpreted and explained the situation which it predicted and which happened to come true. Any development whatever will fit the dialectic scheme; the dialectician need never be afraid of any refutation by future

[13] In *L.Sc.D.* I have tried to show that the scientific content of a theory is the greater the more the theory conveys, the more it risks, the more it is exposed to refutation by future experience. If it takes no such risks, its scientific content is zero—it has no scientific content, it is metaphysical. By this standard we can say that dialectic is unscientific: it is metaphysical.

experience.[13] As mentioned before, it is not just the dialectical approach, it is, rather, the idea of a theory of historical development—the idea that scientific sociology aims at large-scale historical forecasts—which is mistaken. But this does not concern us here.

Apart from the role dialectic plays in Marx's historical method, Marx's anti-dogmatic attitude should be discussed. Marx and Engels strongly insisted that science should not be interpreted as a body of final and well-established knowledge, or of 'eternal truth', but rather as something developing, progressive. The scientist is not the man who knows a lot but rather the man who is determined not to give up the search for truth. Scientific systems develop; and they develop, according to Marx, dialectically.

There is not very much to be said against this point—although personally I think that the dialectical description of scientific development is not always applicable unless it is forced, and that it is better to describe scientific development in a less ambitious and ambiguous way, as for example, in terms of the trial and error theory. But I am prepared to admit that this criticism is not of great importance. It is, however, of real moment that Marx's progressive and anti-dogmatic view of science has never been applied by orthodox Marxists within the field of their own activities. Progressive, anti-dogmatic science is critical—criticism is its very life. But criticism of Marxism, of dialectical materialism, has never been tolerated by Marxists.

Hegel thought that philosophy develops; yet his own system was to remain the last and highest stage of this development and could not be superseded. The Marxists adopted the same attitude towards the Marxian system. Hence, Marx's anti-dogmatic attitude exists only in the theory and not in the practice of orthodox Marxism, and dialectic is used by Marxists, following the example of Engels' *Anti-Dühring*, mainly for the purposes of apologetics—to defend the Marxist system against criticism. As a rule critics are denounced for their failure to understand the dialectic, or proletarian science, or for being traitors. Thanks to dialectic the anti-dogmatic attitude has disappeared, and Marxism has established itself as a dogmatism which is elastic enough, by using its dialectic method, to evade any further attack. It has thus become what I have called a reinforced dogmatism.

Yet there can be no worse obstacle to the growth of science than a reinforced dogmatism. There can be no scientific development without the free competition of thought—this is the essence of the anti-dogmatic attitude once so strongly supported by Marx and Engels; and in general there cannot be free competition in scientific thought without freedom for all thought.

Thus dialectic has played a very unfortunate role not only in the development of philosophy, but also in the development of political theory. A full understanding of this unfortunate role will be easier if we try to see how Marx originally came to develop such a theory. We have to consider the whole situation. Marx, a young man who was progressive, evolutionary and even revolutionary in his thought, came under the influence of Hegel, the most famous German philosopher. Hegel had been a representative of Prussian reaction. He had used his principle of the identity of reason and reality to support the existing powers—for what exists, is reasonable—and to defend the idea of the Absolute State (an idea nowadays called 'Totalitarianism'). Marx, who admired him, but who was of a very different political temperament, needed a philosophy on which to base his own political opinions. We can understand his elation at discovering that Hegel's dialectical philosophy could easily be turned against its own master—that dialectic was in favour of a revolutionary political theory, rather than of a conservative and apologetic one. Besides this, it was excellently adapted to his need for a theory which should be not only revolutionary, but also optimistic—a theory forecasting progress by emphasizing that every new step is a step upwards.

This discovery, although undeniably fascinating for a disciple of Hegel and in an era dominated by Hegel, has now, together with Hegelianism, lost all significance, and can hardly be considered to be more than the clever *tour de force* of a brilliant young student revealing a weakness in the speculations of his undeservedly famous master. But it became the theoretical basis of what is called 'Scientific Marxism'. And it helped to turn Marxism into a dogmatic system by preventing the scientific development of which it might have been capable. So Marxism has for decades kept its dogmatic attitude, repeating against its opponents just the same arguments as were originally used by its founders. It is sad but illuminating to see how orthodox Marxism today officially recommends, as a basis for the study of scientific methodology, the reading of Hegel's *Logic*—which is not merely obsolete but typical of pre-scientific and even pre-logical ways of thinking. It is worse than recommending Archimedes' mechanics as a basis for modern engineering.

The whole development of dialectic should be a warning against the dangers inherent in philosophical system-building. It should remind us that philosophy must not be made a basis for any sort of scientific system and that philosophers should be much more modest in their claims. One task which they can fulfil quite usefully is the study of the critical methods of science.

16

PREDICTION AND PROPHECY IN THE SOCIAL SCIENCES

I

The topic of my address is 'Prediction and Prophecy in the Social Sciences'. My intention is to criticize the doctrine that it is the task of the social sciences to propound historical prophecies, and that historical prophecies are needed if we wish to conduct politics in a rational way.[1] I shall call this doctrine 'historicism'. I consider historicism to be the relic of an ancient superstition, even though the people who believe in it are usually convinced that it is a very new, progressive, revolutionary, and scientific theory.

The tenets of historicism—that it is the task of the social sciences to propound historical prophecies, and that these historical prophecies are needed for any rational theory—are topical today because they form a very important part of that philosophy which likes to call itself

An address delivered to the Plenary Session of the 10th International Congress of Philosophy, Amsterdam 1948, and published in the Library of the 10th International Congress of Philosophy, **1**, *Amsterdam*, 1948; *and in* Theories of History, *ed. P. Gardiner*, 1959.

[1] A fuller discussion of this problem, and of a number of related problems, will be found in my book *The Poverty of Historicism*, 1957, 1959, 1961.

by the name of 'Scientific Socialism' or 'Marxism'. My analysis of the role of prediction and prophecy could therefore be described as a criticism of the historical method of Marxism. But in fact it does not confine itself to that economic variant of historicism which is known as Marxism, for it aims at criticizing the historicist doctrine in general. Nevertheless, I have decided to speak as if Marxism were my main or my only object of attack, since I wish to avoid the accusation that I am attacking Marxism surreptitiously under the name of 'historicism'. But I should be glad if you would remember that whenever I mention Marxism, I always have in mind a number of other philosophies of history also; for I am trying to criticize a certain historical method which has been believed to be valid by many philosophers, ancient and modern, whose political views were very different from those of Marx.

As a critic of Marxism, I shall try to interpret my task in a liberal spirit. I shall feel free not only to criticize Marxism but also to defend certain of its contentions; and I shall feel free to simplify its doctrines radically.

One of the points on which I feel sympathy with Marxists is their insistence that the social problems of our time are urgent, and that philosophers ought to face the issues; that we should not be content to interpret the world but should help to change it. I am very much in sympathy with this attitude, and the choice by the present assembly of the theme 'Man and Society', shows that the need to discuss these problems is widely recognized. The mortal danger into which mankind has floundered—no doubt the gravest danger in its history—must not be ignored by philosophers.

But what kind of contribution can philosophers make—not just as men, not just as citizens, but as philosophers? Some Marxists insist that the problems are too urgent for further contemplation, and that we ought to take sides at once. But if—as philosophers—we can make any contribution at all then, surely, we must refuse to be rushed into blindly accepting ready-made solutions, however great the urgency of the hour; as philosophers we can do no better than bring rational criticism to bear on the problems that face us, and on the solutions advocated by the various parties. To be more specific, I believe that the best I can do as philosopher is to approach the problems armed with the weapons of a *critic of methods*. This is what I propose to do.

II

I may, by way of introduction, say why I have chosen this particular subject. I am a rationalist, and by this I mean that I believe in discussion, and argument. I also believe in the possibility as well as the desirability of applying science to problems arising in the social field. But believing as I do in social science, I can only look with apprehension upon social pseudo-science.

Many of my fellow-rationalists are Marxists; in England, for example, a considerable number of excellent physicists and biologists emphasize their allegiance to the Marxist doctrine. They are attracted to Marxism by its claims: (*a*) that it is a science, (*b*) that it is progressive, and (*c*) that it adopts the methods of prediction which the natural sciences practise. Of course, everything depends upon this third claim. I shall therefore try to show that this claim is not justified, and that the kind of prophecies which Marxism offers are in their logical character more akin to those of the Old Testament than to those of modern physics.

III

I shall begin with a brief statement and criticism of the historical method of the alleged science of Marxism. I shall have to oversimplify matters; this is unavoidable. But my oversimplifications may serve the purpose of bringing the decisive points into focus.

The central ideas of the historicist method, and more especially of Marxism, seem to be these:

(*a*) It is a fact that we can predict solar eclipses with a high degree of precision, and for a long time ahead. Why should we not be able to predict revolutions? Had a social scientist in 1780 known half as much about society as the old Babylonian astrologers knew about astronomy, then he should have been able to predict the French Revolution.

The fundamental idea that it should be possible to predict revolutions just as it is possible to predict solar eclipses gives rise to the following view of the task of the social sciences:

(*b*) The task of the social sciences is fundamentally the same as that of the natural sciences—to make predictions, and, more especially, historical predictions, that is to say, predictions about the social and political development of mankind.

(*c*) Once these predictions have been made, the task of politics can be determined. For it is to lessen the 'birthpangs' (as Marx calls them) unavoidably connected with the political developments which have been predicted as impending.

These simple ideas, especially the one claiming that it is the task of the social sciences to make historical predictions, such as predictions of social revolutions, I shall call the *historicist doctrine of the social sciences*. The idea that it is the task of politics to lessen the birthpangs of impending political developments I shall call the *historicist doctrine of politics*. Both these doctrines may be considered as parts of a wider philosophical scheme which may be called historicism—the view that the story of mankind has a plot, and that if we can succeed in unravelling this plot, we shall hold the key to the future.

IV

I have briefly outlined two historicist doctrines concerning the task of the social sciences and of politics. I have described these doctrines as Marxist. But they are not peculiar to Marxism. On the contrary, they are among the oldest doctrines in the world. In Marx's own time they were held, in exactly the form described, not only by Marx who inherited them from Hegel, but by John Stuart Mill who inherited them from Comte. And they were held in ancient times by Plato, and before him by Heraclitus and Hesiod. They seem to be of oriental origin; indeed, the Jewish idea of the chosen people is a typical historicist idea—that history has a plot whose author is Jahwe, and that the plot can be partly unravelled by the prophets. These ideas express one of the oldest dreams of mankind—the dream of prophecy, the idea that we can know what the future has in store for us, and that we can profit from such knowledge by adjusting our policy to it.

This age-old idea was sustained by the fact that prophecies of eclipses and of the movements of the planets were successful. The close connection between historicist doctrine and astronomical knowledge is clearly exhibited in the ideas and practices of astrology.

These historical points have, of course, no bearing on the question whether or not the historicist doctrine concerning the task of the social sciences is tenable. This question belongs to the methodology of the social sciences.

V

The historicist doctrine which teaches that it is the task of the social sciences to predict historical developments is, I believe, untenable.

Admittedly all theoretical sciences are predicting sciences. Admittedly there are social sciences which are theoretical. But do these admissions imply—as the historicists believe—that the task of the social sciences is historical prophecy? It looks like it: but this impression disappears once we make a clear distinction between what I shall call '*scientific prediction*' on the one side and '*unconditional historical prophecies*' on the other. Historicism fails to make this important distinction.

Ordinary predictions in science are conditional. They assert that certain changes (say, of the temperature of water in a kettle) will be accompanied by other changes (say, the boiling of the water). Or to take a simple example from a social science: Just as we can learn from a physicist that under certain physical conditions a boiler will explode, so we can learn from the economist that under certain social conditions, such as shortage of commodities, controlled prices, and, say, the absence of an effective punitive system, a black market will develop.

Unconditional scientific predictions can sometimes be derived from these conditional scientific predictions, together with historical statements which assert that the conditions in question are fulfilled. (From these premises we can obtain the unconditional prediction by the *modus ponens*.) If a physician has diagnosed scarlet fever then he may, with the help of the conditional predictions of his science, make the unconditional prediction that his patient will develop a rash of a certain kind. But it is possible, of course, to make such unconditional prophecies without any such justification in a theoretical science, or—in other words—in scientific conditional predictions. They may be based, for example, on a dream—and by some accident they may even come true.

My contentions are two.

The first is that the historicist does not, as a matter of fact, derive his historical prophecies from conditional scientific predictions. The second (from which the first follows) is that he cannot possibly do so because long-term prophecies can be derived from scientific conditional predictions only if they apply to systems which can be described as well-isolated, stationary, and recurrent. These systems are very rare in nature; and modern society is surely not one of them.

Let me develop this point a little more fully. Eclipse prophecies, and indeed prophecies based on the regularity of the seasons (perhaps the oldest natural laws consciously understood by man) are possible only because our solar system is a stationary and repetitive system; and this is so because of the accident that it is isolated from the influence of other mechanical systems by immense regions of empty space and is therefore relatively free of interference from outside. Contrary to popular belief the analysis of such repetitive systems is not typical of natural science. These repetitive systems are special cases where scientific prediction becomes particularly impressive—but that is all. Apart from this very exceptional case, the solar system, recurrent or cyclic systems are known especially in the field of biology. The life cycles of organisms are part of a semi-stationary or very slowly changing biological chain of events. Scientific predictions about life cycles of organisms can be made in so far as we abstract from the slow evolutionary changes, that is to say, in so far as we treat the biological system in question as stationary.

No basis can therefore be found in examples such as these for the contention that we can apply the method of long-term unconditional prophecy to human history. Society is changing, developing. This development is not, in the main, repetitive. True, in so far as it is repetitive, we may perhaps make certain prophecies. For example, there is undoubtedly some repetitiveness in the manner in which new religions arise, or new tyrannies; and a student of history may find that he can foresee such developments to a limited degree by comparing them with earlier instances, i.e. by studying the conditions under which they arise. But this application of the method of conditional prediction does not take us very far. For the most striking aspects of historical development are non-repetitive. Conditions are changing, and situations arise (for example, in consequence of new scientific discoveries) which are very different from anything that ever happened before. The fact that we can predict eclipses does not, therefore, provide a valid reason for expecting that we can predict revolutions.

These considerations hold not only for the evolution of man, but also for the evolution of life in general. There exists no law of evolution, only the historical fact that plants and animals change, or more precisely, that they have changed. The idea of a law which determines the direction and the character of evolution is a typical nineteenth-century mistake, arising out of the general tendency to ascribe to the 'Natural Law' the functions traditionally ascribed to God.

VI

The realization that the social sciences cannot prophesy future historical developments has led some modern writers to despair of reason, and to advocate political irrationalism. Identifying predictive power with practical usefulness, they denounce the social sciences as useless. In an attempt to analyse the possibility of forecasting historical developments, one of these modern irrationalists writes[2]: 'The same element of uncertainty from which the natural sciences suffer affects the social sciences, only more so. Because of its quantitative extension, it affects here not only theoretical structure but also *practical usefulness*.'

But there is no need as yet to despair of reason. Only those who do not distinguish between ordinary prediction and historical prophecy, in other words, only historicists—disappointed historicists—are likely to draw such desperate conclusions. The main usefulness of the physical sciences does not lie in the prediction of eclipses; and similarly, the practical usefulness of the social sciences does not depend on their power to prophesy historical or political developments. Only an uncritical historicist, that is to say, one who believes in the historicist doctrine of the task of the social sciences as a matter of course, will be driven to despair of reason by the realization that the social sciences cannot prophesy: and some have in fact been driven even to hatred of reason.

[2] H. Morgenthau, *Scientific Man and Power Politics*, London, 1947, p. 122, italics mine. As indicated in my next paragraph, Morgenthau's anti-rationalism can be understood as resulting from the disillusionment of a historicist who cannot conceive of any form of rationalism except a historicist form.

VII

What then is the task of the social sciences, and how can they be useful?

In order to answer this question, I shall first briefly mention two naïve theories of society which must be disposed of before we can understand the function of the social sciences.

The first is the theory that the social sciences study the behaviour of social wholes, such as groups, nations, classes, societies, civilizations, etc. These social wholes are conceived as the empirical objects which the social sciences study in the same way in which biology studies animals or plants.

This view must be rejected as naïve. It completely overlooks the fact that these so-called social wholes are very largely postulates of popular social theories rather than empirical objects; and that while there are, admittedly, such empirical objects as the crowd of people here assembled, it is quite untrue that names like 'the middle-class' stand for any such empirical groups. What they stand for is a kind of ideal object whose existence depends upon theoretical assumptions. Accordingly, the belief in the empirical existence of social wholes or collectives, which may be described as *naïve collectivism*, has to be replaced by the demand that social phenomena, including collectives, should be analysed in terms of individuals and their actions and relations.

But this demand may easily give rise to another mistaken view, the second and more important of the two views to be disposed of. It may be described as the *conspiracy theory of society*. It is the view that whatever happens in society—including things which people as a rule dislike, such as war, unemployment, poverty, shortages—are the results of direct design by some powerful individuals or groups. This view is very widespread, although it is, I have no doubt, a somewhat primitive kind of superstition. It is older than historicism (which may even be said to be a derivative of the conspiracy theory); and in its modern form, it is the typical result of the secularization of religious superstitions. The belief in the Homeric gods whose conspiracies were responsible for the vicissitudes of the Trojan War is gone. But the place of the gods on Homer's Olympus is now taken by the Learned Elders of Zion, or by the monopolists, or the capitalists, or the imperialists.

Against the conspiracy theory of society I do not, of course, assert that conspiracies never happen. But I assert two things. First, they are not very frequent, and do not change the character of social life. Assuming that conspiracies were to cease, we should still be faced with fundamentally the same problems which have always faced us. Secondly, I assert that conspiracies are very rarely successful. The results achieved differ widely, as a rule, from the results aimed at. (Consider the Nazi conspiracy.)

VIII

Why do the results achieved by a conspiracy as a rule differ widely from the results aimed at? Because this is what usually happens in social life, conspiracy or no conspiracy. And this remark gives us an opportunity to formulate the *main task of the theoretical social sciences. It is to trace the unintended social repercussions of intentional human actions*. I may give a simple example. If a man wishes urgently to buy a house in a certain district, we can safely assume that he does not wish to raise the market price of houses in that district. But the very fact that he appears on the market as a buyer will tend to raise market prices. And analogous remarks hold for the seller. Or to take an example from a very different field, if a man decides to insure his life, he is unlikely to have the intention of encouraging other people to invest their money in insurance shares. But he will do so nevertheless.

We see here clearly that not all consequences of our actions are intended consequences; and accordingly, that the conspiracy theory of society cannot be true because it amounts to the assertion that all events, even those which at first sight do not seem to be intended by anybody, are the intended results of the actions of people who are interested in these results.

It should be mentioned in this connection that Karl Marx himself was one of the first to emphasize the importance, for the social sciences, of these unintended consequences. In his more mature utterances, he says that we are all caught in the net of the social system. The capitalist is not a demoniac conspirator, but a man who is forced by circumstances to act as he does; he is no more responsible for the state of affairs than is the proletarian.

This view of Marx's has been abandoned—perhaps for propagandist reasons, perhaps because people did not understand it—and a Vulgar Marxist Conspiracy theory has very largely replaced it. It is a come-down—the come-down from Marx to Goebbels. But it is clear that the adoption of the conspiracy theory can hardly be avoided by those who believe that they know how to make heaven on earth. The only explanation for their failure to produce this heaven is the malevolence of the devil who has a vested interest in hell.

IX

The view that it is the task of the theoretical social sciences to discover the unintended consequences of our actions brings these sciences very close to the experimental natural sciences. The analogy cannot here be developed in detail, but it may be remarked that both lead us to the formulation of practical technological rules stating *what we cannot do*.

The second law of thermodynamics can be expressed as the technological warning, 'You cannot build a machine which is 100 per cent efficient'. A similar rule of the social sciences would be, 'You cannot, without increasing productivity, raise the real income of the working population' and 'You cannot equalize real incomes and at the same time raise productivity'. An example of a promising hypothesis in this field which is by no means generally accepted—or, in other words, a problem that is still open—is the following: 'You cannot have a full employment policy without inflation.' These examples may show the way in which the social sciences are practically important. They do not allow us to make historical prophecies, but they may give us an idea of what can, and what cannot, be done in the political field.

We have seen that the historicist doctrine is untenable, but this fact does not lead us to lose faith in science or in reason. On the contrary, we now see that it gives rise to a clearer insight into the role of science in social life. Its practical role is the modest one of helping us to understand even the more remote consequences of possible actions, and thus of helping us to choose our actions more wisely.

X

The elimination of the historicist doctrine destroys Marxism completely as far as its scientific pretensions go. But it does not yet destroy the more technical or political claims of Marxism—that only a social revolution, a complete recasting of our social system, can produce social conditions fit for men to live in.

I shall not here discuss the problem of the humanitarian aims of Marxism. I find that there is a very great deal in these aims which I can accept. The hope of reducing misery and violence, and of increasing freedom, is one, I believe, which inspired Marx and many of his followers; it is a hope which inspires most of us.

But I am convinced that these aims cannot be realized by revolutionary methods. On the contrary, I am convinced that revolutionary methods can only make things worse—that they will increase unnecessary suffering; that they will lead to more and more violence; and that they must destroy freedom.

This becomes clear when we realize that a revolution always destroys the institutional and traditional framework of society. It must thereby endanger the very set of values for the realization of which it has been undertaken. Indeed, a set of values can have social significance only in so far as there exists a social tradition which upholds them. This is true of the aims of a revolution as much as of any other values.

But if you begin to revolutionize society and to eradicate its traditions, you cannot stop this process if and when you please. In a revolution, everything is questioned, including the aims of the well-meaning revolutionaries; aims which grow from, and which were necessarily a part of, the society which the revolution destroys.

Some people say that they do not mind this; that it is their greatest wish to clean the canvas thoroughly—to create a social *tabula rasa* and to begin afresh by painting on it a brand new social system. But they should not be surprised if they find that once they destroy tradition, civilization disappears with it. They will find that mankind have returned to the position in which Adam and Eve began—or, using less biblical language, that they have returned to the beasts. All that these revolutionary progressivists will then be able to do is to begin the slow process of human evolution again (and so to arrive in a few thousand years perhaps at another capitalist period, which will lead them to another sweeping revolution, followed by another return to the beasts, and so on, for ever and ever). In other words, there is no earthly reason why a society whose traditional set of values has been destroyed should, of its own accord, become a better society (unless you believe in political miracles,[3] or hope that once the conspiracy of the devilish capitalists is broken up, society will naturally tend to become beautiful and good).

Marxists, of course, will not admit this. But the Marxist view, that is to say, the view that the social revolution will lead to a better world, is only understandable on the *historicist assumptions* of Marxism. If you know, on the basis of historical prophecy, what the result of the social revolution must be, and if you know that the result is all that we hope for, then, but only then, can you consider the revolution with its untold suffering as a means to the end of untold happiness. But with the elimination of the historicist doctrine, the theory of revolution becomes completely untenable.

The view that it will be the task of the revolution to rid us of the capitalist conspiracy, and with it, of opposition to social reform, is widely held; but it is untenable, even if we assume for a moment that such a conspiracy exists. For a revolution is liable to replace old masters by new ones, and who guarantees that the new ones will be better? The theory of revolution overlooks the most important aspect of social life—that what we need is not so much good men as good institutions. Even the best man may be corrupted by power; but institutions which permit the ruled to exert some effective control over the rulers will force even bad rulers to do what the ruled consider to be in their interests. Or to put it another way, we should like to have good rulers, but historical experience shows us that we are not likely to get them. This is why it is of such importance to design institutions which prevent even bad rulers from causing too much damage.

There are only two kinds of governmental institutions, those which provide for a change of the government without bloodshed, and those which do not. But if the government cannot be changed without bloodshed, it cannot, in most cases, be removed at all. We need not

[3] The phrase is due to Julius Kraft.

quarrel about words, and about such pseudo-problems as the true or essential meaning of the word 'democracy'. You can choose whatever name you like for the two types of government. I personally prefer to call the type of government which can be removed without violence 'democracy', and the other 'tyranny'. But, as I said, this is not a quarrel about words, but an important distinction between two types of institutions.

Marxists have been taught to think in terms not of institutions but of classes. Classes, however, never rule, any more than nations. The rulers are always certain persons. And, whatever the class they may have belonged to, when they are rulers they belong to the ruling class.

Marxists nowadays do not think in terms of institutions; they put their faith in certain personalities, or perhaps in the fact that certain persons were once proletarians—a result of their belief in the overruling importance of classes and class loyalties. Rationalists, on the contrary, are more inclined to rely on institutions for controlling men. This is the main difference.

XI

But what ought the rulers to do? In opposition to most historicists, I believe that this question is far from vain; it is one which we ought to discuss. For in a democracy, the rulers will be compelled by the threat of dismissal to do what public opinion wants them to do. And public opinion is a thing which all can influence, and especially philosophers. In democracies, the ideas of philosophers have often influenced future developments—with a very considerable time-lag, to be sure. British social policy is now that of Bentham, and of John Stuart Mill who summed up its aim as that of 'securing full employment at high wages for the whole labouring population'.[4]

I believe that philosophers should continue to discuss the proper aims of social policy in the light of the experience of the last fifty years. Instead of confining themselves to discussing the 'nature' of ethics, or of the greatest good, etc., they should think about such fundamental

[4] In his *Autobiography*, 1873, p. 105. My attention has been drawn to this passage by F. A. Hayek. (For further comments on *public opinion* see also chapter 17, below.)

and difficult ethical and political questions as are raised by the fact that political freedom is impossible without some principle of equality before the law; that, since absolute freedom is impossible, we must, with Kant, demand in its stead equality with respect to those limitations of freedom which are the unavoidable consequences of social life; and that, on the other hand, the pursuit of equality, especially in its economic sense, much as it is desirable in itself, may become a threat to freedom.

And similarly, they should consider the fact that the greatest happiness principle of the Utilitarians can easily be made an excuse for a benevolent dictatorship, and the proposal[5] that we should replace it by a more modest and more realistic principle—the principle that the fight against avoidable misery should be a recognized aim of public policy, while the increase of happiness should be left, in the main, to private initiative.

This modified Utilitarianism could, I believe, lead much more easily to agreement on social reform. For new ways of happiness are theoretical, unreal things, about which it may be difficult to form an opinion. But misery is with us, here and now, and it will be with us for a long time to come. We all know it from experience. Let us make it our task to impress on public opinion the simple thought that it is wise to combat the most urgent and real social evils one by one, here and now, instead of sacrificing generations for a distant and perhaps forever unrealizable greatest good.

XII

The historicist revolution, like most intellectual revolutions, seems to have had little effect on the basically theistic and authoritarian structure of European thought.[6]

The earlier, naturalistic, revolution against God replaced the name 'God' by the name 'Nature'. Almost everything else was left

[5] I am using the term 'proposal' here in the technical sense in which it is advocated by L. J. Russell. (Cp. his paper 'Propositions and Proposals', in the *Proc. of the Tenth Intern. Congress of Philosophy*, Amsterdam, 1948.)

[6] See pp. 14–18 and 25–27 above. (Section xii of the present chapter has not been previously published.)

unchanged. Theology, the Science of God, was replaced by the Science of Nature; God's laws by the laws of Nature; God's will and power by the will and the power of Nature (the natural forces); and later God's design and God's judgment by Natural Selection. Theological determinism was replaced by a naturalistic determinism; that is, God's omnipotence and omniscience were replaced by the omnipotence of Nature[7] and the omniscience of Science.

Hegel and Marx replaced the goddess Nature in its turn by the goddess History. So we get laws of History; powers, forces, tendencies, designs, and plans, of History; and the omnipotence and omniscience of historical determinism. Sinners against God are replaced by 'criminals who vainly resist the march of History'; and we learn that not God but History (the History of 'Nations' or of 'Classes') will be our judge.

It is this deification of history which I am combatting.

But the sequence *God—Nature—History*, and the sequence of the corresponding secularized religions, does not end here. The historicist discovery that all standards are after all only historical facts (in God, standards and facts are one) leads to the deification of *Facts*—of existing or actual Facts of human life and behaviour (including, I am afraid, merely alleged Facts)—and thus to the secularized religions of Nations and of Classes, and of existentialism, positivism, and behaviourism. Since human behaviour includes verbal behaviour, we are led still further to the deification of the Facts of Language.[8] Appeal to the logical and moral authority of these Facts (or alleged Facts) is, it would seem, the ultimate wisdom of philosophy in our time.

[7] See Spinoza's *Ethics*, i, propos. xxix, and pp. 9 and 20, above.

[8] See for example point (13), pp. 84 f., and p. 22, above. For legal positivism see my *Open Society and Its Enemies*, especially vol. i, pp. 71–73, and vol. ii, pp. 392–5; and F. A. Hayek, *The Constitution of Liberty*, 1960, pp. 236 ff. See also F. A. Hayek, *Studies in Philosophy, Politics and Economics*, 1967.

17

PUBLIC OPINION AND LIBERAL PRINCIPLES

The following remarks were designed to provide material for debate at an international conference of liberals (in the English sense of this term: see the end of the Preface). My purpose was simply to lay the foundations for a good general discussion. Because I could assume liberal views in my audience I was largely concerned to challenge, rather than to endorse, popular assumptions favourable to these views.

1. THE MYTH OF PUBLIC OPINION

We should beware of a number of myths concerning 'public opinion' which are often accepted uncritically.

There is, first, the classical myth, *vox populi vox dei*, which attributes to the voice of the people a kind of final authority and unlimited wisdom. Its modern equivalent is faith in the ultimate common-sense rightness of that mythical figure, 'the man in the street', his vote, and his voice. The avoidance of the plural in both cases is characteristic. Yet people

This paper was read before the Sixth Meeting of the Mont Pèlerin Society at their Conference in Venice, September 1954; it was published (in Italian) in Il Politico, **20**, *1955, and (in German) in* Ordo, **8**, *1956; it has not been previously published in English.*

are, thank God, seldom univocal; and the various men in the various streets are as different as any collection of V.I.P.s in a conference-room. And if, on occasion, they do speak more or less in unison, what they say is not necessarily wise. They may be right, or they may be wrong. 'The voice' may be very firm on very doubtful issues. (Example: the nearly unanimous and unquestioning acceptance of the demand for 'unconditional surrender'.) And it may waver on issues over which there is hardly room for doubt. (Example: the question whether to condone political blackmail, and mass-murder.) It may be well-intentioned but imprudent. (Example: the public reaction which destroyed the Hoare-Laval plan.) Or it may be neither well-intentioned nor very prudent. (Example: the approval of the Runciman mission; the approval of the Munich agreement of 1938.)

I believe nevertheless that there is a kernel of truth hidden in the *vox populi* myth. One might put it in this way: In spite of the limited information at their disposal, many simple men are often wiser than their governments; and if not wiser, then inspired by better or more generous intentions. (Examples: the readiness of the people of Czechoslovakia to fight, on the eve of Munich; the Hoare-Laval reaction again.)

One form of the myth—or perhaps of the philosophy behind the myth—which seems to me of particular interest and importance is the doctrine that *truth is manifest*. By this I mean the doctrine that, though error is something that needs to be explained (by lack of good will or by bias or by prejudice), truth will always make itself known, as long as it is not suppressed. Thus arises the belief that liberty, by sweeping away oppression and other obstacles, must of necessity lead to a Reign of Truth and Goodness—to 'an Elysium created by reason and graced by the purest pleasures known to the love of mankind', in the words of the concluding sentence of Condorcet's *Sketch for a Historical Picture of the Progress of the Human Mind*.

I have consciously oversimplified this important myth which also may be formulated: 'Nobody, if presented with the truth, can fail to recognize it.' I propose to call this 'the theory of rationalist optimism'. It is a theory, indeed, which the Enlightenment shares with most of its political offspring and its intellectual forebears. Like the *vox populi* myth, it is another myth of the univocal voice. If humanity is a Being we ought to worship, then the unanimous voice of mankind ought to be our final authority. But we have learned that this is a myth, and we have learned to distrust unanimity.

A reaction to this rationalist and optimistic myth is the romantic version of the *vox populi* theory—the doctrine of the authority and uniqueness of the popular will, of the '*volonté generale*', of the spirit of the people, of the genius of the nation, of the group mind, or of the instinct of the blood. I need hardly repeat here the criticism which Kant and others—among them myself—have levelled against these doctrines of the irrational grasp of truth which culminates in the Hegelian doctrine of the cunning of reason which uses our passions as instruments for the instinctive or intuitive grasp of truth; and which makes it impossible for the people to be wrong, especially if they follow their passions rather than their reason.

An important and still very influential variant of the myth may be described as the myth of the progress of public opinion, which is the myth of public opinion of the nineteenth-century Liberal. It may be illustrated by quoting a passage from Anthony Trollope's *Phineas Finn*, to which Professor E. H. Gombrich has drawn my attention. Trollope describes the fate of a parliamentary motion for Irish tenant rights. The division comes, and the Ministry is beaten by a majority of twenty-three. 'And now', says Mr Monk, M.P., 'the pity is that we are not a bit nearer tenant-rights than we were before.'

> 'But we are nearer to it.'
>
> 'In one sense, yes. Such a debate and such a majority will make men think. But no;—think is too high a word; as a rule men don't think. But it will make them believe that there is something in it. Many who before regarded legislation on the subject as chimerical, will now fancy that it is only dangerous, or perhaps not more than difficult. And so in time it will come to be looked on as among the things possible, then among the things probable;—and so at last it will be ranged in the list of those few measures which the country requires as being absolutely needed. That is the way in which public opinion is made.'
>
> 'It is not loss of time,' said Phineas, 'to have taken the first great step in making it.'
>
> 'The first great step was taken long ago,' said Mr Monk,—'taken by men who were looked upon as revolutionary demagogues, almost as

> traitors, because they took it. But it is a great thing to take any step that leads us onwards.'

The theory here expounded by the radical-liberal Member of Parliament, Mr Monk, may be perhaps called the '*avant-garde theory of public opinion*', or the theory of the leadership of the advanced. It is the theory that there are some leaders or creators of public opinion who, by books and pamphlets and letters to *The Times*, or by parliamentary speeches and motions, manage to get some ideas first rejected and later debated and finally accepted. Public opinion is here conceived as a kind of public response to the thoughts and efforts of those aristocrats of the mind who produce new thoughts, new ideas, new arguments. It is conceived as slow, as somewhat passive and by nature conservative, but nevertheless as capable, in the end, of intuitively discerning the truth of the claims of the reformers—as the slow-moving but final and authoritative umpire of the debates of the elite. This, no doubt, is again another form of our myth, however much of the English reality may at first sight appear to conform to it. No doubt, the claims of reformers have often succeeded in exactly this way. But did only the valid claims succeed? I am inclined to believe that, in Great Britain, it is not so much the truth of an assertion or the wisdom of a proposal that is likely to win for a policy the support of public opinion, as the feeling that injustice is being done which can and must be rectified. It is the characteristic *moral sensitivity* of public opinion, and the way in which it has often been roused, at least in the past, which is described by Trollope; its intuition of injustice rather than its intuition of factual truth. It is debatable how far Trollope's description is applicable to other countries; and it would be dangerous to assume that even in Great Britain public opinion will remain as sensitive as in the past.

2. THE DANGERS OF PUBLIC OPINION

Public opinion (whatever it may be) is very powerful. It may change governments, even non-democratic governments. Liberals ought to regard any such power with some degree of suspicion.

Owing to its anonymity, public opinion is an *irresponsible form of power*, and therefore particularly dangerous from the liberal point of view.

(Example: colour bars and other racial questions.) The remedy in *one* direction is obvious: by minimizing the power of the state, the danger of the influence of public opinion, exerted through the agency of the state, will be reduced. But this does not secure the freedom of the individual's behaviour and thought from the direct pressure of public opinion. Here, the individual needs the powerful protection of the state. These conflicting requirements can be at least partly met by a certain kind of tradition—of which more below.

The doctrine that public opinion is not irresponsible, but somehow 'responsible to itself'—in the sense that its mistakes will rebound upon the public who held the mistaken opinion—is another form of the collectivist myth of public opinion: the mistaken propaganda of one group of citizens may easily harm a very different group.

3. LIBERAL PRINCIPLES: A GROUP OF THESES

(1) The state is a necessary evil: its powers are not to be multiplied beyond what is necessary. One might call this principle the '*Liberal Razor*'. (In analogy to Ockham's Razor, i.e. the famous principle that entities or essences must not be multiplied beyond what is necessary.)

In order to show the necessity of the state I do not appeal to Hobbes' *homo-homini-lupus* view of man. On the contrary, its necessity can be shown even if we assume that *homo homini felis*, or even that *homo homini angelus*—in other words, even if we assume that, because of their gentleness, or angelic goodness, nobody ever harms anybody else. In such a world there would still be weaker and stronger men, and the weaker ones would have *no legal right* to be tolerated by the stronger ones, but would owe them gratitude for their being so kind as to tolerate them. Those (whether strong or weak) who think this an unsatisfactory state of affairs, and who think that every person should have a *right* to live, and that every person should have a *legal claim* to be protected against the power of the strong, will agree that we need a state that protects the rights of all.

It is easy to see that the state must be a constant danger, or (as I have ventured to call it) an evil, though a necessary one. For if the state is to fulfil its function, it must have more power at any rate than any single private citizen or public corporation; and although we might design institutions to minimize the danger that these powers will be misused, we can never eliminate the danger completely. On the contrary, it seems that most men will always have to pay for the protection of the state, not only in the form of taxes but even in the form of humiliation suffered, for example, at the hands of bullying officials. The thing is not to pay too heavily for it.

(2) The difference between a democracy and a tyranny is that under a democracy the government can be got rid of without bloodshed; under a tyranny it cannot.

(3) Democracy as such cannot confer any benefits upon the citizen and it should not be expected to do so. In fact democracy can do nothing—only the citizens of the democracy can act (including, of course, those citizens who comprise the government). Democracy provides no more than a framework within which the citizens may act in a more or less organized and coherent way.

(4) We are democrats, not because the majority is always right, but because democratic traditions are the least evil ones of which we know. If the majority (or 'public opinion') decides in favour of tyranny, a democrat need not therefore suppose that some fatal inconsistency in his views has been revealed. He will realize, rather, that the democratic tradition in his country was not strong enough.

(5) Institutions alone are never sufficient if not tempered by traditions. Institutions are always ambivalent in the sense that, in the absence of a strong tradition, they also may serve the opposite purpose to the one intended. For example, a parliamentary opposition is, roughly speaking, supposed to prevent the majority from stealing the taxpayer's money. But I well remember an affair in a south-eastern European country which illustrates the ambivalence of this institution. There, the opposition shared the spoils with the majority.

To sum up: Traditions are needed to form a kind of link between institutions and the intentions and valuations of individual men.

(6) A Liberal Utopia—that is, a state rationally designed on a traditionless *tabula rasa*—is an impossibility. For the Liberal principle demands that the limitations to the freedom of each which are made necessary by social life should be minimized and equalized as much as possible (Kant). But how can we apply such an *a priori* principle in real life? Should we prevent a pianist from practising, or prevent his neigh-

bour from enjoying a quiet afternoon? All such problems can be solved in practice only by an appeal to existing traditions and customs and to a traditional sense of justice; to common law, as it is called in Britain, and to an impartial judge's appreciation of equity. All laws, being universal principles, have to be interpreted in order to be applied; and an interpretation needs some principles of concrete practice, which can be supplied only by a living tradition. And this holds more especially for the highly abstract and universal principles of Liberalism.

(7) Principles of Liberalism may be described (at least today) as principles of assessing, and if necessary of modifying or changing, existing institutions, rather than of replacing existing institutions. One can express this also by saying that Liberalism is an evolutionary rather than a revolutionary creed (unless it is confronted by a tyrannical regime).

(8) Among the traditions we must count as the most important is what we may call the 'moral framework' (corresponding to the institutional 'legal framework') of a society. This incorporates the society's traditional sense of justice or fairness, or the degree of moral sensitivity it has reached. This moral framework serves as the basis which makes it possible to reach a fair or equitable compromise between conflicting interests where this is necessary. It is, of course, itself not unchangeable, but it changes comparatively slowly. Nothing could be more dangerous than the destruction of this traditional framework, as it was consciously aimed at by Nazism. In the end its destruction will lead to cynicism and nihilism, i.e. to the disregard and the dissolution of all human values.

4. THE LIBERAL THEORY OF FREE DISCUSSION

Freedom of thought, and free discussion, are ultimate Liberal values which do not really need any further justification. Nevertheless, they can also be justified pragmatically in terms of the part they play in the search for truth.

Truth is not manifest; and it is not easy to come by. The search for truth demands at least

(*a*) imagination
(*b*) trial and error

(*c*) the gradual discovery of our prejudices by way of (*a*), of (*b*), and of critical discussion.

The Western rationalist tradition, which derives from the Greeks, is the tradition of critical discussion—of examining and testing propositions or theories by attempting to refute them. This critical rational method must not be mistaken for a method of proof, that is to say, for a method of finally establishing truth; nor is it a method which always secures agreement. Its value lies, rather, in the fact that participants in a discussion will, to some extent, change their minds, and part as wiser men.

It is often asserted that discussion is only possible between people who have a common language and accept common basic assumptions. I think that this is a mistake. All that is needed is a readiness to learn from one's partner in the discussion, which includes a genuine wish to understand what he intends to say. If this readiness is there, the discussion will be the more fruitful the more the partners' backgrounds differ. Thus the value of a discussion depends largely upon the variety of the competing views. Had there been no Tower of Babel, we should invent it. The liberal does not dream of a perfect consensus of opinion; he only hopes for the mutual fertilization of opinions, and the consequent growth of ideas. Even when we solve a problem to universal satisfaction, we create, in solving it, many new problems over which we are bound to disagree. This is not to be regretted.

Although the search for truth through free rational discussion is a public affair, it is not public opinion (whatever this may be) which results from it. Though public opinion may be influenced by science and may judge science, it is not the product of scientific discussion.

But the tradition of rational discussion creates, in the political field, the tradition of government by discussion, and with it the habit of listening to another point of view; the growth of a sense of justice; and the readiness to compromise.

Our hope is thus that traditions, changing and developing under the influence of critical discussion and in response to the challenge of new problems, may replace much of what is usually called 'public opinion', and take over the functions which public opinion is supposed to fulfil.

5. THE FORMS OF PUBLIC OPINION

There are two main forms of public opinion; institutionalized and non-institutionalized.

Examples of institutions serving or influencing public opinion: the press (including Letters to the Editor); political parties; societies like the Mont Pèlerin Society; Universities; book-publishing; broadcasting; theatre; cinema; television.

Examples of non-institutionalized public opinion: what people say in railway carriages and other public places about the latest news, or about foreigners, or about 'coloured men'; or what they say about one another across the dinner table. (This may even become institutionalized.)

6. SOME PRACTICAL PROBLEMS: CENSORSHIP AND MONOPOLIES OF PUBLICITY

No theses are offered in this section—only problems.

How far does the case against censorship depend upon a tradition of self-imposed censorship?

How far do publishers' monopolies establish a kind of censorship? How far are thinkers free to publish their ideas? Can there be complete freedom to publish? And ought there to be complete freedom to publish anything?

The influence and responsibility of the intelligentsia: (*a*) upon the spread of ideas (example: socialism); (*b*) upon the acceptance of often tyrannical fashions (example: abstract art).

The freedom of the Universities: (*a*) state interference; (*b*) private interference; (*c*) interference in the name of public opinion.

The management of (or planning for) public opinion. 'Public relations officers.'

The problem of the propaganda for cruelty in newspapers (especially in 'comics'), cinema, etc.

The problem of *taste*. Standardization and levelling.

The problem of propaganda and advertisement *versus* the spread of information.

7. A SHORT LIST OF POLITICAL ILLUSTRATIONS

This is a list containing cases which should be worthy of careful analysis.

(1) The Hoare-Laval Plan and its defeat by the unreasonable moral enthusiasm of public opinion.
(2) The Abdication of Edward VIII.
(3) Munich.
(4) Unconditional surrender.
(5) The Crichel-Down case.
(6) The British habit of accepting hardship without grumbling.

8. SUMMARY

That intangible and vague entity called public opinion sometimes reveals an unsophisticated shrewdness or, more typically, a moral sensitivity superior to that of the government in power. Nevertheless, it is a danger to freedom if it is not moderated by a strong liberal tradition. It is dangerous as an arbiter of taste, and unacceptable as an arbiter of truth. But it may sometimes assume the role of an enlightened arbiter of justice. (Example: The liberation of slaves in the British colonies.) Unfortunately it can be 'managed'. These dangers can be counteracted only by strengthening the liberal tradition.

Public opinion should be distinguished from the publicity of free and critical discussion which is (or should be) the rule in science, and which includes the discussion of questions of justice and other moral issues. Public opinion is influenced by, but neither the result of, nor under the control of, discussions of this kind. Their beneficial influence will be the greater the more honestly, simply and clearly, these discussions are conducted.

18

UTOPIA AND VIOLENCE

There are many people who hate violence and are convinced that it is one of their foremost and at the same time one of their most hopeful tasks to work for its reduction and, if possible, for its elimination from human life. I am among these hopeful enemies of violence. Not only do I hate violence, but I firmly believe that the fight against it is not at all hopeless. I realize that the task is difficult. I realize that, only too often in the course of history, it has happened that what appeared at first to be a great success in the fight against violence was followed by defeat. I do not overlook the fact that the new age of violence which was opened by the two World wars is by no means at an end. Nazism and Fascism are thoroughly beaten, but I must admit that their defeat does not mean that barbarism and brutality have been defeated. On the contrary, it is no use closing our eyes to the fact that these hateful ideas achieved something like victory in defeat. I have to admit that Hitler succeeded in degrading the moral standards of our Western world, and that in the world of today there is more violence and brutal force than would have been tolerated even in the decade after the first World war. And we must face the possibility that our civilization may ultimately be

An address delivered to the Institut des Arts in Brussels, in June 1947; first published in The Hibbert Journal, **46**, 1948.

destroyed by those new weapons which Hitlerism wished upon us, perhaps even within the first decade[1] after the second World war; for no doubt the spirit of Hitlerism won its greatest victory over us when, after its defeat, we used the weapons which the threat of Nazism had induced us to develop. But in spite of all this I am today no less hopeful than I have ever been that violence can be defeated. It is our only hope; and long stretches in the history of Western as well as of Eastern civilizations prove that it need not be a vain hope—that violence *can* be reduced, and brought under the control of reason.

This is perhaps why I, like many others, believe in reason; why I call myself a rationalist. I am a rationalist because I see in the attitude of reasonableness the only alternative to violence.

When two men disagree, they do so either because their opinions differ, or because their interests differ, or both. There are many kinds of disagreement in social life which must be decided one way or another. The question may be one which must be settled, because failure to settle it may create new difficulties whose cumulative effects may cause an intolerable strain, such as a state of continual and intense preparation for deciding the issue. (An armaments race is an example.) To reach a decision may be a necessity.

How can a decision be reached? There are, in the main, only two possible ways: argument (including arguments submitted to arbitration, for example to some international court of justice) and violence. Or, if it is interests that clash, the two alternatives are a reasonable compromise or an attempt to destroy the opposing interest.

A rationalist, as I use the word, is a man who attempts to reach decisions by argument and perhaps, in certain cases, by compromise, rather than by violence. He is a man who would rather be unsuccessful in convincing another man by argument than successful in crushing him by force, by intimidation and threats, or even by persuasive propaganda.

We shall understand better what I mean by reasonableness if we consider the difference between trying to convince a man by argument and trying to persuade him by propaganda.

[1] This was written in 1947. Today I should alter this passage merely by replacing 'first' by 'second'.

The difference does not lie so much in the use of argument. Propaganda often uses argument too. Nor does the difference lie in our conviction that our arguments are conclusive, and must be admitted to be conclusive by any reasonable man. It lies rather in an attitude of give and take, in a readiness not only to convince the other man but also possibly to be convinced by him. What I call the attitude of reasonableness may be characterized by a remark like this: 'I think I am right, but I may be wrong and you may be right, and in any case let us discuss it, for in this way we are likely to get nearer to a true understanding than if we each merely insist that we are right.'

It will be realized that what I call the attitude of reasonableness or the rationalistic attitude presupposes a certain amount of intellectual humility. Perhaps only those can take it up who are aware that they are sometimes wrong, and who do not habitually forget their mistakes. It is born of the realization that we are not omniscient, and that we owe most of our knowledge to others. It is an attitude which tries as far as possible to transfer to the field of opinions in general the two fundamental rules of every legal proceeding: first, that one should always hear both sides, and secondly, that one does not make a good judge if one is a party to the case.

I believe that we can avoid violence only in so far as we practise this attitude of reasonableness when dealing with one another in social life; and that any other attitude is likely to produce violence—even a one-sided attempt to deal with others by gentle persuasion, and to convince them by argument and example of those insights we are proud of possessing, and of whose truth we are absolutely certain. We all remember how many religious wars were fought for a religion of love and gentleness; how many bodies were burned alive with the genuinely kind intention of saving souls from the eternal fire of hell. Only if we give up our authoritarian attitude in the realm of opinion, only if we establish the attitude of give and take, of readiness to learn from other people, can we hope to control acts of violence inspired by piety and duty.

There are many difficulties impeding the rapid spread of reasonableness. One of the main difficulties is that it always takes two to make a discussion reasonable. Each of the parties must be ready to learn from the other. You cannot have a rational discussion with a man who prefers shooting you to being convinced by you. In other words, there are limits to the attitude of reasonableness. It is the same with tolerance. You must not, without qualification, accept the principle of tolerating all those who are intolerant; if you do, you will destroy not only yourself, but also the attitude of tolerance. (All this is indicated in the remark I made before—that reasonableness must be an attitude of *give and take.*)

An important consequence of all this is that we must not allow the distinction between attack and defence to become blurred. We must insist upon this distinction, and support and develop social institutions (national as well as international) whose function it is to discriminate between aggression and resistance to aggression.

I think I have said enough to make clear what I intend to convey by calling myself a rationalist. My rationalism is not dogmatic. I fully admit that I cannot rationally prove it. I frankly confess that I choose rationalism because I hate violence, and I do not deceive myself into believing that this hatred has any rational grounds. Or to put it another way, my rationalism is not self-contained, but rests on an irrational faith in the attitude of reasonableness. I do not see that we can go beyond this. One could say, perhaps, that my irrational faith in equal and reciprocal rights to convince others and be convinced by them is a faith in human reason; or simply, that I believe in man.

If I say that I believe in man, I mean in man as he is; and I should never dream of saying that he is wholly rational. I do not think that a question such as whether man is more rational than emotional or *vice versa* should be asked: there are no ways of assessing or comparing such things. I admit that I feel inclined to protest against certain exaggerations (arising largely from a vulgarization of psycho-analysis) of the irrationality of man and of human society. But I am aware not only of the power of emotions in human life, but also of their value. I should never demand that the attainment of an attitude of reasonableness should become the one dominant aim of our lives. All I wish to assert is that this attitude can become one that is never wholly absent—not even in relationships which are dominated by great passions, such as love.[2]

[2] The existentialist Jaspers writes 'This is why love is cruel, ruthless; and why it is believed in, by the genuine lover, only if it is so'. This attitude, to my mind, reveals weakness rather than the strength it wishes to show; it is not so much plain barbarism as an hysterical attempt to play the barbarian. (Cf. my *Open Society*, 4th edn., vol. II, p. 317.)

My fundamental attitude towards the problem of reason and violence will by now be understood; and I hope I share it with some of my readers and with many other people everywhere. It is on this basis that I now propose to discuss the problem of Utopianism.

I think we can describe Utopianism as a result of a form of rationalism, and I shall try to show that this is a form of rationalism very different from the form in which I and many others believe. So I shall try to show that there exist at least two forms of rationalism, one of which I believe is right and the other wrong; and that the wrong kind of rationalism is the one which leads to Utopianism.

As far as I can see, Utopianism is the result of a way of reasoning which is accepted by many who would be astonished to hear that this apparently quite inescapable and self-evident way of reasoning leads to Utopian results. This specious reasoning can perhaps be presented in the following manner.

An action, it may be argued, is rational if it makes the best use of the available means in order to achieve a certain end. The end, admittedly, may be incapable of being determined rationally. However this may be, we can judge an action rationally, and describe it as rational or adequate, only relative to some given end. Only if we have an end in mind, and only relative to such an end, can we say that we are acting rationally.

Now let us apply this argument to politics. All politics consists of actions; and these actions will be rational only if they pursue some end. The end of a man's political actions may be the increase of his own power or wealth. Or it may perhaps be the improvement of some of the laws of the state, thus leading to a change in the structure of the state or of society.

In the latter case political action will be rational only if we first determine the final ends of the political changes which we intend to bring about. It will be rational only relative to certain ideas of what a state ought to be like. Thus it appears that as a preliminary to any rational political action we must first attempt to become as clear as possible about our ultimate political ends; for example the kind of state which we should consider the best; and only afterwards can we begin to determine the means which may best help us to realize this state, or to move slowly towards it, taking it as the aim of a historical process which we may to some extent influence and steer towards the goal selected.

Now it is precisely this view which I call Utopianism. Any rational and non-selfish political action, on this view, must be preceded by a determination of our ultimate ends, not merely of intermediate or partial aims which are only steps towards our ultimate end, and which therefore should be considered as means rather than as ends; therefore rational political action must be based upon a more or less clear and detailed description or blueprint of our ideal state, and also upon a plan or blueprint of the historical path that leads towards this goal.

I consider what I call Utopianism an attractive and, indeed, an all too attractive theory; for I also consider it dangerous and pernicious. It is, I believe, self-defeating, and it leads to violence.

That it is self-defeating is connected with the fact that it is impossible to determine ends scientifically. There is no scientific way of choosing between two ends. Some people, for example, love and venerate violence. For them a life without violence would be shallow and trivial. Many others, of whom I am one, hate violence. This is a quarrel about ends. It cannot be decided by science. This does not mean that the attempt to argue against violence is necessarily a waste of time. It only means that you may not be able to argue with the admirer of violence. He has a way of answering an argument with a bullet if he is not kept under control by the threat of counter-violence. If he is willing to listen to your arguments without shooting you, then he is at least infected by rationalism, and you may, perhaps, win him over. This is why arguing is no waste of time—as long as people listen to you. But you cannot, by means of argument, make people listen to argument; you cannot, by means of argument, convert those who suspect all argument, and who prefer violent decisions to rational decisions. You cannot prove to them that they are wrong. And this is only a particular case, which can be generalized. No decision about aims can be established by *purely* rational or scientific means. Nevertheless argument may prove extremely helpful in reaching a decision about aims.

Applying all this to the problem of Utopianism, we must first be quite clear that the problem of constructing a Utopian blueprint cannot possibly be solved by science alone. Its aims, at least, must be given before the social scientist can begin to sketch his blueprint. We find the same situation in the natural sciences. No amount of physics will tell a scientist that it is the right thing for him to construct a plough, or an aeroplane, or an atomic bomb. Ends must be adopted by him, or given to him; and what he does *qua* scientist is only to construct means by which these ends can be realized.

In emphasizing the difficulty of deciding, by way of rational argument, between different Utopian ideals, I do not wish to create the impression that there is a realm—such as the realm of ends—which goes altogether beyond the power of rational criticism (even though I certainly wish to say that the realm of ends goes largely beyond the power of *scientific* argument). For I myself try to argue about this realm; and by pointing out the difficulty of deciding between competing Utopian blueprints, I try to argue rationally against choosing ideal ends of this kind. Similarly, my attempt to point out that this difficulty is likely to produce violence is meant as a rational argument, although it will appeal only to those who hate violence.

That the Utopian method, which chooses an ideal state of society as the aim which all our political actions should serve, is likely to produce violence can be shown thus. Since we cannot determine the ultimate ends of political actions scientifically, or by purely rational methods, differences of opinion concerning what the ideal state should be like cannot always be smoothed out by the method of argument. They will at least partly have the character of religious differences. And there can hardly be tolerance between these different Utopian religions. Utopian aims are designed to serve as a basis for rational political action and discussion, and such action appears to be possible only if the aim is definitely decided upon. Thus the Utopianist must win over, or else crush, his Utopianist competitors who do not share his own Utopian aims and who do not profess his own Utopianist religion.

But he has to do more. He has to be very thorough in eliminating and stamping out all heretical competing views. For the way to the Utopian goal is long. Thus the rationality of his political action demands constancy of aim for a long time ahead; and this can only be achieved if he not merely crushes competing Utopian religions, but as far as possible stamps out all memory of them.

The use of violent methods for the suppression of competing aims becomes even more urgent. For unavoidably, the period of Utopian construction is liable to be one of social change. In such a time ideas are liable to change also. Thus what may have appeared to many as desirable at the time when the Utopian blueprint was decided upon may appear less desirable at a later date. If this is so, the whole approach is in danger of breaking down. For if we change our ultimate political aims while attempting to move towards them we may soon discover that we are moving in circles. The whole method of first establishing an ultimate political aim and then preparing to move towards it must be futile if the aim may be changed during the process of its realization. It may easily turn out that the steps so far taken lead in fact away from the new aim. And if we then change direction in accordance with our new aim we expose ourselves to the same risk. In spite of all the sacrifices which we may have made in order to make sure that we are acting rationally, we may get exactly nowhere—although not exactly to that 'nowhere' which is meant by the word 'Utopia'.

Again, the only way to avoid such changes of our aims seems to be to use violence, which includes propaganda, the suppression of criticism, and the annihilation of all opposition. With it goes the affirmation of the wisdom and foresight of the Utopian planners, of the Utopian engineers who design and execute the Utopian blueprint. The Utopian engineers must in this way become omniscient as well as omnipotent. They become gods. Thou shalt have no other Gods before them.

Utopian rationalism is a self-defeating rationalism. However benevolent its ends, it does not bring happiness, but only the familiar misery of being condemned to live under a tyrannical government.

It is important to understand this criticism fully. I do not criticize political ideals as such, nor do I assert that a political ideal can never be realized. This would not be a valid criticism. Many ideals have been realized which were once dogmatically declared to be unrealizable, for example, the establishment of workable and untyrannical institutions for securing civil peace, that is, for the suppression of crime within the state. Again, I see no reason why an international judicature and an international police force should be less successful in suppressing international crime, that is, national aggression and the ill-treatment of minorities or perhaps majorities. I do not object to the attempt to realize such ideals.

Wherein, then, lies the difference between those benevolent Utopian plans to which I object because they lead to violence, and those other important and far-reaching political reforms which I am inclined to recommend?

If I were to give a simple formula or recipe for distinguishing between what I consider to be admissible plans for social reform and inadmissible Utopian blueprints, I might say:

Work for the elimination of concrete evils rather than for the realization of abstract goods. Do not aim at establishing happiness by political means. Rather aim at the elimination of concrete miseries. Or, in more practical terms: fight for the elimination of poverty by direct means—for example, by making sure that everybody has a minimum income. Or fight against epidemics and disease by erecting hospitals and schools of medicine. Fight illiteracy as you fight criminality. But do all this by direct means. Choose what you consider the most urgent evil of the society in which you live, and try patiently to convince people that we can get rid of it.

But do not try to realize these aims indirectly by designing and working for a distant ideal of a society which is wholly good. However deeply you may feel indebted to its inspiring vision, do not think that you are obliged to work for its realization, or that it is your mission to open the eyes of others to its beauty. Do not allow your dreams of a beautiful world to lure you away from the claims of men who suffer here and now. Our fellow men have a claim to our help; no generation must be sacrificed for the sake of future generations, for the sake of an ideal of happiness that may never be realized. In brief, it is my thesis that human misery is the most urgent problem of a rational public policy and that happiness is not such a problem. The attainment of happiness should be left to our private endeavours.

It is a fact, and not a very strange fact, that it is not so very difficult to reach agreement by discussion on what are the most intolerable evils of our society, and on what are the most urgent social reforms. Such an agreement can be reached much more easily than an agreement concerning some ideal form of social life. For the evils are with us here and now. They can be experienced, and are being experienced every day, by many people who have been and are being made miserable by poverty, unemployment, national oppression, war and disease. Those of us who do not suffer from these miseries meet every day others who can describe them to us. This is what makes the evils concrete. This is why we can get somewhere in arguing about them; why we can profit here from the attitude of reasonableness. We can learn by listening to concrete claims, by patiently trying to assess them as impartially as we can, and by considering ways of meeting them without creating worse evils.

With ideal goods it is different. These we know only from our dreams and from the dreams of our poets and prophets. They cannot be discussed, only proclaimed from the housetops. They do not call for the rational attitude of the impartial judge, but for the emotional attitude of the impassioned preacher.

The Utopianist attitude, therefore, is opposed to the attitude of reasonableness. Utopianism, even though it may often appear in a rationalist disguise, cannot be more than a pseudo-rationalism.

What, then, is wrong with the apparently rational argument which I outlined when presenting the Utopianist case? I believe that it is quite true that we can judge the rationality of an action only in relation to some aims or ends. But this does not necessarily mean that the rationality of a political action can be judged only in relation to an *historical* end. And it surely does not mean that we must consider every social or political situation merely from the point of view of some preconceived historical ideal, from the point of view of an alleged ultimate aim of the development of history. On the contrary, if among our aims and ends there is anything conceived in terms of human happiness and misery, then we are bound to judge our actions in terms not only of possible contributions to the happiness of man in a distant future, but also of their more immediate effects. We must not argue that a certain social situation is a mere means to an end on the grounds that it is merely a transient historical situation. For all situations are transient. Similarly we must not argue that the misery of one generation may be considered as a mere means to the end of securing the lasting happiness of some later generation or generations; and this argument is improved neither by a high degree of promised happiness nor by a large number of generations profiting by it. All generations are transient. All have an equal right to be considered, but our immediate duties are undoubtedly to the present generation and to the next. Besides, we should never attempt to balance anybody's misery against somebody else's happiness.

With this the apparently rational arguments of Utopianism dissolve into nothing. The fascination which the future exerts upon the Utopianist has nothing to do with rational foresight. Considered in this light the violence which Utopianism breeds looks very much like the running amok of an evolutionist metaphysics, of an hysterical philosophy of history, eager to sacrifice the present for the splendours of the future, and unaware that its principle would lead to sacrificing each particular future period for one which comes after it; and likewise unaware of the trivial truth that the ultimate future of man—whatever fate may have in store for him—can be nothing more splendid than his ultimate extinction.

The appeal of Utopianism arises from the failure to realize that we cannot make heaven on earth. What I believe we can do instead is to make life a little less terrible and a little less unjust in each generation. A good deal can be achieved in this way. Much has been achieved in the last hundred years. More could be achieved by our own generation. There are many pressing problems which we might solve, at least partially, such as helping the weak and the sick, and those who suffer under oppression and injustice; stamping out unemployment; equalizing opportunities; and preventing international crime, such as blackmail and war instigated by men like gods, by omnipotent and omniscient leaders. All this we might achieve if only we could give up dreaming about distant ideals and fighting over our Utopian blueprints for a new world and a new man. Those of us who believe in man as he is, and who have therefore not given up the hope of defeating violence and unreason, must demand instead that every man should be given the right to arrange his life himself so far as this is compatible with the equal rights of others.

We can see here that the problem of the true and the false rationalisms is part of a larger problem. Ultimately it is the problem of a sane attitude towards our own existence and its limitations—that very problem of which so much is made now by those who call themselves 'Existentialists', the expounders of a new theology without God. There is, I believe, a neurotic and even an hysterical element in this exaggerated emphasis upon the fundamental loneliness of man in a godless world, and upon the resulting tension between the self and the world. I have little doubt that this hysteria is closely akin to Utopian romanticism, and also to the ethic of hero-worship, to an ethic that can comprehend life only in terms of 'dominate or prostrate yourself'. And I do not doubt that this hysteria is the secret of its strong appeal. That our problem is part of a larger one can be seen from the fact that we can find a clear parallel to the split between true and false rationalism even in a sphere apparently so far removed from rationalism as that of religion. Christian thinkers have interpreted the relationship between man and God in at least two very different ways. The sane one may be expressed by: 'Never forget that men are not Gods; but remember that there is a divine spark in them.' The other exaggerates the tension between man and God, and the baseness of man as well as the heights to which men may aspire. It introduces the ethic of 'dominate or prostrate yourself' into the relationship of man and God. Whether there are always either conscious or unconscious dreams of godlikeness and of omnipotence at the roots of this attitude, I do not know. But I think it is hard to deny that the emphasis on this tension can arise only from an unbalanced attitude towards the problem of power.

This unbalanced (and immature) attitude is obsessed with the problem of power, not only over other men, but also over our natural environment—over the world as a whole. What I might call, by analogy, the 'false religion', is obsessed not only by God's power over men but also by His power to create a world; similarly, false rationalism is fascinated by the idea of creating huge machines and Utopian social worlds. Bacon's 'knowledge is power' and Plato's 'rule of the wise' are different expressions of this attitude which, at bottom, is one of claiming power on the basis of one's superior intellectual gifts. The true rationalist, by contrast, will always know how little he knows, and he will be aware of the simple fact that whatever critical faculty or reason he may possess he owes to intellectual intercourse with others. He will be inclined, therefore, to consider men as fundamentally equal, and human reason as a bond which unites them. Reason for him is the precise opposite of an instrument of power and violence: he sees it as a means whereby these may be tamed.

19

THE HISTORY OF OUR TIME: AN OPTIMIST'S VIEW

In a series of lectures instituted to keep alive the memory of that inspired and successful social reformer, Eleanor Rathbone, it is perhaps not out of place to devote a lecture to a general though tentative assessment of the problem of social reform in our time. What have we achieved, if anything? How does our western society compare with others? These are the questions which I propose to discuss.

I have chosen as the title of my lecture 'The History of Our Time: An Optimist's View', and I feel that I should begin by explaining this title.

When I say 'History', I wish to refer particularly to our social and political history, but also to our moral and intellectual history. By the word 'our', I mean the free world of the Atlantic Community—especially England, the United States, the Scandinavian countries and Switzerland, and the outposts of this world in the Pacific, Australia and New Zealand. By 'our time' I mean, in particular, the period since 1914. But I also mean the last fifty or sixty years—that is to say the time since the Boer War, or the age of Winston Churchill, as one might call it; the last hundred years—that is, in the main, the time since the

The Sixth Eleanor Rathbone Memorial lecture, delivered at the University of Bristol on October 12th 1956. (Not previously published.)

abolition of slavery and since John Stuart Mill; the last two hundred years—that is, in the main, the time since the American Revolution, since Hume, Voltaire, Kant, and Burke; and to a lesser extent, the last three hundred years—the time since the Reformation; since Locke, and since Newton. So much for the phrase 'The History of Our Time'.

Now I come to the word 'Optimist'. First let me make it quite clear that if I call myself an optimist, I do not wish to suggest that I know anything about the future. I do not wish to pose as a prophet, least of all as a historical prophet. On the contrary, I have for many years tried to defend the view that historical prophecy is a kind of quackery.[1] I do not believe in historical laws, and I disbelieve especially in anything like a law of progress. In fact, I believe that it is much easier for us to regress than to progress.

Though I believe all this, I think that I may fairly describe myself as an optimist. For my optimism lies entirely in my interpretation of the present and the immediate past. It lies in my strongly appreciative view of our own time. And whatever you might think about this optimism you will have to admit that it has a scarcity value. In fact the wailings of the pessimists have become somewhat monotonous. No doubt there is much in our world about which we can rightly complain if only we give our mind to it; and no doubt it is sometimes most important to find out what is wrong with us. But I think that the other side of the story might also get a hearing.

Thus it is with respect to the immediate past and to our own time that I hold optimistic views. And this brings me finally to the word 'view' which is the last word of my title. What I shall be aiming at in this lecture is to sketch, in a few strokes, a kind of bird's-eye view of our time. It will no doubt be a very personal view—an interpretation rather than a description. But I shall try to support it by argument. And although pessimists will feel that my view is superficial, I shall at least try to present it in a way that may challenge them.

And so I begin with a challenge. I will challenge a certain belief which seems to be widely held, and held in widely different quarters; not only by many Churchmen whose sincerity is beyond doubt, but

[1] See my *Poverty of Historicism*, 1957; and ch. 16.

also by some rationalists such as Bertrand Russell, whom I greatly admire as a man and as a philosopher.

Russell has more than once expressed the belief I wish to challenge. He has complained that our intellectual development has outrun our moral development.

We have become very clever, according to Russell, indeed too clever. We can make lots of wonderful gadgets, including television, high-speed rockets, and an atom bomb, or a thermonuclear bomb, if you prefer. But we have not been able to achieve that moral and political growth and maturity which alone could safely direct and control the uses to which we put our tremendous intellectual powers. This is why we now find ourselves in mortal danger. Our evil national pride has prevented us from achieving the world-state in time.

To put this view in a nutshell: we are clever, perhaps too clever, but we are also wicked; and this mixture of cleverness and wickedness lies at the root of our troubles.

As against this, I shall maintain precisely the opposite. My *first thesis* is this.

We are good, perhaps a little too good, but we are also a little stupid; and it is this mixture of goodness and stupidity which lies at the root of our troubles.

To avoid misunderstandings, I should stress that when I use the word 'we', in this thesis, I include myself.

You may perhaps ask me why my first thesis should be part of an optimist's view. There are various reasons. One is that wickedness is even more difficult to combat than a limited measure of stupidity, because good men who are not very clever are usually very anxious to learn.

Another reason is that I do not think that we are hopelessly stupid, and this is surely an optimist's view. What is wrong with us is that we so easily mislead ourselves, and that we are so easily 'led by the nose' by others, as Samuel Butler says in *Erewhon*. I hope you will let me quote from one of my favourite passages: 'It will be seen', Butler writes, '. . . that the Erewhonians are a meek and long-suffering people, easily led by the nose, and quick to offer up common sense at the shrine of logic, when a philosopher arises among them, who carries them away . . . by convincing them that their existing institutions are not based on the strictest principles of morality.'

You see that my first thesis, although it is directly opposed to such an authority as Bertrand Russell, is far from original. Samuel Butler seems to have thought along similar lines.

Both Butler's formulation of this thesis and my own are somewhat flippant in form. But the thesis might be put more seriously in this way.

The main troubles of our time—and I do not deny that we live in troubled times—are not due to our moral wickedness, but, on the contrary, to our often misguided moral enthusiasm: to our anxiety to better the world we live in. Our wars are fundamentally religious wars; they are wars between competing theories of how to establish a better world. And our moral enthusiasm is often misguided, because we fail to realize that our moral principles, which are sure to be over-simple, are often difficult to apply to the complex human and political situations to which we feel bound to apply them.

I certainly do not expect you to agree at once, either with my thesis or with Butler's. And even if you sympathize with Butler's, you are hardly likely to sympathize with mine. Butler, you might say, was a Victorian. But how can I hold the view that we do not live in a world of wickedness? Have I forgotten Hitler and Stalin? I have not. But I do not allow myself to be over-impressed by them. In spite of them, and with my eyes open, I remain an optimist. They, and their immediate helpers, may be set aside in this context. What is more interesting is the fact that the great dictators had a very large following. But I contend that my first thesis or, if you like, Butler's thesis, does apply to most of their followers. Most of those who followed Hitler and Stalin did so precisely because, to use Butler's phrase, they were 'easily led by the nose'. Admittedly, the great dictators did appeal to all sorts of fears and hopes, to prejudices and to envy, and even to hatred. But their main appeal was an appeal to a kind of morality—no doubt a dubious morality. They had a message; and they demanded sacrifices. It is sad to see how easily an appeal to morality can be misused. But it is simply a fact that the great dictators were always trying to convince their people that they knew the way to a higher morality.

To illustrate my point, I may remind you of a remarkable pamphlet, published as recently as 1942. In this pamphlet the then Bishop of Bradford attacked a certain form of society which he described as 'immoral' and 'un-Christian', and of which he said: 'when something is so plainly the work of the devil, . . . nothing can excuse a minister of the Church from working for its destruction'. The society which, in the Bishop's opinion, was the work of the devil was not Hitler's Germany or Stalin's Russia; it was our own Western society, the free world of the Atlantic Community. And the Bishop said these things in a pamphlet which was written in order to support the truly satanic system of Stalin. I am absolutely convinced that the Bishop's moral condemnation was sincere. But moral fervour blinded him, and many like him, to facts which others could easily see; for example, to the fact that countless innocent people were being tortured in Stalin's prisons.[2]

Here, I am afraid, you have an example of a typical refusal to face facts, even if they are obvious facts; of a typical lack of criticism; of a typical readiness to be 'led by the nose' (to use Butler's words again); to be led by the nose by anybody who claims that our 'existing institutions are not based on the strictest principles of morality'. You have here an example of how dangerous goodness can be if too much of it is combined with too little rational criticism.

But the Bishop does not stand alone. Some of you may remember an uncontradicted report from Prague in *The Times*, about four or five years ago, in which a famous British physicist was said to have described Stalin as the greatest of all scientists. One wonders what this famous physicist will say now that the doctrine of Stalin's satanism has become, if only for the time being, an essential component of the party line itself. It all shows how astonishingly liable we are to be led by the nose if anybody arises who claims to know the way to a higher morality.

The believers in Stalin offer a sad spectacle today. But if we admire the martyrs of Christianity, we cannot completely withhold a reluctant admiration from those who retained their faith in Stalin while being tortured in Russian prisons. Theirs was a faith in a cause we know to be

[2] The pamphlet is *Christians in the Class Struggle*, by Gilbert Cope, with a Foreword by the Bishop of Bradford, 1942. Cp. my *Open Society and its Enemies* (1950 and later editions), notes 3 to ch. 1 and 12 to ch. 9.

bad; today even party members know it. But they believed in it in all sincerity.

We see how important this aspect of our troubles is if we remember that the great dictators were all forced to pay homage to the goodness of man. They were forced to pay lip-service to a morality in which they did not believe. Communism and nationalism are both believed in as moralities and religions. This is their only strength. Intellectually they border on absurdity.

The absurdity of the communist faith is manifest. Appealing to the belief in human freedom, it has produced a system of oppression without parallel in history.

But the nationalist faith is equally absurd. I am not alluding here to Hitler's racial myth. What I have in mind is, rather, an alleged natural right of man—*the alleged right of a nation to self-determination*. That even a great humanitarian and liberal like Masaryk could uphold this absurdity as one of the natural rights of man is a sobering thought. It suffices to shake one's faith in the wisdom of philosopher kings, and it should be contemplated by all who think that we are clever but wicked rather than good but stupid. For the utter absurdity of the principle of national self-determination must be plain to anybody who devotes a moment's effort to criticizing it. The principle amounts to the demand that each state should be a nation-state: that it should be confined within a natural border, and that this border should coincide with the location of an ethnic group; so that it should be the ethnic group, the 'nation', which should determine and protect the natural limits of the state.

But nation-states of this kind do not exist. Even Iceland—the only exception I can think of—is only an apparent exception to this rule. For its limits are determined, not by its ethnic group, but by the North Atlantic—just as they are protected, not by the Icelandic nation, but by the North Atlantic Treaty. Nation-states do not exist, simply because the so-called 'nations' or 'peoples' of which the nationalists dream do not exist. There are no, or hardly any, homogenous ethnic groups long settled in countries with natural borders. Ethnic and linguistic groups (dialects often amount to linguistic barriers) are closely intermingled everywhere. Masaryk's Czechoslovakia was founded upon the principle of national self-determination. But as soon as it was founded, the

Slovaks demanded, in the name of this principle, to be free from Czech domination; and ultimately it was destroyed by its German minority, in the name of the same principle. Similar situations have arisen in practically every case in which the principle of national self-determination has been applied to fixing the borders of a new state: in Ireland, in India, in Israel, in Yugoslavia. There are ethnic minorities everywhere. The proper aim cannot be to 'liberate' all of them; rather, it must be *to protect* all of them. *The oppression of national groups is a great evil; but national self-determination is not a feasible remedy*. Moreover, Britain, the United States, Canada, and Switzerland, are four obvious examples of states which in many ways violate the nationality principle. Instead of having its borders determined by one settled group, each of them has managed to unite a variety of ethnic groups. So the problem does not seem insoluble.

Yet, in the face of all these obvious facts, the principle of national self-determination continues to be widely accepted as an article of our moral faith; and it is rarely challenged outright. A Cypriot appealed recently, in a letter to *The Times*, to this principle. He described it as a universally accepted principle of morality. The defenders of this principle, he proudly claimed, were defending the sacred human values and the natural rights of man (apparently even when terrorizing their own dissenting countrymen). The fact that this letter did not mention the ethnic minority of Cyprus; the fact that it was printed; and the fact that its moral doctrines remained completely unanswered in a long sequence of letters on this subject, all go a long way towards proving my first thesis. Indeed, it seems to me possible that more people are killed out of righteous stupidity than out of wickedness.

The nationalist religion is strong. Many are ready to die for it, fervently believing that it is morally good, and factually true. But they are mistaken; just as mistaken as their communist bedfellows. Few creeds have created more hatred, cruelty, and senseless suffering than the belief in the righteousness of the nationality principle; and yet it is still widely believed that this principle will help to alleviate the misery of national oppression. My optimism is a little shaken, I admit, when I look at the near-unanimity with which this principle is still accepted, even today, without any hesitation, without any doubt—even by those whose political interests are clearly opposed to it. But I refuse to abandon the hope that the absurdity and cruelty of this alleged moral principle will one day be recognized by all thinking men.

But let us now leave all these sad stories of misguided moral enthusiasm, and turn to our own free world. Resisting the temptation to offer further arguments in support of my first thesis, I will now proceed to my second.

I have said that I am an optimist. Optimism as a philosophical creed is best known as the famous doctrine, elaborately defended by Leibniz, that this world of ours is the best of all possible words. I do not believe that this thesis of Leibniz is true. But I am sure you will concede me the happy title of optimist when you hear my second thesis which refers to our free world—the Society of the Atlantic Community. My *second thesis* is this.

In spite of our great and serious troubles, and in spite of the fact that ours is surely not the best possible society, I assert that our own free world is by far the best society which has come into existence during the course of human history.

Thus I do not say, with Leibniz, that our world is the best of all possible worlds. Nor do I say that our social world is the best of all possible social worlds. My thesis is merely that our own social world is the best that has ever been—the best, at least, of which we have any historical knowledge.

I suppose you will by now concede me the right to call myself an optimist. But you may perhaps suspect me of being a materialist—of calling our society the best because it is the wealthiest which history has ever seen.

But I can assure you that this is not the reason why I call our society the best. Admittedly, I believe it to be a great thing to have succeeded, or very nearly succeeded, in abolishing hunger and poverty. But it is neither nylons nor nutrition, neither terylene nor television, which I chiefly admire. When I call our social world 'the best', I have in mind the very same values which led the former Bishop of Bradford to brand it as the work of the devil, only fourteen years ago: I have in mind the standards and values which have come down to us through Christianity from Greece and from the Holy Land; from Socrates, and from the Old and New Testaments.

At no other time, and nowhere else, have men been more respected, as men, than in our society. Never before have their human rights, and their human dignity, been so respected, and never before have so many been ready to bring great sacrifices for others, especially for those less fortunate than themselves.

I believe that these are facts.

But before examining these facts more closely, I wish to stress that I am very much alive to other facts also. Power still corrupts, even in our world. Civil servants still behave at times like uncivil masters. Pocket dictators still abound; and a normally intelligent man seeking medical advice must be prepared to be treated as a rather tiresome type of imbecile, if he betrays an intelligent interest—that is, a critical interest—in his physical condition.

But all this is not so much due to lack of good intentions as to clumsiness and sheer incompetence. And there is much to balance it. For example, in some countries belonging to the free world (I am thinking of Belgium), hospital services are being most successfully reorganized with the obvious aim of making them pleasant rather than depressing places, with due consideration for the sensitive, and for those whose self-respect may be wounded by practices now prevailing. And it is realized there how important it is to establish a genuine and intelligent co-operation between doctor and patient, and to ensure that a man, even a sick man, should never be encouraged to surrender his final responsibility for himself.

But let us turn to larger problems. Our free world has very nearly, if not completely, succeeded in abolishing the greatest evils which have hitherto beset the social life of man.

Let me give you a list of what I believe to be some of the greatest of those evils which can be remedied, or relieved, by social co-operation: They are:

Poverty
Unemployment and some similar forms of Social Insecurity
Sickness and Pain
Penal Cruelty
Slavery and other forms of Serfdom
Religious and Racial Discrimination
Lack of Educational Opportunities

Rigid Class Differences
War

Let us see what has been achieved; not only here in Great Britain, through the Welfare State, but by one method or another everywhere in the free world.

Abject poverty has been practically abolished. Instead of being a mass phenomenon, the problem has almost become one of detecting the isolated cases which still persist.

The problems of unemployment and of some other forms of insecurity have changed completely. We are now faced with new problems brought into being by the fact that the problem of mass-unemployment has largely been solved.

Fairly continuous progress is being made in dealing with the problems of sickness and pain.

Penal reform has largely abolished cruelty in this field.

The story of the successful fight against slavery has become the everlasting pride of this country and of the United States.

Religious discrimination has practically disappeared. Racial discrimination has diminished to an extent surpassing the hopes of the most hopeful. What makes these two achievements even more astonishing is the fact that religious prejudices, and even more so racial prejudices, are probably as widespread as they were fifty years ago, or very nearly so.

The problem of educational opportunities is still very serious, but it is being tackled sincerely and with energy.

Class differences have diminished enormously everywhere. In Scandinavia, the United States, Canada, Australia and New Zealand, we have, in fact, something approaching classless societies.

My eighth point was war. This point I must discuss more fully. It may be best to formulate what I have to say here as my *third thesis*.

My third thesis is that since the time of the Boer War, none of the democratic governments of the free world has been in a position to wage a war of aggression. No democratic government would be united upon the issue, because they would not have the nation united behind them. Aggressive war has become almost a moral impossibility.[3]

[3] This lecture was delivered before the Suez adventure. It seems to me that the sad history of this adventure supports my first three theses.

The Boer War led to a revulsion of feeling in Great Britain, amounting to a moral conversion in favour of peace. It was because of this attitude that Great Britain hesitated to resist the Kaiser, and that it entered the first world war only after the violation of Belgium. It was under its influence that Britain was ready to make allowances for Hitler. When Hitler's army entered the Rhineland, this was undeniably an act of aggression on his part. Yet public opinion in this country made it impossible for the Government to meet the challenge—although it would have been the most reasonable course to take, under the circumstances. On the other hand, Mussolini's open attack on Ethiopia so much outraged British public opinion that the Hoare-Laval plan, which wisely tried to keep Mussolini and Hitler apart, was swept away by an outburst of public indignation.

But a still stronger example is the public attitude towards the issue of preventive war against Russia. You may remember that, around 1950, even Bertrand Russell advocated a preventive war. And it must be admitted that there were strong reasons in favour of it. Russia was not yet in possession of an atomic arsenal; and it was the last opportunity of preventing Russia from acquiring the hydrogen bomb.

I do not envy the American President his power to decide between such terrible alternatives. The one alternative was to begin a war. The other was to allow Stalin to acquire the power to destroy the world; a power with which he certainly ought not to have been entrusted. Bertrand Russell was no doubt right in maintaining that from a purely rational point of view the second alternative was even worse than the first. But the decision went the other way. An aggressive war, even in these crucial circumstances, and with the then practical certainty of victory, had become morally impossible.

The free world is still ready to go to war. It is ready to go to war against heavy odds, as it has done more than once in the past. But it will do this only if faced with unambiguous aggression. Thus as far as the free world itself is concerned war has been conquered.

I have briefly discussed my list of eight great social evils.

I believe that it is most important to say what the free world has achieved. For we have become unduly sceptical about ourselves. We are suspicious of anything like self-righteousness, and we find self-praise unpalatable. One of the great things we have learned is not only to be tolerant of others, but to ask ourselves seriously whether the other fellow is not perhaps in the right, and altogether the better man. We have learned the fundamental moral truth that nobody should be judge in his own cause. This, no doubt, is a symptom of a certain moral maturity; yet one may learn a lesson too well. Having discovered the sin of self-righteousness, we have fallen into its stereotyped inversion: into a stereotyped pose of self-depreciation, of inverted smugness. Having learned that one should not be judge in one's own cause, we are tempted to become advocates for our opponents. Thus we become blind to our own achievements. But this tendency must be resisted.

When Mr Krushchev on his Indian tour indicted British colonialism, he was no doubt convinced of the truth of all he said. I do not know whether he was aware that his accusations were derived, *via* Lenin, largely from British sources. Had he known it, he would probably have taken it as an additional reason for believing in what he was saying. But he would have been mistaken; for this kind of self-accusation is a peculiarly British virtue as well as a peculiarly British vice. The truth is that the idea of India's freedom was born in Great Britain; as was the general idea of political freedom in modern times. And those Britishers who provided Lenin and Mr Krushchev with their moral ammunition were closely connected, or even identical, with those Britishers who gave India the idea of freedom.

I shall always regret that the great British statesman who answered Mr Krushchev had so little to say for himself, and for our different way of life. I am quite sure that he made no impression at all on Mr Krushchev. But I think he could have done so. Had he pointed to the difference between our free world and the communist world by way of the following example, I am sure Mr Krushchev would have understood him. Our statesman might have spoken thus:

'The difference between your country and mine can be explained as follows. Imagine that my chief, Sir Anthony, suddenly dies tomorrow. I can assure you that in our country nobody in his senses would even for a moment consider the possibility that I had murdered Sir Anthony. Not even a British communist would think so. This illustrates the simple difference between our respective ways of conducting our affairs. It is not a racial difference, to be sure, for we may learn from

Shakespeare that not so very long ago we too conducted our affairs in that other manner.'

I believe in the importance of answering all those absurd but terrible accusations against Great Britain, often originating from British sources, which are current in the world today. For I believe in the power of ideas, including the power of false and pernicious ideas. And I believe in what I might call the war of ideas.

The war of ideas is a Greek invention. It is one of the most important inventions ever made. Indeed, the possibility of fighting with words instead of fighting with swords is the very basis of our civilization, and especially of all its legal and parliamentary institutions. And this habit of fighting with words and ideas is one of the few things which still unite the worlds on the two sides of the Iron Curtain (although on the other side, words have only inadequately replaced swords, and are sometimes used to prepare for the kill). To see how powerful ideas have become since the days of the Greeks, we only need to remember that all religious wars were wars of ideas, and that all revolutions were revolutions of ideas. Although these ideas were more often false and pernicious than true and beneficial there is perhaps a certain tendency for some of the better ones to survive, provided they find sufficiently powerful and intelligent support.

All this may be formulated in my *fourth thesis*. It is as follows.

The power of ideas, and especially of moral and religious ideas, is at least as important as that of physical resources.

I am well aware of the fact that some students of politics are strongly opposed to this thesis; that there is an influential school of so-called political realists who declare that 'ideologies', as they call them, have little influence upon political reality, and that whatever influence they have must be pernicious. But I do not think that this is a tenable view. Were it true, Christianity would have had no influence on history; and the United States would be inexplicable, or merely the result of a pernicious mistake.

My fourth thesis, the doctrine of the power of ideas, is characteristic of the liberal and rationalist thought of the eighteenth and nineteenth centuries.

But the liberal movement did not believe only in the power of ideas. It also upheld a view which I consider mistaken. It believed that there was little need for competing ideas to join battle. This was because it supposed that truth, once put forward, would always be recognized. It believed in the theory that truth is manifest—that it cannot be missed once the powers which are interested in its suppression and perversion are destroyed.

This important and influential idea—that truth is manifest—is one form of optimism which I cannot support. I am convinced that it is mistaken, and that, on the contrary, truth is hard, and often painful, to come by. This, then, is my *fifth thesis*.

Truth is hard to come by.

This thesis explains to some extent the wars of religion. And although it is a piece of epistemology, it can throw much light upon the history of Europe since the Renaissance, and even since classical antiquity.

Let me now, in the time that remains, try to give a brief glimpse of this history—of the history of our time, especially since the Renaissance and the Reformation.

The Renaissance, and the Reformation, may be considered as the conflict between the idea that truth is manifest—that it is an open book, there to be read by anybody of good will—and the idea that truth is hidden: that it is discernible only by the elect; that the book must be deciphered only by the ministry of the Church, and interpreted only by its authority.

Although 'the book' meant, in the first instance, the Bible, it subsequently came to mean the book of nature. This book of nature, Bacon believed, was an open book. Those who misread it were misled by prejudice, impatience, and 'anticipation'. If only you will read it without prejudice, patiently, and without anticipating the text, you will not err. Error is always your own fault. It is your own perverse and sinful refusal to see the truth which is manifest before you.

This naïve and, I believe, mistaken view that truth is manifest became the inspiration for the advancement of learning in modern times. It became the basis of modern rationalism, as opposed to the more sceptical classical rationalism of the Greeks.

In the field of social ideas, the doctrine that truth is manifest leads to the doctrines of individual moral and intellectual responsibility and of freedom; it leads to individualism, and to a rationalist liberalism.

This doctrine makes the spiritual authority of the Church and its interpretation of the truth superfluous, and even pernicious.

A more sceptical attitude towards truth, on the other hand, leads to an emphasis upon the authority of the Church, and to other forms of authoritarianism. For if the truth is not manifest, then you cannot leave it to each individual to interpret it; for this would of necessity lead to chaos, to social disintegration, to religious schisms, and to religious wars. Thus the book must be interpreted by an over-riding authority.

The issue here can be described as one between individualistic rationalism and authoritarian traditionalism.

The issue between rationalism and authoritarian traditionalism can also be described as that between, on the one hand, faith in man, in human goodness as well as in human reason, and, on the other hand, distrust of man, of his goodness and of his reason.

I may confess that in the issue between faith in man and distrust of man, my feelings are all on the side of the naïve liberal optimists, even though my reason tells me that their epistemology was all wrong, and that truth is in fact hard to come by. I am repelled by the idea of keeping men under tutelage and authority. But I must admit, on the other hand, that the pessimists who feared the decline of authority and tradition were wise men. The terrible experience of the great religious wars, and of the French and Russian revolutions, prove their wisdom and foresight.

But although these wars and revolutions prove that the cautious pessimists were wise, they do not prove that they were right. On the contrary, the verdict of history—I mean, of course, the history of our time—seems, by and large, to favour those who had faith in man and in human reason.

For the society of our free world since the Reformation has indeed seen a decline of authority without parallel in any other epoch. It is a society without authority, or, as one might call it, a fatherless society.

The Reformation, by stressing the conscience of the individual, has dethroned God as the responsible ruler of Man's world: God can only rule in our hearts, and through our hearts. The Protestant believes that it is through his own human conscience that God rules the world. The responsibility for the world is mine and yours: this is the Protestant faith; and the Bishop of Bradford spoke as a good Protestant when he appealed to his ministers to destroy a social world which was the work of the devil.

But the authoritarians and traditionalists were convinced that a non-authoritarian or fatherless society must spell the destruction of all human values. They were wise, I have said, and in a way they were the better epistemologists. And yet, they were wrong. For there were other revolutions, the Glorious Revolution, and the American Revolution. And there is our present free world, our Atlantic Community. It is a fatherless society ruled by the interplay of our own individual consciences. And, as I have tried to convince you, it is the best society that has ever existed.

What was the mistake of the authoritarians? Why must their wisdom be rejected? I believe that there are three elements in our free world which have successfully replaced the dethroned authority.

The first is our respect for the authority of truth: of an impersonal, interpersonal, objective truth which it is our task to find, and which it is not in our power to change, or to interpret to our liking.

The second is a lesson learnt in the religious wars. For I think that in these wars we did learn our lesson: we did learn from our mistakes (though in the social and political field this seems a rare and difficult thing). We learnt that religious faith and other convictions can only be of value when they are freely and sincerely held, and that the attempt to force men to conform was pointless because those who resisted were the best, and indeed the only ones whose assent was worth having. Thus we learnt not only to tolerate beliefs that differ from ours, but to respect them and the men who sincerely held them. But this means that we slowly began to differentiate between sincerity and dogmatic stubbornness or laziness, and to recognize the great truth that truth is not manifest, not plainly visible to all who ardently want to see it, but hard to come by. And we learnt that we must not draw authoritarian conclusions from this great truth but, on the contrary, suspect all those who claim that they are authorized to teach the truth.

The third is that we have also learnt that by listening to one another, and criticizing one another, we may get nearer to the truth.

I believe that this critical form of rationalism and, above all, this belief in the authority of objective truth is indispensable for a free society based on mutual respect. (This is why it is important not to let our thoughts be seriously influenced by such intellectual misunderstandings as relativism and irrationalism, the understandable results of disappointment with dogmatism and authoritarianism.)

But this critical approach makes room, at the same time, for a reconciliation between rationalism and traditionalism. The critical rationalist can appreciate traditions, for although he believes in truth, he does not believe that he himself is in certain possession of it. He can appreciate every step, every approach towards it, as valuable, indeed as invaluable; and he can see that our traditions often help to encourage such steps, and also that without an intellectual tradition the individual could hardly take a single step towards the truth. It is thus the critical approach to rationalism, the compromise between rationalism and scepticism, which for a long time has been the basis of the British middle way: the respect for traditions, and at the same time the recognition of the need to reform them.

What the future will bring us, we do not know. But the achievements of the past and of our own time show us what is humanly possible. And they can teach us that although ideas are dangerous we may learn from our mistakes how to handle them; how to approach them critically, how to tame them, and how to use them in our struggles, including our struggle to get a little nearer to the hidden truth.

20

HUMANISM AND REASON

The first of a series of books, *Studia Humanitatis*, published in Switzerland, is written in German by two friends, Ernesto Grassi, an Italian scholar interested in the 'Humanist' writers of the Renaissance, and Thure von Uexküll, son of the German biologist Jakob von Uexküll, famous for his *Theoretical Biology*. The book[1] which deals with *The Origin and the Limits of the Moral and the Natural Sciences*, is part of a movement of considerable interest that aims at re-awakening the spirit of the humanists. This neo-humanist movement is characteristically Central European, born of the disasters suffered by the Continent during this century; and although the book under review is not only scholarly but also serene, some of its moods, and some of the conclusions drawn, may not easily be appreciated by those who have no personal knowledge of the shattering experience of social disintegration through which it was the lot of these European thinkers to live. The neo-humanist movement is inspired by the conviction (shared by a number of other movements) that it knows both the causes and the cure of the

This review, written in 1951, appeared first (with considerable cuts, made by the Editor, to save space) in The Philosophical Quarterly, **2**, 1952.

[1] *Von Ursprung und Grenzen der Geisteswissenschaften und Naturwissenschaften*, by E. Grassi and T. von Uexküll, Berne, 1950.

widespread depravity and perversion of everything human which Central Europe has had to witness. Its message is that only the understanding of man and his 'essential nature'—his cultural creativity—can bring relief to our ills; and it tries, as is made clear by Grassi's 'Introductory Remark', to take up again the task of developing a philosophy of man and of that important human activity, science. Science, according to this philosophy, is to be reinterpreted as a part of 'humanism'; consequently a meaning of 'humanism' and of 'humanistic' which confines humanism to the 'humanities'—that is, to historical, philological and literary studies—is rejected as too narrow.

The book may thus be said to aim at a new philosophy of man which puts both the humanities and the natural sciences in their proper place. It consists of two parts—*On the Origin and the Limits of the Humanities* (*Geisteswissenschaften*[2]) by Grassi, and *On the Origin of the Natural Sciences*, by Uexküll. The two parts are loosely connected by a vague relativistic pragmatism (reminiscent of F. C. S. Schiller, who also called himself a humanist) combined with a repudiation of pragmatist views. No doubt the authors will disagree with this opinion which they may take as proof that the reviewer is incapable of seeing their main point; but their various attempts to stress the identity of their views appear somewhat forced. This, however, does not diminish the value or the interest either of the whole or of its two parts.

The first part, Grassi's contribution, is a philosophical essay on the essence of humanism. Its main topic is indicated by the German word *Bildung* (often translated by 'culture'), which is here understood as the growth, the development, or the self-formation of the human mind; and it attempts to reestablish an educational ideal of mental growth designed to meet the criticisms raised against the old *humanistische Bildungsideal* (the educational aim of the humanities) which, according to Grassi, has become pointless owing to the disappearance of the social

[2] The term '*die Geisteswissenschaften*' ('the humanities') has become a typical German term, and almost untranslatable, even though it can be literally translated as 'mental sciences' (or 'moral and mental sciences'), and even though it seems to have reached Germany, ironically enough, through Theodor Gomperz's translation of J. S. Mill's expression 'the Moral Sciences'. (I say 'ironically enough' because the term has, in its present German usage, a strong irrationalist and even anti-rationalist and anti-empiricist flavour; but Gomperz and Mill were rationalists and empiricists.)

and cultural traditions in which it was rooted. The text on which Grassi's neo-humanistic sermon is based is a disputation concerning the relative merits of legal and medical science, C. Salutati's *De nobilitate legum et medicinae*. (Written in 1390, it was published in the middle of the fifteenth century; a critical edition by E. Garin was published in 1947 by the *Istituto di studi filosofici* in Florence. Together with Petrarch's famous attack on medical men it is perhaps the earliest ancestor of Kant's *Streit der Fakultäten*.) Grassi takes this as a discussion of the relative merits of the humanities and the natural sciences, and as a vindication of the claim of the humanities to superiority. This superiority, he says, was much better understood at the time when the natural sciences were founded than it is today.

The superiority claimed is threefold. First, it is claimed that the various natural sciences have the character of 'arts' (in the sense of *artes* = *technai*) rather than of science or knowledge (*scientia* or *epistēmē*); this means, in Salutati's view, that they have to take their 'principles' (corresponding to Bacon's 'middle principles') from elsewhere, i.e. from philosophical knowledge, and that they are therefore logically inferior to those disciplines which establish their own principles. (This view derives from Aristotle and was shared by contemporaries of Salutati as well as by later thinkers such as Leonardo.) Secondly, it is claimed (with Francis Bacon) that the natural sciences are arts (*artes*) in the sense of techniques or rather technologies—that they give us power; but such power is not, as Bacon thought, knowledge, for true knowledge springs from first principles rather than from secondary or middle principles. Thirdly, although these technologies may be the servants of man, and although they may be of some help to him in his ultimate and essential task of furthering his mental growth, they cannot carry him on to the fulfilment of this task; for they inquire into reality only within the narrow limits of their particular secondary principles without which their efforts would be pointless.

As opposed to all this, legal science, which is political science, is the science of right and wrong. As such it is not only useful to man ('*ius . . . a iuvando*', says Salutati), but useful in an essential sense, for it 'saves his humanity', it 'aims at his completion'. Only by leaving the primitive jungle or bush (*hūlē*) and settling in ordered political communities do men transcend the beasts, as Protagoras taught. This is the first step in their mental growth (*Bildung*), and the basis of all others; and 'human history is nothing but the success or failure of man-designed norms, enabling community life in the political and social spheres to proceed' (p. 106).

This is by no means a complete outline of Grassi's contribution, which deals at length with such problems as the Aristotelian doctrine that all poetry is *imitation*, with problems of the theory of tragedy, especially that of *katharsis*, and with the philosophy of time. Yet the discussions of these latter topics suffer severely from insufficient clarity and coherence; they do not, in my opinion, shed new light on the problems discussed, even though they contain some interesting asides. Outstanding among these are Grassi's emphasis on *imaginative power* (*Phantasie*) as an essential element in human nature and mental growth; but his hint (pp. 102–3) that its role in the natural sciences is confined to that of tracing out their framework does not appear to me to do justice to them. One of the most interesting remarks from the educational or self-educational point of view is contained in Grassi's analysis of the 'humanistic conception of mental growth' (*Bildung*). In trying to interpret a literary passage we may discover that in the context in question the words have an unusual and even a new meaning. 'This leads us to something new and unexpected. An unsuspected world opens itself before us—and thus we "grow" (*und dabei "bilden" wir uns*).'

Grassi very fairly concedes that the natural scientist's mind can 'grow' in precisely the same way when he finds himself compelled to adopt a new 'interpretation' of a natural phenomenon; but this concession seems to me to destroy his attempt to make use of Salutati's arguments to establish the educational priority of the humanities.

Returning to Grassi's central claim—the threefold superiority of the humanities—I admit that the natural sciences are in danger of stifling mental growth, instead of furthering it, if they are taught as technologies (the same is probably true of painting and of poetry); and that they should be treated (like painting and poetry) as human achievements, as great adventures of the human mind, as chapters in the history of human ideas, of the making of myths (as I have explained elsewhere[3]), and of their criticism. Neither the possibility of such a

[3] See chs. 4 f. of this volume. Cf. also note 6 to ch. 11 of my *Open Society* (revised editions).

humanistic approach to science, nor the need for it, is mentioned by Grassi; on the contrary, he seems to believe that salvation lies in the realization and explicit recognition of the inferior technological character of the natural sciences—in other words, in making them keep their place. But while I am ready to admit the educational priority of a 'humanist' approach, I cannot admit the validity of the Grassi-Salutati theory of the natural sciences—a theory which, of course, is directly derived from Aristotle. That the natural sciences have blindly to accept their principles from First Philosophy is a doctrine whose truth I cannot admit in any sense. Grassi tries to meet this criticism (p. 52) by conceding that the natural sciences *may* question, criticize, and replace their 'principles' (an admission which seems to me tantamount to abandoning Salutati and Aristotle), and by asserting that it is (*a*) the *aims* of science, and (*b*) the *conception* of a 'principle' (rather than their various principles) which the various natural sciences must blindly presuppose. But this position, although not incompatible with the Aristotelian view on which Salutati's argument is based, is nevertheless completely different from it.

The truth of the matter seems to be this. Although medicine happens to be an 'art', a technology, it is a mistake to conclude that it may be taken as representing the natural sciences; for it is an applied rather than a pure science. As to the latter, I agree that natural science—as opposed to pure mathematics—is not *scientia* or *epistēmē*; not, however, because it is a *technē*, but because it belongs to the realm of *doxa*—just like the myths which Grassi rightly values so highly. (The realization that natural science belongs to the realm of *doxa*, but that it was usually mistaken, until fairly recently, for *epistēmē* is, I believe, fertile for understanding the history of ideas.) Thus Grassi's central claim that we ought to return to Salutati's superior understanding of the status and significance of the natural sciences seems to me unfounded. Moreover, in Britain at least, the (Aristotelian) view of the matter which Grassi tries to reestablish never lost its hold, and it is therefore hardly in need of a restatement—not even of a restatement that uses valid arguments.

The second part of the book, written by Thure von Uexküll, is an excitingly original attempt to develop a new theory of science—a biologically orientated epistemology. A beautifully clear piece of writing, perhaps the best piece of contemporary German prose I can recall, it introduces us to a new approach to biology, a new development of ideas which originated with the author's father, Jakob von Uexküll.

The fundamental category (p. 248) of this approach is that of a biological *action* (*Handlung*). To explain it, we may perhaps start from the obvious fact that the natural sciences try to describe and explain the behaviour of things under various conditions, and especially any order or regularity which may be discovered in this behaviour. This is true for physics, chemistry and biology. In the biological sciences we are interested in the behaviour of organs, tissues, cells, and, of course, whole organisms. The central idea of Uexküll's biology is that the most successful way of describing the behaviour of a whole organism is in terms of *actions* which follow certain schematic patterns or 'schemata', and that these 'schemata of action' and 'rules of the game' may be understood as elaborations and modifications of a small number of fundamental schemata and rules. This idea appears at first sight attractive if not very surprising, although one may be inclined to suspend judgment until it has proved its fruitfulness. But the fruitfulness of the idea is shown, I believe, by Uexküll's brilliant application of it to the problem of the behaviour of the *parts* of the organism (organs, tissues, etc.), and to a most interesting and truly revolutionary analysis of 'the significance of physical and chemical methods within biology' (p. 166).

According to Uexküll's theory, there exists for each kind of organism a definite number of action schemata, each of which is released by a certain 'release-signal' (*Auslöser*), whose nature can be found by experiment, by constructing an *imitative contraption* (*Attrappe*, dummy). These, in most cases, can be reduced to astonishingly simple schematic representation. The Viennese biologist Konrad Lorenz found, for example, that (p. 162) certain species of geese follow, as if it were their mother, the first moving object they encounter upon breaking their shells, and that they continue to do so even when they are confronted by their real mother.[4] For certain other fledglings (p. 169) the imitative contraption which may replace the parent by operating as a release signal for normal actions (opening their beaks) consists simply of two

[4] See K. Z. Lorenz, *King Solomon's Ring* (published in English in 1952, after the present review was first published).

round pieces of cardboard or sheet metal giving something like a generalized silhouette of the head and body of the parent bird. 'With the help of such imitative contraptions, we can make our entry into the scene of life of some animals. It is a moving and even a shattering experience for a sensitive mind to realize the strangeness of this world. The magical and threatening character of this reality creates an impression before which all our old ideas and conception of nature must fail' (p. 169). Uexküll's extension of this approach to the problem of tissue-reactions, and of the use of physical and chemical methods, is, I can only repeat, of the greatest interest. He suggests that what we actually do in biochemistry is to construct imitative contraptions (dummies) serviceable as release signals for the actions of organs or tissues. This, I believe, is an idea with a great future, likely to throw much light on some vexed questions. (I have in mind, for example, the question of the functional equivalence of certain chemical and electrical stimuli in some neuro-muscular reactions in the face of even such subtle tests as the measurement of 'end-plate potentials'. Another of the many cases which, I think, might be used to illustrate Uexküll's point is a well-known hypothesis which has been used to explain bacteriostasis: the bacteria, it is suggested, absorb a certain chemical which they cannot assimilate, mistaking it for food; that is, the chemical acts, and is acted upon, like a dummy.)

All that Uexküll has to say about the application of his ideas to biology is beyond praise. I do not know whether his theories are true, but they are strikingly original. They not only have great explanatory power, but also the power to put familiar things in an entirely new light; and one day they may well open a new era in biological thinking, especially in the fields of physiology and biochemistry—provided, of course, that the experimentalists take notice of these new ideas and their countless applications in almost all fields of biology.

Yet Uexküll speaks in this book not only as a biologist (and methodologist of biology) but also as a philosopher.

Encouraged, perhaps, by his biological applications, Uexküll tries to apply his fundamental categories to the whole problem of the theory of knowledge. Starting from the Kantian question whether it is possible to know things 'in themselves', he discusses the old aspirations of physics to discover the innermost secret of nature itself (*das Innere der*

Natur), and the failure of these aspirations; and after an elaborate (but I do not think successful) attempt to determine the role of physics in a world of biological actions, he ultimately arrives at a biological ontology—the doctrine that reality (which can only be *our* world, a reality-for-us[5]), is a structure of actions; of 'actions of various kinds and various extension' (p. 248); and he replaces the problem of our *knowledge* of the world-in-itself by that of our *participation* in the structure of actions which is the world.

Although much of this is reminiscent of certain forms of pragmatism, operationalism, and instrumentalism, it is nevertheless one of the most original attempts since Schopenhauer and Bergson to erect a new metaphysical world, and one capable of accommodating modern science. This new attempt commands respect; but it does not carry conviction. On the contrary, it seems to me clear that Uexküll's theory of knowledge and his ontology are founded upon a mistake. Anybody acquainted with the pitfalls of idealistic epistemology will have no difficulty in seeing that the mistake made must be akin to that of identifying what is with what is *known*; or *esse* = *sciri*. This led to Berkeley's *esse* = *percipi* as well as to Hegel's *esse* = *concipi*, and it now leads a biologist for whom knowledge is, rightly, a kind of action, to *esse* = *agi*, i.e. to the doctrine that 'reality' is the thing acted upon, or the object in the way of action, or a factor—the situational factor—of the schemata of our biological actions.

To be more specific, three mistakes may be pointed out in Uexküll's argument. The first can be found in his analysis describing the failure of the aspirations of physics. This analysis appears to me to exhibit some typical and popular misunderstandings of the theory of relativity. (It is a mistake to maintain that the relativist universe does not know continuous time or continuous space, but only 'islandlike space-time connections'; and it is a mistake to infer from the principle of the equivalence of reference systems the relativization of reality: on the contrary, relativity teaches both the reality and the invariance of

[5] Compare the following remarks made by the older von Uexküll in 1920 in his *Theoretical Biology* (see the English translation, 1920, p. xv; the second set of italics is mine): '*All reality is subjective appearance*: this must be the great fundamental admission even of biology. . . . We always *come up against objects* that owe their construction [and so, presumably, their existence] to the subject.'

spatio-temporal intervals.) Modern physics (*pace* Heisenberg) does attempt to give us a picture of the universe; whether this is drawn well or badly is, of course, a very different question. If we realize this, the suggestion that we must replace an allegedly dissolving world-view of physics by a new world-view of biology loses much of its force.

The second mistake is an extremely interesting one. It is made at a point (pp. 201 ff.) where Uexküll blames Lorenz for reasoning in a circle, and for failing to realize the full consequences of his own (and Uexküll's) new biological attitude. Lorenz, he tells us, believes that the action schemata (including those of 'biological experience') have developed by adapting themselves to the external world by the method of trial and error. This view is rejected by Uexküll. Lorenz, he claims, 'fails to grasp the new attitude which is the result of the discovery' (due partly to Lorenz himself) 'that the world around us, as it is given to our senses, is only the sum total of the biological release signals, and that it exists therefore only as a factor of the schemata of our biological actions' (p. 202). Uexküll asserts that Lorenz's circular argument is due to his failure 'to rid himself of the objectivist assumptions upon which the picture of the universe of classical physics rests' (p. 203).

I have no doubt that the accusation of arguing in a circle falls back on Uexküll, and that his faulty reasoning is at least partly due to his untenable subjectivist interpretation of modern physics. For Uexküll overlooks the fact that his whole biological analysis presupposes the possibility of a (more or less) objectivist approach. It is only such an approach which enables us to speak, for example, of an 'imitative contraption' taking over the functions of a bird's mother. It is only because we know—in our 'objective' world, which goes beyond the bird's 'subjective' world—what its real mother is, and what a contraption is, that we can say that, if animal *A* differentiates by its actions between its real mother and an imitative contraption of a certain kind while animal B does not, then *A* has, to that extent, the greater powers of discrimination or differentiation, and is, to the same extent, better adapted to certain possible environmental situations.

Lorenz's view (which I have shared for many years[6]) is not only

[6] Cf. chs. 1 and 15, above.

defensible, but necessary for understanding the peculiar human situation—the phenomenon, based on the argumentative use of the human language,[7] of *critical* knowledge, as opposed to the uncritical and, as it were, accidental adaptations of the animal's 'knowledge'.

And this brings me to the third mistake in Uexküll's argument; a mistake which is very hard to understand in one who admires Kant. It is the gravest mistake of the book, and one which both authors share. It is their complete (and it seems, almost hostile) neglect of human reason—of man's power to grow, to transcend himself, not only by the imaginative invention of myths (whose importance is so well emphasized by Grassi), but also by the rational criticism of his own imaginative inventions. These inventions, if *formulated in some language*, are from the start somewhat different from other biological actions; this may be seen from the fact that each of two schemata of biological actions which otherwise are indistinguishable may contain a myth (concerning, say, the origin of the world) which is contradictory to the other. For although some of our beliefs may be immediately relevant to practice, others are only remotely relevant to it, if at all. Their differences may make it possible for them to clash, and their comparative remoteness may make it possible for them to be argued about. In this way, rational *criticism* may develop, and standards of rationality—some of the first inter-subjective standards—and the idea of an objective truth. And this criticism may, in time, develop into systematic attempts to discover what is weak and untrue in other people's theories and beliefs, and also in one's own. It is by this mutual criticism that man, if only by degrees, can break through the subjectivity of a world of biological release signals, and, beyond this, through the subjectivity of his own imaginative inventions, and the subjectivity of the historical accidents upon which these inventions may in part depend. For these standards of rational criticism and of objective truth make his knowledge structurally different from its evolutionary antecedents (even though it will always remain possible to subsume it under some biological or anthropological schema of action). It is the acceptance of these standards which creates the dignity of the individual man; which makes him responsible, morally as well as intellectually; which enables

[7] Cf. chs. 4 and 12, above.

him not only to act rationally, but also to contemplate and adjudicate, and to discriminate between, competing theories.

These standards of objective truth and criticism may teach him to try again, and to think again; to challenge his own conclusions, and to use his imagination in trying to find whether and where his own conclusions are at fault. They may teach him to apply the method of trial and error in every field, and especially in science; and thus they may teach him how to learn from his mistakes, and how to search for them. These standards may help him to discover how little he knows, and how much there is that he does not know. They may help him to grow in knowledge, and also to realize that he is growing. They may help him to become aware of the fact that he owes his growth to other people's criticisms, and that reasonableness is readiness to listen to criticism. And in this way they may even help him to transcend his animal past, and with it that subjectivism and voluntarism in which romantic and irrationalist philosophies may try to hold him captive.

This is the way in which our mind grows and transcends itself. If humanism is concerned with the growth of the human mind, what then is the tradition of humanism if not a tradition of criticism and reasonableness?

ADDENDA
SOME TECHNICAL NOTES

1. Empirical Content

We arrive at the idea of empirical content as follows. By the logical content (or the consequence class) of *a* we mean the class of all statements which follow from *a*. Thus we may first, and tentatively, consider calling the *empirical* content of *a* the class of all observational statements (or 'basic statements', see below) which follow from *a*.

But this tentative idea does not work. For what interests us most is the empirical content of an explanatory universal theory; yet from such a theory alone no observational statement follows. (From 'All ravens are black' we cannot derive any observational statement like 'There is a black raven here now'; although we can indeed derive the statement 'There is no white raven here now'.)

This is the reason why, in defining empirical content, I fell back on the idea that *a theory tells us the more about observable facts the more such facts it forbids*—that is to say, the more observable facts are incompatible with

The technical Addenda, 1–6, are relevant to ch. 10 of this volume. Addenda 6–9 (especially sections 6–10 of Addendum 8) are relevant to chs. 3–5. Addendum 10 is relevant to the Introduction.

it.[1] We then can say that the empirical content of a theory is determined by (and equal to) the class of those observational statements, or basic statements, which *contradict* the theory.

A basic statement which contradicts a theory t may be called a 'potential falsifier' of t. Using this terminology we can say that the empirical content of t consists of the class of its potential falsifiers.

That the name 'empirical content' is justifiably applied to this class is seen from the fact that whenever the measures of the empirical contents, $ECt(t_1)$ and $ECt(t_2)$, of two *empirical* (i.e. non-metaphysical) theories, t_1 and t_2, are so related that

$$(1) \qquad ECt(t_1) \leqslant ECt(t_2)$$

holds, the measures of their logical contents will also be so related that

$$(2) \qquad Ct(t_1) \leqslant Ct(t_2)$$

will hold; and similar relations will hold for the equality of contents.

Proceeding now to the notion of '*basic statements*', there is a point in which I wish to improve upon my discussion of what I have called '*basic statements*' in *The Logic of Scientific Discovery* (see especially sections 28 and 29). I introduced the term 'basic statement' in order to denote a class of statements (true or false) which, in our discussion, we can assume to be of *unquestioned empirical character*. 'Unquestioned' means here that we are prepared to confine the class of basic statements in accordance with the requirements of the most scrupulous and exacting empiricist we may be confronted with, provided that these requirements are not less exacting than our own (objectivist) minimum requirements. These are: (i) basic statements state (truly or falsely) the existence of observable facts (occurrences) within some sufficiently narrow spatio-temporal region. (ii) The negation of a basic statement will not be in general basic. In some simple cases of basic statements (example: 'There is now a full grown Great Dane in my study') their negations *may* be acceptable as basic; in most cases of basic statements (example:

[1] See *L.Sc.D.*, sections 31, 34. This idea has been accepted by Carnap; see especially his *Logical Foundations of Probability*, 1950, p. 406, and also his *Symbolische Logik*, 2nd edn., 1960, p. 21.

'There is now a mosquito in my study') their negations will not be acceptable as basic, for obvious reasons. (iii) The conjunction of two basic statements is always basic if (and only if) it is logically consistent. (Thus whenever a statement and its negation are both basic, their conjunction will not be basic.) We *may* single out from a class of otherwise acceptable basic statements those which are not compound ('relative atomic' statements; cp. *L.Sc.D.*, section 38). We can then, if we like, start with these, and construct a new class of basic statements from them, as follows. (*a*) We do not admit as basic any of the negations of the relative atomic basic statements. (*b*) We admit as basic all conjunctions of basic statements so far as they are consistent. (Consistency seems intuitively a necessary requirement, and its adoption greatly simplifies various formulations of the ensuing theory, but we might dispense with it, as long as we exclude inconsistent statements from the class of falsifiers.) (*c*) We do *not* admit any negation of any compound basic statement, or any compounds other than conjunctions of basic statements.

These last exclusions may appear somewhat severe; but it is not our purpose to admit *all* empirical statements as basic—not even all statements of observable facts: I do not mind excluding such compound observation statements as 'There is either a full grown Great Dane or a full grown Shetland Pony in my study' from the class of basic statements, though I should not like to exclude it from the class of empirical statements. For although it is our intention to ensure that all basic statements are obviously empirical, we do not intend to ensure the converse—that all obviously empirical statements (or even all observation statements) are 'basic'.

The purpose of the exclusion of negations of basic statements (or of negations of almost all basic statements) from the class of basic statements, and of the exclusion, from this class, of disjunctions and conditionals of basic statements is this: we do not wish to admit conditional statements such as 'If there is a raven in this room then it is black', or 'If there is a mosquito in this room then it is an *anopheles*'. These are no doubt empirical statements; but they are not of the character of *test statements* of theories but rather of *instantiation statements*, and therefore less interesting, and less 'basic', from the point of view of the theory of knowledge here expounded; a theory of knowledge which holds the empirical basis of all theories to be tests; or in other words, attempted refutations.

It might be worth mentioning, in this context, that the word 'basic' in the term 'basic statement' seems to have misled some of my readers. My use of the term has a history which is as follows.

Before using the terms 'basic' and 'basic statement', I made use of the term 'empirical basis', meaning by it the class of all those statements which may function as tests of empirical theories (that is, as potential falsifiers). In introducing the term 'empirical basis' my intention was, partly, to give an ironical emphasis to my thesis that the empirical basis of our theories is far from firm; that it should be compared to a swamp rather than to solid ground.[2]

Empiricists usually believed that the empirical basis consisted of absolutely 'given' perceptions or observations, of '*data*', and that science could build on these *data* as if on rock. In opposition, I pointed out that the apparent 'data' of experience were always interpretations in the light of theories, and therefore affected by the hypothetical or conjectural character of all theories.

That those experiences which we call 'perceptions' are interpretations—interpretations, I suggest, of the total situation in which we find ourselves when 'perceiving'—is an insight due to Kant. It has often been formulated, somewhat awkwardly, by saying that perceptions are interpretations of what is given to us by our senses; and from this formulation sprang the belief that there must be present some ultimate 'data', some ultimate material which must be uninterpreted (since interpretation must be *of* something, and since there cannot be an infinite regress). But this argument does not take into account that (as already suggested by Kant) the process of interpretation is at least partly physiological, so that there are never any uninterpreted data experienced by us: the existence of these uninterpreted 'data' is therefore a *theory*, not a fact of experience, and least of all an ultimate, or 'basic' fact.

Thus there is no uninterpreted empirical basis; and the test statements which form the empirical basis cannot be statements expressing uninterpreted 'data' (since no such data exist) but are, simply,

[2] See especially the last paragraph of section 30 of my L.Sc.D.

statements which state observable simple facts about our physical environment. They are, of course, facts interpreted in the light of theories; they are soaked in theory, as it were.

As I pointed out in my *Logic of Scientific Discovery* (end of section 25), the statement 'Here is a glass of water', cannot be verified by any observational experience. The reason is that the *universal terms* which occur in this statement ('glass', 'water') are dispositional: they 'denote physical bodies which exhibit a certain *law-like behaviour*'.[3]

What has been said here about 'glass' and 'water' holds for all descriptive universals. The famous cat on the mat so much beloved by empiricists (I too find cats endearing) is an entity even more highly theoretical than is either glass or water. *All terms are theoretical terms, though some are more theoretical than others.* ('Breakable' is more highly theoretical, or more highly dispositional, than 'broken', but the latter term is also theoretical or dispositional, as mentioned for example at the end of chapter 3, above.)

This view of the matter makes it possible for us to include into our 'empirical basis' statements containing highly theoretical terms, provided they are singular statements about observable facts; for example, statements like 'Here is a potentiometer which reads 145' or 'This clock reads 30 minutes past 3'. That the instrument is in fact a potentiometer cannot be finally established or verified—no more than that the glass before us contains water. But it is a *testable* hypothesis, and we can easily *test* it in any physical laboratory.

Thus every statement (or 'basic statement') remains essentially conjectural; but it is a conjecture which can be easily tested. These tests, in their turn, involve new conjectural and testable statements, and so on, *ad infinitum*; and should we try to *establish* anything with our tests, we should be involved in an infinite regress. But as I explained in my *Logic of Scientific Discovery* (especially section 29), we do not *establish* anything by this procedure: we do not wish to 'justify' the 'acceptance' of anything, we only test our theories critically, in order to see whether or not we can bring a case against them.

Thus our 'basic statements' are anything but 'basic' in the sense of

[3] *L.Sc.D.*, section 25, p. 95; new appendix *x, (1) to (4), pp. 422–6. See also for example chs. 1 (sections iv and v) and 3 (section 6, the last six paragraphs) of the present volume.

'final'; they are 'basic' only in the sense that they belong to that class of statements which are used in testing our theories.

2. Probability and the Severity of Tests

The severity of our tests can be objectively compared; and if we like, we can define a measure of their severity.

In this definition, and also in later discussions in this *addendum*, I shall make use of the idea of *probability*, in the sense of the *calculus of probability*; or more precisely, of the idea of relative probability,

$$p(x,y),$$

to be read 'the probability of x, given y'. The idea of absolute probability

$$p(x),$$

to be read 'the absolute probability of x', will be here taken as defined in terms of relative probability, by the explicit definition

$$\mathrm{D(AP)}\ p(a) = p(a,b) \leftrightarrow (c)(((d)(p(c,d) \geqslant p(d,c))) \rightarrow p(a,b) = p(a,c)).$$

Here '(a)' abbreviates 'for every a'; '(Ea)' abbreviates 'there exists an a'; '$\leftrightarrow$' abbreviates 'if and only if'; and '. . . $\rightarrow$ . . .' abbreviates 'if . . . then . . .'. I shall also use '&' as an abbreviation for 'and'. D(AP) stipulates that $p(a) = p(a,b)$ provided b has maximal (relative) probability.

The idea of relative probability, $p(x,y)$, will be used here mainly as a definiens, as in D(AP). It can in its turn be defined implicitly through an *axiom system*, as in my *L.Sc.D.* (new appendices *iv and *v). The six axioms there given may be reduced to three, one of them, A, an existential axiom, and two, B and C, axioms in the form of ('creative'[4]) definitions:

[4] For a discussion of 'creative' and 'non-creative' definitions see, for example, P. Suppes, *Introduction to Logic*, 1957, pp. 153 f., and also my paper 'Creative and Non-Creative Definitions in the Calculus of Probability', *Synthese*, **15**, 1963, No. 2, pp. 167–86.

A $$(Ea)(Eb)p(a,b) \neq p(b,b)$$

that is, there are at least two different probabilities.

B $$((d)p(ab,d) = p(c,d)) \leftrightarrow (e)(f)(p(a,b) \leqslant p(c,b) \;\&\; p(a,e) \geqslant p(c,e) \leqslant p(b,c) \;\&\; ((p(b,e) \leqslant p(f,e) \;\&\; p(b,f) \geqslant p(f,f) \leqslant p(e,f)) \rightarrow p(a,f)p(b,e) = p(c,e)))$$

Axiom B defines the product ab (read 'a-and-b') in terms of $p(x,y)$.

C $$p(-a,b) = p(b,b) - p(a,b) \leftrightarrow (Ec)p(b,b) \neq p(c,b)$$

Axiom C defines the complement $-a$ (read 'non-a') in terms of $p(x,y)$.

To these three axioms we may add three (non-creative, or ordinary) definitions: of absolute probability, $p(a)$ defined above by D(AP); of the Boolean identity, $a = b$; and of n-termed independence relative to b.

Identity is defined as follows:

D(=) $$a = b \leftrightarrow (c)p(a,c) = p(b,c)$$

We say that a set of n elements or an n-termed sequence, $A_n = a_1, \ldots, a_n$, is 'n-termed independent (relative to b)' if the so-called 'special multiplication theorem' (relative to b) applies to every one of the $2^n - 1$ non-empty sub-sets of the set A_n. Let $a_i, \ldots, a_m$ be the elements of *any* such sub-set (or sub-sequence); then if A_n is n-independent, we have

(m) $$p(a_i \ldots a_m,b) = p(a_i,b).p(a_{i+1},b) \ldots p(a_m,b)$$

where the right-hand side is a product of $m - i$ probabilities. Among these $2^n - 1$ equations, corresponding to the $2^n - 1$ non-empty sub-sets of A_n, there will be n trivial ones (for the unit sub-sets), since for $m = i$, our equation (m) degenerates to

(i) $$p(a_i,b) = p(a_i,b);$$

that is, every single element is, trivially, 1-termed independent relative to every b. Thus the n-termed independence of A_n is defined by $2^n - n - 1$ *non-trivial* equations.[5]

This somewhat clumsy definition which uses $2^n - n - 1$ equations can be simplified by introducing a recursive definition of '$\text{Ind}_n(\{a_1, \ldots, a_n\}; b)$', to be read '$a_1, \ldots, a_n$, are n-independent relative to b':

For this purpose, I consider a set S of elements, $a_1 \in S$, $b \in S$, etc.; and I use the following notation: I write '$\{a_1, \ldots, a_n\}$' to denote the sub-set of S whose elements are $a_1, \ldots, a_n$; and I write '$\{a_1, \ldots, a_n\} - \{a_i\}$' to denote the same sub-set *with the exclusion* of the element a_i. I now define n-independence relative to b as follows.

D(Ind) (1) $\text{Ind}_1 (\{a_1\};b)$ for every a_i and b in S.
(2) $\text{Ind}_{n+1}(\{a_1, \ldots, a_{n+1}\};b)$ if and only if
(a) $\text{Ind}_n(\{a_1, \ldots, a_{n+1}\} - \{a_i\};b)$ for every i, $1 \leqslant i \leqslant n + 1$
(b) $p(a_1 \ldots a_{n+1},b) = p(a_1 \ldots a_n,b)p(a_{n+1},b)$.

There are various related concepts. A weaker one is that of serial independence, $\text{Sind}_n(a_1, \ldots, a_n;b)$. The definition is like that of Ind_n, except that we may omit the {}-brackets and replace (2)(a) by the formula

$$\text{Sind}_n(a_1, \ldots, a_n;b).$$

Now we can turn to the definition of the severity of tests.

Let h be the hypothesis to be tested; let e be the test statement (the evidence), and b the 'background knowledge', that is to say, all those things which we accept (tentatively) as unproblematic while we are testing the theory. (b may also contain statements of the character of initial conditions.) Let us assume, to start with, that e is a logical consequence of h and b (this assumption will later be relaxed), so that $p(e,hb) = 1$. For example, e may be a statement of a predicted position of

[5] Cp. for example W. Feller, *An Introduction to Probability Theory and its Applications*, vol. i, 2nd ed., 1957, p. 117. Incidentally, we may identify the empty sub-set with that unit-set whose only element is $-(a. - a)$, since this element is (relative to any b) *absolutely independent*, i.e. independent with respect to any set A_n. We so obtain 2^n equations of which $n + 1$ refer to unit-classes and are trivial.

the planet Mars, derived from Newton's theory h and our knowledge of past positions which forms part of b.

We then can say that, if we take e as a test of h, then the severity of this test interpreted as supporting evidence, will be the greater the less probable is e, given b alone (without h); that is to say, the smaller is $p(e,b)$, the probability of e given b.

There are in the main two methods[6] of defining the severity

$$S(e,b)$$

of the test e, given b. Both start from the *measure of content*, Ct. The first takes the complement of probability as a measure of content Ct:

$$\text{(1)} \qquad \mathrm{Ct}(a) = 1 - p(a);$$

the second takes the reciprocal of probability as a measure of content:

$$\text{(2)} \qquad \mathrm{Ct}'(a) = 1 \,/\, p(a)$$

The first suggests a definition like $S(e,b) = 1 - p(e,b)$ or, better,

$$\text{(3)} \qquad S(e,b) = (1 - p(e,b)) \,/\, (1 + p(e,b))$$

that is to say, it suggests that we measure the severity of the test by Ct or, better, by something like a 'normalized' Ct (using $1 \,/\, (1 + p(e,b))$ as a normalizing factor). The second suggests that we measure the severity of the test simply by its content Ct′:

$$\text{(4)} \qquad S'(e,b) = \mathrm{Ct}'(e,b) = 1/p(e,b).$$

We may now generalize these definitions by relaxing the demand that e logically follows from h and b, or even the weaker demand that

$$p(e,hb) = 1$$

[6] See *L.Sc.D.*, note *2 to section 83 (p. 270).

Instead we now assume that there is some probability, $p(e,hb)$, which may or may not be equal to 1.

This suggests that, in order to obtain a generalization of (3) and (4) we substitute in both these formulae the more general term '$p(e,hb)$' for '1'. We thus arrive at the following generalized definitions of the severity of the test e interpreted as *supporting evidence* of the theory h, given the background knowledge b.

$$S(e,h,b) = (p(e,hb) - p(e,b)) \,/\, (p(e,hb) + p(e,b)) \tag{5}$$

$$S'(e,h,b) = p(e,hb) \,/\, p(e,b) \tag{6}$$

These are our measures of the severity of tests, *qua* supporting evidence. There is little to choose between them since the transition from the one to the other is order-preserving;[7] that is to say, the two are topologically invariant. (The same holds, if we replace the measures Ct' and S' by their logarithms[8]— for example by $\log_2 Ct'$ and by $\log_2 S'$—in order to make these measures additive.)

Having defined a measure of the severity of our tests, we can now use the same method to define the explanatory power of the theory h, $E(h,e,b)$ (and if we like, in a somewhat similar way, the degree of corroboration[9] of h) with respect to e, in the presence of b:

$$E(h,e,b) = S(e,h,b) \tag{7}$$

$$E'(h,e,b) = S'(e,h,b) \tag{8}$$

These definitions indicate that the explanatory power of a theory h (with respect to some explicandum e) is the greater the more severe is e if taken as test of the theory h.

It can now be shown quite easily that the maximum degree of the explanatory power of a theory, or of the severity of its tests, depends upon the (informative or empirical) content of the theory.

Thus our criterion of progress or of the potential growth of knowledge, will be the increase of the informative content, or the empirical

[7] See *L. Sc.D.*, p. 404.
[8] Ibid., pp. 402–6.
[9] Ibid., pp. 400–2.

content, of our theories; and, at the same time, the increase of their testability; and also their explanatory power with respect to (known *and* as yet unknown) evidence.

3. Verisimilitude

In this section, the ideas of sections X and XI of chapter 10 (which are here taken as read) are further discussed and developed.

In Tarski's theory of truth, 'truth' is a property of statements. We may take 'T' to denote the class of all true statements of some more or less artificial language (object language; see section 5, below). And we may express by

$$a \varepsilon \mathrm{T}$$

the assertion (of some metalanguage) that the statement a is a member of the class of true statements; or in other words, that a is true.

Our first task here is to define the idea of the *truth-content* of a statement a, which we denote by '$\mathrm{Ct_T}(a)$'. It will have to be defined in such a way that a false statement as well as a true one has a truth-content.

If a is true, then $\mathrm{Ct_T}(a)$, the truth-content of a (or rather its *measure*) will be simply the measure of the content of a; that is:

$$(1) \qquad a \varepsilon \mathrm{T} \rightarrow \mathrm{Ct_T}(a) = \mathrm{Ct}(a)$$

where we may, as in section 2, (1), put

$$(2) \qquad \mathrm{Ct}(a) = 1 - p(a).$$

If a is false, it may still, as suggested, have a truth-content. For assume that today is Monday. Then the statement 'Today is Tuesday' will be false. But this false statement will entail a number of true statements, such as 'Today is not Wednesday' or 'Today is either Monday or Tuesday'; and the class of all those true statement which it entails will be its (logical) truth-content. In other words, the fact that every false statement entails a class of true statements is the basis for ascribing a truth-content to every false statement.

We therefore shall define the (logical) *truth-content of the statement* a as the class of statements which belong to both, the (logical) content of a, and to T; and we interpret the *measure* of its *truth content*, $Ct_T(a)$, accordingly.

In order to give a definition of the idea of $Ct_T(a)$ within the theory of Ct or of p (where $Ct(a) = 1 - p(a)$), various methods are open to us.

The simplest method is perhaps to agree that in expressions like $p(a)$ or $p(a,b)$, the letters 'a', 'b', etc., may not only be names of statements (and thus, for example, of conjunctions of a finite number of statements) but also names of classes of statements (or of the finite or infinite conjunctions of all statements which are members of these classes). We then agree to use, in place of 'T' the symbol 't'[10] in contexts like $p(t)$ or $p(a,t)$ or $p(t,b)$, and to operate with t exactly as if it were the (finite or infinite) conjunction of all true statements of the language system (or system of statements) under consideration. In other words, we use the symbol 't' as one of the constant values which may be taken up by our variables 'a', 'b', etc., and agree to use it in such a way that

(3) The consequence class or logical content of t is T.

Next we define a new symbol, 'a_T', by the definition

(4) $$a_T = a \vee t$$

As a result of this definition we have (using '$\vdash$' for 'entails' or 'from . . . follows . . .')

(5) $$a \vdash a_T$$

and therefore also

[10] Note that 't' is *not* used for 'tautology', for which we shall later introduce the symbol '*tautol*'. (Since T may well be non-axiomatizable, this method of using 't' might be said to amount to interpreting $a, b, \ldots, t, \ldots,$ as *deductive systems* (rather than as statements); see Tarski, *Logic, Semantics, Metamathematics*, pp. 342 ff., and the reference to S. Mazurkiewicz on p. 383.)

(6) $$p(aa_T) = p(a),$$
(7) $$p(a,a_T)p(a_T) = p(aa_T) = p(a).$$

We also have

(8) $$a_T \vdash x \text{ if, and only if, } a \vdash x \;\&\; x \,\varepsilon\, T,$$

where '$a \vdash b$' again reads 'b is deducible from (or entailed by) a'. Thus (8) means that a_T is the logically strongest *true* statement (or deductive system) entailed by a. Thus *we can now define the truth-content of a as the content of a_T*; and its measure, $Ct_T(a)$, can now be defined as follows:

(9) $$Ct_T(a) = Ct(a_T) = 1 - p(a_T)$$

It follows from (9) and (5) that

(10) $$Ct_T(a) \leqslant Ct(a)$$

and

(11) $$\text{If } a \,\varepsilon\, T, \text{ then } a_T = a, \text{ and } Ct_T(a) = Ct(a)$$

In order to define '$Vs(a)$'—that is (a measure of) the verisimilitude of a—we need not only the truth-content of a but also its falsity-content—or a measure of it—since we wish to define $Vs(a)$ as something like the difference of the truth-content and the falsity-content of a. But the definition of a falsity-content of a, or something to serve in its place, is not quite simple, owing to the fundamental fact that, while T can be said to form a consequence class or content (the content of t, see (3) above), the class F of *all false statements of our system* is not a consequence class. For while T contains all the logical consequences of T—since the logical consequence of anything true must also be true—F does not contain all its logical consequences: while from a true statement only true statements follow, from a false statement follow not only false statements but always true statements also.

As a result of this, a definition of 'falsity-content' on lines analogous to 'truth-content' appears not to be workable.

In order to arrive at a satisfactory definition of $Ct_F(a)$, the measure of the falsity-content of a, it will be useful to lay down a number of *desiderata*:

$$\text{(i)} \quad a \,\varepsilon\, T \rightarrow Ct_F(a) = 0$$
$$\text{(ii)} \quad a \,\varepsilon\, F \rightarrow Ct_F(a) \leqslant Ct(a)$$
$$\text{(iii)} \quad 0 \leqslant Ct_F(a) \leqslant Ct(a) \leqslant 1$$
$$\text{(iv)} \quad Ct_F(\mathit{contrad}) = Ct(\mathit{contrad}) = 1$$

where '*contrad*' is a name of a self-contradictory statement. *Desideratum* (iv) should be compared and contrasted with the theorem

$$Ct_T(\mathit{tautol}) = Ct(\mathit{tautol}) = 0$$

where '*tautol*' is a name of a tautological statement.

$$\text{(v)} \quad Ct_T(a) = 0 \rightarrow Ct_F(a) = Ct(a)$$
$$\text{(vi)} \quad Ct_F(a) = 0 \rightarrow Ct_T(a) = Ct(a)$$
$$\text{(vii)} \quad Ct_T(a) + Ct_F(a) \geqslant Ct(a)$$

(the reason for putting here '$\geqslant$' rather than '$=$' will be seen if we take 'a' to be, for example, '*contrad*'; for in this case, we obtain

$$Ct_F(a) = Ct(a) = 1, \qquad \text{by (iv)}$$

and

$$Ct_T(a) = Ct(t);$$

but $Ct(t)$ is *the maximum truth-content*, which will in general be different from zero. In an infinite universe, $Ct(t) = 1 - p(t)$ will as a rule be equal to 1.) (viii) Ct_F and Ct_T are symmetrical with respect to Ct in the following sense: there exist two functions, f_1 and f_2, such that

$$\text{(a)} \quad \begin{aligned} Ct_T(a) + Ct_F(a) &= Ct(a) + f_1(Ct_T(a), Ct_F(a)) \\ &= Ct(a) + f_1(Ct_F(a), Ct_T(a)) \end{aligned}$$

that is to say, f_1 is symmetrical with respect to Ct_T and Ct_F; so that, as a consequence we get

$$\text{(b)} \qquad Ct_T(a) = f_2(Ct(a), Ct_F(a))$$
$$\text{(c)} \qquad Ct_F(a) = f_2(Ct(a), Ct_T(a))$$

Among the various possibilities of defining '$Ct_F(a)$' on these lines, the following definition recommends itself, and will be adopted here:

$$\text{(12)} \qquad Ct_F(a) = 1 - p(a,a_T) = Ct(a,a_T)$$

This definition satisfies our *desiderata*. This is obvious for the *desiderata* (i) and (ii); and it becomes clear for the rest if we consider the following theorems:

$$\text{(13)} \qquad \begin{aligned} Ct_F(a)\, p(a_T) &= p(a_T) - (p(a,a_T)\, p(a_T)) \\ &= p(a_T) - p(a) \qquad \text{see (7)} \\ &= Ct(a) - Ct_T(a) \end{aligned}$$

so that

$$\text{(14)} \qquad Ct_T(a) = Ct(a) - (Ct_F(a)\, p(a_T)) \leqslant Ct(a).$$
$$\text{(15)} \qquad \begin{aligned} Ct_F(a) &= (Ct(a) - Ct_T(a)) \,/\, p(a_T) \\ &= (Ct(a) - Ct_T(a)) \,/\, (1 - Ct_T(a)) \end{aligned}$$
$$\text{(16)} \qquad \begin{aligned} Ct_T(a)\, p(a,a_T) &= p(a,a_T) - (p(a_T)\, p(a,a_T)) \\ &= p(a,a_T) - (p(a) \\ &= Ct(a) - Ct_F(a) \end{aligned}$$

Thus we obtain

$$\text{(17)} \qquad Ct_F(a) = Ct(a) - (Ct_T(a)\, p(a,a_T)) \leqslant Ct(a)$$
$$\text{(18)} \qquad \begin{aligned} Ct_T(a) &= (Ct(a) - Ct_F(a)) \,/\, p(a,a_T) \qquad \text{see (iii)} \\ &= (Ct(a) - Ct_F(a)) \,/\, (1 - Ct_F(a)) \qquad \text{see (15)} \end{aligned}$$

We also obtain from (15)

$$\text{(19)} \qquad Ct_F(a) - Ct_T(a)\, Ct_F(a) = Ct(a) - Ct_T(a)$$

and thus

$$(20) \qquad Ct_T(a) + Ct_F(a) = Ct(a) + Ct_T(a)\, Ct_F(a)$$

Thus (17) shows that (iii) is satisfied, and (20) shows that (v), (vi), (vii) and (viii) are satisfied. The satisfaction of (iv) follows from $p(\mathit{contrad}, t) = 0$.

This shows that the proposed definition, (12), of $Ct_F(a)$ satisfies all our *desiderata*. Yet one of our *desiderata*, (vii), may perhaps appear unsatisfactory: it may perhaps appear—in spite of our comment on (vii)—that we should have postulated that

$$(-) \qquad Ct_T(a) + Ct_F(a) = Ct(a)$$

It can be shown that the equation (–) would indeed determine Ct_F: it would lead to the definition (which we shall not adopt)

$$Ct_F(a) = Ct(a_T \rightarrow a) = 1 - p(a_T \rightarrow a),$$

where '$a_T \rightarrow a$' (or, as we can also write, '$a \leftarrow a_T$'), is the conditional statement 'if a_T then a', or 'a if a_T'.

It is of interest to compare this definition with our (12), or in other words, to compare $Ct(a \leftarrow a_T)$ with $Ct(a, a_T)$ (the latter being our $Ct_F(a)$), or to compare $p(a \leftarrow a_T)$ with $p(a, a_T)$.

We have, to be sure,

$$Ct_T(a) + Ct(a \leftarrow a_T) = Ct(a),$$

and this appears, at first sight, satisfactory.

But let us substitute '*contrad*' for a.

$$Ct_T(\mathit{contrad}) = Ct(t) = 1 - p(t)$$

which, as we have seen, is the maximal truth-content obtainable in our system; and since $Ct(\mathit{contrad}) = 1$, we obtain for $Ct(a \leftarrow a_T) = Ct(\mathit{contrad} \leftarrow t) = 1 - p(\mathit{contrad} \vee -t) = p(t)$. Now while $Ct_T(\mathit{contrad}) = Ct(t)$ would be quite unobjectionable—it is a clear consequence of a satisfactory definition of $Ct_T(a)$ and of the fact that everything, and therefore t, follows from a self-contradictory statement—this is not so with $Ct_F(\mathit{contrad}) = p(t)$; for this would allow, in most cases, the falsity-content of a contradiction to be smaller than its truth-content, while we should expect the falsity-content of a contradiction to be *at least* equal to its truth-content.

To take an example, let our universe of discourse be a throw of a die; let t be 'three turned up'; and let $p(t)$ be 1/6. The proposed (but here rejected) definition of $Ct_F(a) = Ct(a \leftarrow a_T)$ would lead in this universe to the result that the falsity-content of a contradictory statement (such as 'six will turn up and will not turn up'), $Ct_F(\mathit{contrad})$, would be equal to 1/6, while its truth-content, $Ct_T(\mathit{contrad})$, would be equal to 5/6. Thus the truth-content of a contradictory statement would greatly exceed the falsity-content, which is clearly counter-intuitive. This is the reason for adopting our *desideratum* (iv); and this *desideratum* leads to cases in which $Ct_T(a) + Ct_F(a) > Ct(a)$.

It will be seen from all this that our *desideratum* (iv) might be replaced by the following two highly intuitive ones:

$$\text{(iv, } a\text{)} \qquad Ct_F(\mathit{contrad}) = \text{constant}$$

$$\text{(iv, } b\text{)} \qquad Ct_F(\mathit{contrad}) \geqslant Ct_T(\mathit{contrad}).$$

Incidentally, the fact that we have, quite generally,

$$(21) \qquad Ct_F(a) - Ct(a \leftarrow a_T) = Ct_F(a)Ct_T(a)$$

may appear somewhat surprising. Yet it is an immediate consequence of the following more general formula

$$(22) \qquad p(a \leftarrow b) - p(a,b) = Ct(a,b)Ct(b),$$

a formula which I derived many years ago in order to show that the 'conditional probability' $p(a \leftarrow b)$, i.e. the absolute probability of the *one* conditional statement 'a if b' (or of the statement 'if b then a') exceeds in general the relative probability of the statement a, given the statement b.

(Formula (22) thus compares, as it were, the arrow to the left '←' with the comma ',' and calculates the never negative *excess*,

$$Exc(a,b) = p(a \leftarrow b) - p(a,b),$$

of the conditional probability over the relative probability.)

Having defined the measures of truth-content and of falsity-content we may now proceed to define the $Vs(a)$, the verisimilitude of a. As long as we are merely interested in comparative values, we could use

$$Ct_T(a) - Ct_F(a) = p(a,a_T) - p(a_T)$$

as definiens. If we are interested in numerical values, then it becomes preferable to multiply this by a normalizing factor, and to use $(p(a,a_T) - p(a_T)) \,/\, (p(a,a_T) + p(a_T))$ as definiens. For we wish the following *desiderata* to be satisfied.

(i) $Vs(a) \gtreqless Vs(b) \leftrightarrow Ct_T(a) - Ct_F(a) \gtreqless Ct_T(b) - Ct_F(b)$

(ii) $-1 \leqslant Vs(a) \leqslant Vs(t) \leqslant 1$

(iii) $Vs(tautol) = 0$

(iv) $Vs(contrad) = -1$

so that we get

(v) $-1 = Vs(contrad) \leqslant Vs(a) \leqslant +1$

(vi) In an infinite universe in which $Ct(t)$ may become 1, $Vs(t)$ should be able to become 1 also.

It should be noted here that $Ct(t) = 1 - p(t)$ will depend upon the choice of our universe of discourse. Even in a potentially infinite universe it may be less than 1, as the following example shows: let our universe contain a denumerably infinite set of exclusive possibilities, a_1, a_2, . . . and let $p(a_1) = 1/2$, $p(a_2) = 1/4$, $p(a_3) = 1/8$, $p(a_n) = 1/2^n$; let, moreover, just one of these possibilities be realized: $t = a_1$; then $Ct(t) = 1/2$.

It is thus preferable to replace, for purposes of numerical calculations, $p(a,a_T) - p(a_T)$ by a normalized form; we choose the normalizing factor $1 \,/\, (p(a,a_T) + p(a_T))$; that is to say, we define, as indicated:

(23) $$Vs(a) = (p(a,a_T) - p(a_T))/(p(a,a_T) + p(a_T)).$$

We obtain:

(24) $$\text{If } a \; \varepsilon \; T \text{ then } Vs(a) = Ct_T(a) \; / \; (1 + p(a_T)) = Ct(a) \; / \; (1 + p(a))$$

(25) $$Vs(tautol) = 0,$$

and

(26) $$Vs(contrad) = -1.$$

There are various other possible definitions. For example, we might introduce other normalizing factors, such as $Ct_T(a)$, or $Ct(a)$, or $Ct_T(a) + Ct_F(a)$. These would not, I think, lead to adequate definitions of $Vs(a)$, but rather to definitions of such ideas as, say, 'degree of truth-value'.

4. Numerical Examples

Before discussing some numerical examples—which have to be taken from theories which apply probability to games of chance, or from statistical theories—I wish to make some general remarks about *numerical values in pure theories of content and probability*.

Apart from those applications of probability theory in which we can measure probabilities in the usual way (with the help of either the assumption of equal probabilities as in dicing or with the help of statistical hypotheses) I see no possibility of attaching numerical values (other than 0 and 1) to our measures of probability or content. Pure probability theory and pure content theory are, in this respect, like Euclidean geometry: there is no actual unit defined in Euclidean geometry. (The definition of the Paris unit-meter is decidedly extra-geometrical.) There is no need to worry if pure probability theory or content theory do not supply us with actual numerical values (except 0 and 1). Their status is thus, in many respects, more like topology than metrical geometry.[11]

[11] The theory of probability here presupposed is developed in *L. Sc. D.*, appendices *iv and *v; see also the second section of the present *addendum*, above.

Turning now to *numerical examples*, I shall distinguish two kinds.

(i) Examples of the type of ordinary dicing. Here if, say, 4 turns up, while our guess was that 5 would turn up, we consider this as no better or worse a guess than, say, the guess that 6 will turn up. (Better or worse are here used in the sense of nearer to, or further from, the truth.)

(ii) Examples in which we have a kind of measure of the *distance* of our guesses from the truth. We can represent this by the assumption that, if *in fact 4 turns up*, the guess or the proposition that 6 will turn up (or that 2 will turn up) is separated from the truth by the proposition that 5 will turn up (or that 3 will turn up); and that, for this reason, if $a = 6$, a_T will be 6 v 5 v 4, rather than 6 v 4 (or alternatively, $a_T = 2 \text{ v } 3 \text{ v } 4$).[12]

Here and in what follows, '$a = 6$' or '$a = 6 \text{ v } 4$' is used to express 'a = 6 will turn up' or 'a = 6 or 4 will turn up', etc.

We assume homogeneous dice.

I shall first calculate three examples of type (i).

(1) $$a = 6; b = 4; b = t$$

We have: $$a_T = 6 \text{ v } 4; p(a,a_T) = 1/2; p(a_T) = 1/3$$
$$Vs(a) = 1/5$$

(2) $$a = 5; b = 4; b = t$$

We have $a_T = 5 \text{ v } 4$. The calculation and the result are the same as in case (1).

(3) $$a = 6 \text{ v } 5; b = 4; b = t$$

We have $$a_T = 6 \text{ v } 5 \text{ v } 4; p(a,a_T) = 2/3; p(a_T) = 1/2$$
$$Vs(a) = 1/7$$

We can now compare these with three corresponding examples of type (ii). The difference lies in the calculation of a_T.

[12] '6 v 5 v 4', and '6 v 4', is here shorthand for 'either 6 or 5 or 4 will turn up', and 'either 6 or 4 will turn up'.

(1′) $a = 6; b = 4; b = t$

We have: $a_T = 6 \vee 5 \vee 4; p(a,a_T) = 1/3; p(a_T) = 1/2$

$$Vs(a) = -1/5$$

(2′) $a = 5; b = 4; b = t$

We have: $a_T = 5 \vee 4; p(a,a_T) = 1/2; p(a_T) = 1/3$

$$Vs(a) = 1/5$$

(3′) $a = 6 \vee 5; b = 4; b = t$

We have: $a_T = 6 \vee 5 \vee 4; p(a,a_T) = 2/3; p(a_T) = 1/2$

$$Vs(a) = 1/7$$

I add two examples of true guesses:

(1″) $a = 6; b = 6; b = t$

$$Vs(a) = 5/7$$

(2″) $a = 6 \vee 5; b = 6; b = t$

$$Vs(a) = 1/2$$

Thus we see that verisimilitude can increase with the content of a, and decrease with the probability of a.

5. Artificial vs. Formalized Languages

It has often been said that Tarski's theory of truth is applicable only to formalized language systems. I do not believe that this is correct. Admittedly it needs a language—an object-language—with a certain degree of artificiality; and it needs a distinction between an object-language and a meta-language—a distinction which is somewhat artificial. But although by introducing certain precautions into ordinary language we rob it of its 'natural' character and make it artificial, we do not necessarily formalize it: although every formalized language is artificial, not every language which is subject to some stated rules, or based on more or less clearly formulated rules (and which is therefore 'artificial') need be a fully formalized language. The recognition of the existence of a whole range of more or less artificial though not formalized languages seems to me a point of considerable importance, and specially important for the philosophical evaluation of the theory of truth.

6. A Historical Note on Verisimilitude (1964)

Some further remarks on the early history of the confusion between verisimilitude and probability (in addition to those in chapter 10, section xiv) will be given here.

1. In brief, my thesis is this. The earliest sayings at our disposal unambiguously use the idea of truthlikeness or verisimilitude. In time, 'like the truth' becomes ambiguous: it acquires additional meanings such as 'likely' or 'likely to be true' or 'probable' or 'possible', so that in some cases it is not clear which meaning is intended.

This ambiguity becomes significant in Plato because of his crucially important theory of imitation or *mimesis*: just as the empirical world *imitates* the (true) world of ideas, so the accounts or theories or myths of the empirical world (of seeming) 'imitate' the truth, and thus are merely '*like the truth*'; or, translating the same expressions in their other meanings, these theories are not provable, or necessary, or true, but merely probable, or possible, or (more or less) seemingly true.

In this way Plato's theory of *mimesis* furnishes something like a philosophical basis for the (then already current) mistaken and misleading equation of '*truthlike*' and '*probable*'.

With Aristotle an additional meaning becomes fairly prominent: 'probable' = 'frequently occurring'.

2. To give a few details, we have first a passage in the *Odyssey* 19,203: wily Odysseus tells Penelope (who does not recognize him) a story which is false, but which contains quite a few elements of truth; or as Homer puts it, 'he made the many lies similar to the truth' ('*etymoisin homoia*'). The phrase is repeated in the *Theogony*, 27 f: the Muses of Olympus, daughters of Zeus, say to Hesiod: 'we know how to speak many lies similar to the truth; but we also know, if we want to, how to speak the truth (*alētheia*).'

The passage is interesting also because in it *etymos* and *alēthēs* occur as synonyms for 'true'.

A third passage in which the phrase '*etymoisin homoia*' occurs is *Theogony*, 713, where cunning is extolled (as in the *Odyssey*) and the power of making lies sound like truth is described as divine (perhaps an allusion to the Muses in the *Theogony*): 'you would make lies similar to the truth with the good tongue of godlike Nestor'.

Now, one thing about these passages is that they are all related to what we call today 'literary criticism'. For the issue is the *telling of stories* which are (and which sound) *like the truth*.

A very similar passage is to be found in Xenophanes, himself a poet and perhaps the first literary critic. He introduces (DK B35) the term '*eoikota*' in place of '*homoia*'. Referring perhaps to his own theological theories, he says: 'these things, we may conjecture, are similar to the truth' (*eoikota tois etymoist*; see also above, p. 205, and Plato's *Phaedrus* 272 D/E, 273 B and D).

Here we have again a phrase which expresses unambiguously the idea of *verisimilitude* (*not* probability) in conjunction with a term (I have translated it by 'we may conjecture') which is derived from the term *doxa* ('opinion'), which plays so important a role in and after Parmenides. (The same term occurs also in the last line of Xenophanes, B34, quoted above on pp. 26 and 153, and is there used in contrast to '*saphes*', that is, '*certain truth*'.)

The next step is important. Parmenides B8: 60 uses *eoikota* ('similar' or 'like') without explicitly mentioning 'truth'. I suggest that it means nevertheless, as in Xenophanes, 'like the truth', and I have translated the passage accordingly ('wholly like truth'; see p. 11 above). My main argument is the similarity between the passage and Xenophanes B35. Both passages speak of the conjectures (*doxa*) of mortal men, and both say something relatively favourable about them; and both clearly imply that a relatively 'good' conjecture is not really a true story. In spite of these similarities, the phrase of Parmenides has often been translated by 'probable and plausible' (see note 19 on p. 320 above).

This passage is interesting because of its close relation to an important passage in Plato's *Timaeus* (27e–30c). In this passage, Plato starts (27e–28a) from the Parmenidean distinction between 'That which always Is and has no Becoming' on the one hand, and 'That which is always Becoming and never Is' on the other; and he says with Parmenides that the first of these can be known by reason, while the second 'is an object of opinion and unreasoning sensation' (compare also pp. 222–3 above).

From this he proceeds to explain that the changing and becoming world (*ouranos* or *kosmos*: 28b) was made by the Creator as a copy or likeness (*eikon*) whose original or paradigm is the eternally unchanging Being that Is.

The transition from the paradigm to the copy (*eikon*) corresponds to the transition, in Parmenides, from the Way of Truth to the Way of Seeming. I have quoted the latter transition above (p. 15), and it contains the term '*eoikota*', which is related to Plato's '*eikon*', i.e. the *likeness of Truth*, or *of What Is*; from which we may perhaps conclude that Plato read '*eoikota*' as 'like (the truth)' rather than 'Probable' or 'likely'.

However, Plato also says that the copy, in being like the truth, cannot be known with certainty, but that we can only have *opinions* of it which are *uncertain* or '*likely*', or '*probable*'. For he says that accounts of the paradigm will be 'abiding, unshakable, irrefutable, and invincible' (29 b–c) while 'accounts of that which is (merely) a copy's likeness of the paradigm will . . . possess (mere) likelihood; for as Being is to Becoming, so is Truth to (mere) Belief'. (See also *Phaedrus* 259 E to 260 B–E, 266 E–267 A.)

This is the passage which introduces likelihood or probability (*eikota*) in the sense of imperfectly certain belief or partial belief, at the same time relating it to verisimilitude.

The passage concludes with yet another echo from the transition to the Way of Seeming: just as the goddess promised Parmenides an account so 'wholly like truth' that no better could be given (p. 11 above), so we read in the *Timaeus* (29d): 'we should be content if we can give an account which is inferior to none in likelihood (*eikota*), remembering that [we] . . . are human creatures and that it becomes us to accept a likely story (*eikota muthon*) . . .'. (To this, 'Socrates' replies: 'Excellent, Timaeus!'.)

It is very interesting to note that this introduction of a systematic ambiguity of 'truthlikeness' and 'likelihood' (i.e. 'probability') does not prevent Plato from using the term '*eikota*' later, in the *Critias* (107d/e), in the sense of 'truth-like account'. For considering what precedes it, that passage should be read: 'in respect to matters celestial and divine, we should be satisfied with an account which has a small degree of truthlikeness, while we should check carefully the accuracy of accounts that pertain to mortal men'.

3. Apart from this systematic and no doubt conscious ambiguity in Plato's use of '*eikota*' (and kindred terms), and apart from a wide range

of differing usages in which its meaning is definite, there is also a wide range of usages in which its meaning is simply vague. Examples of different usages in Plato (and Aristotle) are: its use in opposition to 'demonstrable' and to 'necessary'; its use to express 'the next best to certainty'. It is also often used as a synonym for 'to be sure' or 'certainly', or 'this seems all right to me', especially by way of interjections in the dialogues. It is used in the sense of 'perhaps'; and it is even used in the sense of 'occurring frequently'; for example, in Aristotle's *Rhetoric* 1402b22: '. . . the probable (*eikos*) is that which occurs not invariably but only in most cases . . .'

4. I should like to end with another passage of literary criticism, one that occurs twice in Aristotle's *Poetics* (1456a22–25, and 1461b12–15) and which on its first occurrence he attributes to the poet Agathon. 'It is likely that the unlikely should happen.' Or less vaguely, though less elegantly: '*It is like the truth that improbable things should happen.*'

7. Some Further Hints on Verisimilitude (1968)

1. Since my interest in the distinction between verisimilitude on the one hand, and probability (in its many meanings) on the other, seems to be open to misinterpretation, I will first stress that I am not at all interested in words and their meanings, but only in *problems*. Least of all am I interested in making the meanings of words 'precise', or in 'defining' or 'explicating' them.

There is an analogy between words or concepts and the question of their meaning on the one hand, and statements or theories and the question of their truth on the other, as I pointed out in the table on p. 25, above. *Yet I regard only statements or theories and the question of their truth or falsity as important.*

The mistaken ('essentialist') doctrine that we can 'define' (or 'explicate') a word or term or concept, that we can make its meaning 'definite' or 'precise', is in every way analogous to the mistaken doctrine that we can prove or establish or justify the truth of a theory; in fact, it is part of the latter ('justificationist') doctrine.

While words and their precise meanings are never important, the clearing up of confusions may be important for solving problems; problems concerning theories, of course. *We cannot define, but we must often distinguish*. For confusions, or merely the lack of distinctions, may prevent us from solving our problems.

2. In connection with verisimilitude, the main problem at stake is *the realist's problem of truth*—the correspondence of a theory with the facts, or with reality. (See pp. 302 ff. and 156 above.)

The dangerous confusion or muddle which has to be cleared up is that between truth in the realist's sense—the 'objective' or 'absolute' truth—and truth in the subjectivist sense as that in which I (or we) 'believe'.

This distinction is of fundamental importance, especially for the theory of knowledge. The only important problem of knowledge concerns the problem of truth in the objective sense. My thesis is, simply, that the theory of subjective belief is utterly irrelevant to the philosophical theory of knowledge. Indeed, it is destructive of the latter if the two are mixed up (as they still are, in accordance with tradition).

3. Now it is decisively important that the need to distinguish sharply between objective truth and subjective belief remains as urgent as ever if we bring *approximation to truth* (or truthlikeness or verisimilitude) into the picture: verisimilitude as an objective idea must be sharply distinguished from all such subjective ideas as degrees of belief, or conviction, or persuasion; or of apparent or seeming truth, or plausibility, or of probability in any one of its subjective meanings. (Incidentally, it so happens that even if we take probability in some of its objective meanings, such as propensity, or perhaps frequency, it should still be distinguished from verisimilitude; and the degree of objective verisimilitude should be also sharply distinguished from the degree of corroboration, even though this is an objective notion; for the degree of verisimilitude of a theory, like the idea of truth itself, is timeless, even though it differs from the idea of truth in being a relative concept; while the degree of corroboration of a theory is essentially time-dependent—as pointed out in section 84 of my *Logic of Scientific Discovery*—and thus essentially a historical concept.)

The confusion between verisimilitude and subjective notions like degrees of belief, or of plausibility, or of the appearance of truth, or of subjective probability, is traditional.

The history of this tradition ought to be written. It will turn out to be more or less identical with the history of the theory of knowledge.

In the preceding *Addendum* I sketched, very superficially, this history so far as it was connected with the early philosophical use of the words 'like truth' (words connected with the Greek root *eikō*, such as *eikon*, a likeness, a picture, *eoika*, to be like, to seem like, etc.). That is to say, with words which have at least at times (at any rate in Xenophanes or in Parmenides) been used in connection with a *realist* or *objectivist* idea of truth (whether as 'approximation to truth', as in Xenophanes, B35, or in the sense of a deceptive likeness to truth as in Parmenides, B8: 60).

4. In the present *Addendum* I will just add a few brief remarks on the use of certain words which had from the start a *subjective* meaning. I will refer to two main Greek roots. One is *dokeō* (*dokē*, etc.) to think, to expect, to believe, to have in mind, to hold an opinion, with *doxa*, opinion. (Related are also *dekomai*, to accept, to expect, with *dokimos*, accepted, approved, and *dokeuō*, to expect, to watch closely, to lie in wait.) The second is *peithō*, to persuade (also the power, or the goddess, Persuasion), with the meaning to win over, to make things appear plausible or probable—*subjectively* probable, of course; and with the forms *pithanoō*, to make probable; *pithanos*, persuasive, plausible, probable, even specious; *pistis*, faith, belief (with *kata pistin*, according to belief, according to probability); *pistos*, faithful, believed, deserving belief, probable; *pisteuō*, to trust, to believe; *pistoō*, to make trustworthy, to confirm, to make probable, etc.

There is never a doubt about the fundamentally subjective meaning of these words. They play an important role in philosophy from the earliest times. *Dokos*, for example, occurs in Xenophanes, DK B34, in the beautiful fragment quoted on pp. 34 and 205 above, where I translated the term *dokos* by 'guess' ('guesses'), since it clearly means '*mere* opinion' or '*mere* conjecture'. (Cp. Xenophanes B35; and B14, where *dokeousi* means 'believe wrongly' or 'imagine wrongly'.) One might say that this disparaging usage of *dokein* is the birth of scepticism. It may be perhaps contrasted with the more neutral usage in Heraclitus B5 ('one would think that . . .') or B27: 'When men die, there waits for them what they do not expect or *imagine* (*dokousin*).' But Heraclitus seems to use the term also in the sense of 'mere opinion', as in B17, or in B28: '[For] it is mere opinion what even the most trustworthy [of men] defend [or preserve, or cling to] as knowledge.'

In Parmenides, *doxa*, opinion, is used in direct opposition to truth (*alētheia*); and it is, more than once (B1: 30; B8 : 51), associated with a disparaging reference to 'the mortals'. (Cp. Xenophanes B14, and Heraclitus B27.)

At any rate, *dokei moi* means 'it seems to me', 'it appears to me', and thus comes very near to 'it seems to me plausible, or acceptable' (*dokimos einai*, 'acceptable as real'; cp. p. 15 above, Parmenides, E1 : 32).

5. The term 'probable' itself (*probabilis*) seems to have been invented by Cicero as a translation of the Stoic and Sceptic terms *pithanos, pithanē, pistin*, etc. (*kata pistin kai apistian*—as to probability and improbability, Sextus, *Outline of Pyrrhonism* i, 10, and i, 232). 250 years after Cicero, Sextus, *Against the Logicians* i, 174, distinguishes three 'Academic' senses of the term probability (*to pithanon*, the probable): (1) What appears true and is in fact true. (2) What appears true and is in fact false. (3) What is both true and false.

Under (3), appearance is not specially mentioned: it seems that approximation to truth or verisimilitude in our sense is intended. Elsewhere, appearance is sharply distinguished from objective truth; yet appearance is all we can attain. 'Probable' is, in Sextus' use, that which induces belief. Incidentally Sextus says (*Pyrrhonism* i, 231) with a reference to Carneades and Cleitomachus, that 'the men who . . . use *probability as the guide of life*' are dogmatists: by contrast, 'we [the new sceptics] live in an undogmatic way by following laws, customs, and our natural affections'. At times, Sextus uses 'probability' (or 'apparent probabilities' which seems almost a pleonasm; cp. *Pyrrhonism* ii, 229) in the sense of 'specious'. Cicero's usage is different.

6. 'Such', says Cicero, 'are those things which I felt I should call probable (*probabilia*) or similar to the truth (*veri similia*). I do not mind if you prefer another name.' (*Academica*, Fragm. 19.)

Elsewhere he writes of the sceptics: 'For them something is probable (*probabile*) or resembling the truth (*veri simile*), and this [characteristic] provides them with a rule in the conduct of life, and in philosophical investigations.' (*Academica* ii, 32; in 33 Cicero refers to Carneades, as does Sextus in the same context; cp. *Academica* ii, 104: 'guided by probability'.) In *De Natura Deorum*, probabilities enter *because* falsity may be deceptively similar to truth; yet in *Tusc.* i, 17, and ii, 5, the two terms are synonyms.

7. There is thus no doubt that the terms 'probability' and 'verisimilitude' were introduced by Cicero as synonyms, and in a subjectivist sense. There is also no doubt that Sextus, who uses a subjectivist sense of 'probable', thought of truth and falsity in an objectivist sense, and did clearly distinguish between the subjective appearance of truth—seeming truth—and something like partial truth or approximation to truth.

My proposal is to use, *pace* Cicero, his originally subjectivist term 'verisimilitude' in the objectivist sense of 'like the truth'.

8. As to the terms 'probable', and 'probability', the situation has changed radically since the invention of the *calculus of probability*.

It seems now essential to realize that *there are many interpretations of the calculus of probability* (as I stressed in 1934 in section 48 of my *Logic of Scientific Discovery*), and among them *subjective and objective interpretations* (later called by Carnap 'probability$_1$' and 'probability$_2$').

Some of the objective interpretations, especially the *propensity interpretation*, have been briefly mentioned above (pp. 78 and 160), and in my *Logic of Scientific Discovery*. See also my paper 'The Propensity Interpretation of Probability', in *The British Journal for the Philosophy of Science* **10**, 1959, no. 37, pp. 25–42, and 'Quantum Mechanics Without "The Observer"', in *Quantum Theory and Reality*, edited by Mario Bunge, 1967, pp. 7–44.

8. Further Remarks on the Presocratics, especially on Parmenides (1968)

A few remarks are added here in support of certain views stated in the *Introduction* to the present book, and also in Chapter 5.

1. On p. 12 of the *Introduction*, I stated without argument that Parmenides describes the goddess Dikē 'as the guardian and keeper of the keys of truth, and as the source of all his knowledge'.

In doing so I identified the goddess of DK B1 : 22 with the goddess Dikē of B1 : 14 to 17. This identification (which goes back to Sextus, *Against the Logicians* i, 113) is rejected by some of the main authorities, such as W.K.C. Guthrie, *A History of Greek Philosophy* ii, 1965, p. 10 ('an unnamed goddess'), or Tarán, *Parmenides*, 1965, pp. 15, 31, 230, who both hold that Parmenides leaves his goddess (of B1 : 22) nameless, and who support this view by subtle arguments.

I find this view unconvincing, although the mentioning by the goddess of dikē (or Dikē) in line 28 (B1 : 28) is, admittedly, somewhat strange if we take it as a self-reference. This strangeness, however, is hardly relieved if the reference is taken to be to her own turnkey: it seems best to read, with Tarán, 'themis' and 'dikē' in line 28, rather than the corresponding proper names.

Tarán argues that Parmenides leaves his goddess unnamed in order 'to emphasize the objectivity of his method'. But why, then, does Parmenides name, eight lines before, the goddess Dikē?

There are two arguments for identifying Dikē of lines 14 to 17 with the goddess who reveals the truth about the existing world (and about the origin of error) to Parmenides.

(a) The whole balance of B1, down to line 23, and especially 11 to 22, suggests the identification, as the following details show: Dikē (though on the other view she would be no more than a turnkey) is introduced elaborately, in keeping with the style of the whole passage; she is the main person acting from line 14 down to line 20 (*arērote*); also, the sentence does not seem to stop here—not indeed until the end of line 21, just before the 'goddess' comes in. Moreover, between line 20 and the end of line 21 no more is said than: 'Straight on the road through the gates did the maidens steady the horses.' This in no way implies that Parmenides's journey, elaborately described up to this point, continues any further; rather I find here a strong suggestion that, upon passing through the gates, where he must encounter Dikē, Parmenides's journey ends. And how can we believe that the highest authority and main speaker of the poem enters not only unnamed, but without any introduction or any further ado—even without an epithet? And why should the maidens have to introduce Parmenides to Dikē (and 'appease' her), though on the view here combated she is the inferior person, but not to the superior one?

(b) If, as I do, we believe with Guthrie (*op. cit.*, ii, p. 32, see also pp. 23 f., and Tarán, *op. cit.*, pp. 5 and 61 f.) that the evidence is cumulative 'that Parmenides, in his criticism of earlier thought, had Heraclitus especially in mind', then the role played by Dikē in the *logos* of Heraclitus (see DK Heraclitus B28 which may have influenced Parmenides: its terminology is in more than one way similar to his) would make it understandable why Parmenides in his *antilogia* cites her now as his authority for his own *logos*. (Incidentally, there seems to me no great difficulty in assuming that in the important passage Parmenides B8, line 14, Dikē is speaking about herself but great difficulty in assuming that the 'goddess' speaks in such terms about her own gate keeper.)

2. Of much greater importance than the problem of Dikē seems to me the problem of the early development of epistemology, discussed in chapter 5 of the present book, and in the appendix to chapter 5; a problem on which *Addenda* 6 and 7 have some bearing. What I intend to discuss here is, especially, the early history of the problem of rationalism versus empiricism, particularly in its sensualistic form.

It will be clear to anybody who has glanced at my Introduction or at chapter 5 that my own views are anti-sensualistic. I am an empiricist of sorts, in so far as I hold that 'most of our theories are false anyway' (see point 8 on p. 37 above), and that we learn from experience—that is, from our mistakes—how to correct them. But I also hold that our senses are not sources of knowledge, in any authoritative sense. There is no such thing as *pure* observation or *pure* sense-experience: all perception is interpretation in the light of experience: in the light of expectations, of theories. The structure and the working of our very eyes and ears are the result of trial and error, and they have expectations (and thus theories, or something analogous to them) built into their anatomy and physiology; and so has our nervous system. Thus there does not exist anything like a sense-datum, anything 'given' or uninterpreted which is the given material of that interpretation which leads to perception: everything is interpreted, selected, on some level or other, by our very senses themselves.

On the animal level, this selection is the result of natural selection. On the highest level, it is the result of conscious criticism—of exposing our theories to a critical process of scrutiny, aiming at the elimination of error, by way of critical debate *and* experimental test.

I have more recently come to present the process of selection in the form of a somewhat oversimplified diagram[13]:

[13] I published this diagram (and more elaborate versions of it) in 1966 in my Compton Memorial Lecture *Of Clouds and Clocks* (delivered 1965). See also my lecture 'Epistemology Without a Knowing Subject', published in 1968 in the *Proceedings of the Third International Congress for Logic, Methodology and Philosophy of Science*, pp. 333–373. Both are now in *Objective Knowledge*, Oxford University Press 1972; revised edn 1979.

$$P_1 \rightarrow TT \rightarrow EE \rightarrow P_2$$

P_1 is the *problem* we start from; *TT* are the tentative theories by which we try to solve the problem; *EE* is the process of error elimination to which our theories are exposed (natural selection on the pre-scientific level; critical examination, including experiment, on the scientific level) and P_2 is the new problem which emerges from exposing the errors of our tentative theories.

The whole schema shows that science begins with problems, and ends with problems; and that it grows through the bold invention of theories, and through criticism of the various competing theories.

This tetradic schema may be regarded as a kind of improvement of the dialectical triad discussed on pp. 421–2 above. Like the former, it sums up both prescientific evolution and the evolution of science.

3. The Presocratics seem to represent the breakthrough to the conscious critical debate of science. What is so utterly astonishing about them is that they not only progress, through criticizing each other (a process which leads within a few generations to atomism; to the theory of the spherical shape of the earth, the borrowed light of the moon, the eclipses, Aristarchus's anticipation of the Copernican system): they also begin to reflect on their own critical method, and become conscious of it, as early as Xenophanes.

Anti-sensualist and pro-intellectualist remarks can be found in Heraclitus, for example in B46 and 54 (cp. B8 and 51), and in B123 (cp. B56), all mentioned in chapter 5, but also in B107: 'eyes and ears are false witnesses. . . .' (False witnesses are also alluded to in B28; see also B101a, in the light of which both 107 and 101a may only mean: 'Eye witnesses are better than hearsay.') See also B41. 'Wisdom is knowing the thought [that is, the *logos*, the story, the theory, the law; cp. *pantōn kata ton logon* in B1: "everything happens in accordance with that [story, or theory, or] law"] that steers everything through everything'.

4. But the most decisive step was perhaps Parmenides's defiance of experience and his theory of critical refutation, of which some passages are quoted on p. 15 and especially on pp. 222–3. Although it is especially the latter ones which I want to discuss further, I want to say first a few words about the former.

Parmenides is one of the greatest and most astonishing thinkers of all times. He is a revolutionary thinker, and as conscious of this fact as was Heraclitus; yet his revolution consisted, partly, in trying to prove a doctrine of the immobility or invariance of reality, the non-existence of change.

Some of his other revolutionary innovations were: his discovery of the distinction between *appearance* and the *reality* behind the appearances; and his onslaught upon common sense, empiricism, and traditional belief which he thought was based upon mere convention (name-giving[14]) rather than truth: upon *doxa*, the mere opinion of the mortals. In all this he had, of course, predecessors, but he went far beyond them.

5. This is why I translate B8 : 60–61, where the goddess speaks of the world of *doxa*, of deceptive appearance (cp. p. 15 above):

> Now of this world thus arranged to seem wholly like truth I shall tell you
> Then you will be nevermore led astray by the notions of mortals.

The usual translation of *parelassēi*, which I now translate by 'led astray', is 'outstrip'. For example, Kirk and Raven translate: 'The whole ordering of these I tell thee as it seems likely, that so no thought of mortal men shall ever outstrip thee.' I think that the term 'outstrip' (or similar terms which suggest that the aim of the goddess is to help Parmenides to win in case he gets into a verbal competition or tournament with other 'mortals') is not only misleading, but also destructive to the seriousness of the message of the goddess whose first purpose is to reveal the truth, and whose second purpose is to give Parmenides the intellectual armament needed for avoiding the mistakes of traditional belief, and for avoiding being led astray by it.[15]

[14] It is the giving of names to what is non-existing (to non-existing opposites, light and night) which creates the untruth of the *doxa*. Cp. B8 : 53: 'for they decided to give names. . . .'

[15] I previously translated *parelaunō* by 'overawe', having in mind the weight of traditional religious belief which Parmenides would have to bear, and to resist, and Homer's meaning 'run away with you', or 'carry you off' of *elaunō*. Charles Kahn objected that this ignores the '*para*' in *parelaunō*. My present translation attempts to meet this objection.

6. Most important in connection with Parmenides's attack upon empiricism is his fragment B16, translated on p. 223 above. On that page will be found not only a translation, but also a kind of commentary. I see in this passage both an attack upon and an anticipatory formulation of the sensualist doctrine according to which *nothing can be in the intellect that was not before in the sense-organs*.

Parmenides launches his attack by using the word 'much-erring' (*polyplanktos*) to characterize the sense-organs, and by implying that intellect or 'thought', *so far as it depends upon the senses*, should be equally understood as 'erring': this is indeed stated in B6 : 6, where *plakton noon* clearly means 'erring thought' (or 'erring mind', as Guthrie puts it on p. 21 of volume ii of his *History of Greek Philosophy*, 1965). And it is enforced by the obvious opposition between this 'erring thought' and that critical 'reason' or 'argument' of B7 : 5 to which the goddess appeals against the claim of sense experience. (Cp. my translation of B7 on p. 222.)

The following two terms seem to be crucial for my proposed interpretation of the epistemological fragment B16, the fragment on the much-erring sense-organs quoted on p. 223: (a) The translation 'much-erring' for *polyplanktos*, and (b) the translation 'sense-organs' for *melea*. If these two translations are correct, then the interpretation of the rest of the passage follows almost by necessity.

However, before discussing the translation of these two expressions (a) and (b), let me say that I have two main arguments for my interpretation of the passage: that it fits in with the philosophic tradition (especially with Empedocles and Theophrastus), and that my translation does not only make sense, but philosophically most important sense, while the other interpretations seem to make little sense, if any.[16]

[16] I may perhaps for the convenience of the reader repeat my translation of the epistemological fragments B16 from p. 223, and compare it with some others; mine is:

> For as, at any one time, is the much-erring sense-organs' mixture,
> So does knowledge appear in men. For these two are the same thing:
> That which thinks, and the mixture which makes up the sense-organs' nature.
> What in this mixture prevails becomes thought, in each man and all.

Kirk and Raven translate (*The Presocratic Philosophers*, 1957, 1960, p. 282: 'According to

7. I will now discuss the two crucial expressions (a) and (b).

(a) Much-erring for *polyplanktos*. Parmenides uses the terms *plazō*, *plassō* (in the form *plattō*), which he seems to associate closely with *plazō* (cp. Diels, *Lehrgedicht*, p. 72), and *planaō*, and *always* with the meaning of wandering or straying *away from the truth*. See B6 : 6 (*plakton noon*, 'erring thought'); B8 : 28 (*eplachthēsan*: 'they have been driven far away'); B6 : 5 (*plattontai*: 'they stray', 'they wander about', 'they err'); and B8 : 54 (*peplanemenoi*, 'they have gone astray', 'they err').

In all cases, except perhaps B8 : 28 with the meaning 'they have been driven away *by true conviction*', the words bear the meaning of *erroneous belief or opinion*. There is thus every reason to translate, in a passage like our B16 which is generally admitted to be essentially epistemological, *polyplanktos* as 'much-erring', rather than 'wandering' (Kirk and Raven) or 'much wandering'. (Tarán, who adds on p. 170 that *polyplanktos* 'is supposed to convey *the notion of change*'. [The italics are mine.] This seems strange, since even where *plazō* means 'wandering' or 'straying' it is usually in the sense of 'helplessly erring' or 'roaming about', with the connotation of '*not knowing* where to go'. All the forms seem to be connected with *plagiazo*, to turn and twist; also to deceive.)

(b) Sense-organs for *melea*. As I mentioned before, this was Diel's translation in 1897. This interpretation is strongly suggested by the context in which Parmenides's fragment B16 is transmitted to us, and discussed by Aristotle (*Metaphysics*, in a passage beginning 1009b 13 in which Aristotle discusses the doctrine that 'knowledge is

the mixture that each man has in his wandering limbs, so thought is forthcoming to mankind; for that which thinks is the same thing, namely, the substance of their limbs, in each and all men; for that of which there is more is thought.'

I find this incomprehensible. Yet even stranger is Kirk and Raven's comment on the fragment (p. 282): '. . . the equation of perception and thought comes strangely from [Parmenides].' Since Kirk and Raven speak of 'limbs'—where I think they should speak of 'sense-organs'—it seems strange that they interpret the passage as equating *perception* (sense-perception) and thought. Besides, the equation of *much-erring* perception and *erring* thought (B6 : 6) fits very well with Parmenides's rationalism. H. Diels (*Parmenides's Lehrgedicht*, 1897, p. 112) took *meleōn* to mean 'sense-organs'; and so did K. Reinhardt, *Parmenides*, p. 77. Present authors seem either to take it to mean 'limbs' or 'body'. Thus Tarán, *Parmenides*, 1965, p. 169, translates: 'For as at any time the mixture of the much wandering body is, so does mind come to men. For the same thing is that the nature of the body thinks in each and in all men; for the full is thought.'

sense-perception') and Theophrastus (*De sensu*, 1 ff.). However, this interpretation is argued against by various authors, among them Guthrie (*History of Greek Philosophy*, ii, p. 67) who writes: '*melea* lit. "limbs", i.e. the body, for which no collective word was yet in common use.' Tarán (*Parmenides*, p. 170) does not use this argument, but recommends the same interpretation.

I cannot understand Guthrie's argument. Let us assume that *sōma* was not yet in general use for the living body (Homer uses it for the corpse), although I find it quoted as denoting 'the *living body*, only of men' in my edition of Liddell and Scott from Hesiod, *Works and Days* 538; in Theognis; and in Pindar ('body, as opposed to spirit (*eidōlon*), Pindar. Fr. 96').[17]

Yet even if we admit that *sōma* was not yet established in common use, there was another word: *demas* is found for '*the body*, i.e. *the frame* or *stature of man*, often in Homer'. (Liddell and Scott.) At any rate, it occurs in a passage whose terminology is used by Parmenides, i.e. in Xenophanes B14: 'But the mortals hold the opinion that the gods . . . have bodies like they themselves.' See also Xenophanes B23: 'One single god, the greatest among gods and men, and neither in body or in thought similar to them.' Admittedly, *demas* is also used in the sense of 'frame' or 'stature' or 'shape' or 'form', for example in Parmenides B8 : 55. But a few lines later, in B8 : 59, it does mean 'body' ('dense in body, and heavy', is Tarán's translation; Guthrie translates 'form').

Thus while the sense 'body' for *melea* can hardly be defended by the argument that no other word was available, the meaning of *melea* for 'sense-organs' can be defended on precisely such grounds.

In fact, I do not find in any Presocratic fragment accepted as genuine any general term which may possibly mean 'sense-organ', prior to Parmenides's term *melea* and Empedocles's synonymous term *gyia* (and

[17] *Sōma* was also used by some Presocratics. According to Plato's *Cratylus*, (400C), where the etymology of *sōma* is discussed, it was used by the Orphic poets who suggested that 'the body (*sōma*) is the keeper of the soul (*psychē*)'; cp. DK, Orpheus B3. See also DK Epicharmus B26: 'If your mind (*nous*) is clean, your whole body (*sōma*) is also clean.' (This anticipates 'To the pure all things are pure.') Xenophanes B15 : 4 (quoted above on p. 205, '. . . each would then shape bodies of gods in the likeness, each kind, of its own') contains both *sōma* and *demas*.

his term *palamai*, lit. 'hands', see below). Instead, the various sense-organs are enumerated, as in Parmenides's fragment B7, quoted above on p. 222. There is not even a term for 'sense'. (Strangely enough, I find in the accepted *early* fragments no more than one single occurrence of a general term for 'perceiving' or 'sense-perception': it occurs in a fragment of outstanding interest, Alcmaeon B1a; and it seems possible that Alcmaeon actually wrote 'see' or 'see and hear' where Theophrastus wrote *aisthanetai*, i.e. 'perceive'.) This fact is the more striking as the general idea of sense and of sense-organs (and of sense-perception) obviously existed: eyes and ears are mentioned together, for example by Heraclitus; Parmenides mentions them together with the tongue, and Empedocles together with the hands; and the latter also together with the limbs (*gyia*).

Now *melea*, the plural of *melos*, basically means *limbs*, like *gyia*. (Cp. *kata melea*, 'limb by limb'; *meleïzō* or *melizō*, 'dismember'.) *Melos* also means a song; no doubt originally a *stanza*, a verse, *an organic part*, a *limb* of a song: as Liddell and Scott say, the word involves 'the notion of symmetry of parts, as in German, *Glied, Lied*'.[18] Eyes, ears, and the limbs are all symmetrically arranged, even the nose; and there is every reason to suppose that Parmenides thought also of the hands and feet; see below.

It was not only Parmenides who was in need of a word for sense-organ: Empedocles struggled with the same problem. Interestingly enough, Tarán, who rejects Diels's suggested meaning 'sense-organs', implies (p. 170) that *melea* and *gyia* are synonyms. He writes that both terms are 'used to refer to the whole living body', giving references which, incidentally, I find inconclusive and quite unconvincing: in every case 'limbs' would fit at least as well as 'body'; in Homer even better, as born out by the translations which I consulted. At any rate, since Tarán indicates that *melea* and *gyia* are synonyms, it is most interesting that Empedocles makes use of '*gyia*' (and of such limbs as 'hands') in his search for an acceptable general description of the sense-organs.

[18] It is interesting that in German 'gegliedert' (lit. 'limbed') means 'highly articulated' or 'with organically related and balanced parts'; while 'limbed' or 'possessing limbs' means in English nothing of the kind.

Empedocles is, I think, alluding to Parmenides B7, to the 'blinded eye', and the 'deafened ear';[19] but he tries to defend the sense-organs as highly imperfect yet indispensable sources of knowledge. Thus he writes in B2 : 1 : 'For narrow are the pathways [the openings[20]] of our sense-organs [*palamai*, lit. 'hands'] which are distributed [perhaps 'like soft mounds'] over the limbs (*gyia*); and much of poor significance is bursting upon them which dulls our attention (*merimna*: care, attention, thought, mind).'

That Empedocles means by *palamai* and also by *gyia* 'sense-organs' becomes very clear in B3 : 9–13, where he specifies them as giving us sight, hearing, and the tongue (like Parmenides in B7), and warns us not to suppress the evidence (*pistis*) of '*the other limbs*' (*gyiōn*).

Nothing can be clearer. But there is further evidence: Cicero, in a passage which refers to Empedocles, translates Empedocles's 'narrow sense-organs' (*steinopoi palamai*) by '*angustos sensus*' (*Academica*, i, 12 : 44). There he speaks of the 'obscurity of the facts, which led Socrates to a confession of ignorance, and before Socrates even Democritus, Anaxagoras, and Empedocles . . . who said that nothing can be known, owing to our *narrow senses* . . .'. (Cicero exaggerates Empedocles's mistrust of the senses.)

Since *gyia* means in Empedocles B3 : 13 undoubtedly 'sense-organs', and since, as indicated by Tarán himself, *melea* and *gyia* are synonyms (though the musical sense of *melea* may indicate a special hint towards symmetry and articulation missing in *gyia*) there seems to me no ground left for rejecting Diels's suggestion that *melea* means sense-organs.[21]

8. But once we interpret our epistemological fragment, Parmenides B16, as an attack upon the *much-erring sense-organs*, it becomes an attack

[19] Translated above on p. 222.

[20] Or 'the pathways' (*poros*: B3 : 12) to knowledge.

[21] (Added 1971.) After the third edition of the present book came out in 1968 it occurred to me that evidence might be found in Aristotle's *Parts of Animals* to show that those parts which we call sense-organs (such as the nose or the eyes) may have once been called *melea*. I found the following passage (Aristotle, *De part. an.*, 645b36 to 646a1; I am quoting the Loeb edition, pp. 104 f.): 'Examples of parts are: Nose, Eye, Face; each of these is named a "limb" or "member".' The word translated here by 'limb' or 'member' is *melos*, the (extremely rare) singular of *melea*. This seems to clinch the argument.

upon the sensualistic theory according to which the sense-organs produce true knowledge. This theory, like everything in the *doxa*, forms part of the mistaken opinion of the mortals, and is attributed to convention and habit.[22]

Already with Empedocles, there begins, as we have seen, a reaction against this rationalist revolution. The 'narrowness' and general weakness of the senses is admitted by Empedocles; yet they are defended as sources of knowledge, if we use them all for mutual corroboration.

It took quite a time, however, before Parmenides's criticism of sensualism was ousted by a sensualistic dogma which omitted the word 'erring' (twice) from his sarcastic formula that *there is nothing in the (erring) intellect that was not previously in the (erring) senses.*

9. If Plato is right, a major attack upon Parmenidean anti-sensualism was made by Protagoras who, by his famous proposition 'man is the measure of all things', tried to turn the tables upon Parmenides: since we are mortal men we are forced to accept what Parmenides had contemptuously described as delusive opinion, and as mere sense-appearance. (This would make Protagoras a defender of Parmenides's *doxa*.)

Thus the 'dark saying' (as Plato calls it, *Theaetetus* 152c) of Protagoras may perhaps be best understood as his summary of the following argument. Let us assume that Parmenides is right and that true knowledge of reality—of what really exists—is confined to the gods, while mortal men rely in general upon their much-erring senses and upon human conventions. Then, since we *are* men and have no other standard (or 'measure') than the human standard for deciding upon 'the existence of things that are, and the non-existence of things that are not' (*Theaetetus* 152a), we have to subscribe (not half-heartedly, as did

[22] There are two more points in Parmenides B16 on which I might comment.

(*a*) Although, under the influence of Tarán's argument, I replaced in the third edition of this book, 'prevails' by 'contains' (see also my discussion in *Studies in Philosophy*, edited by J. N. Findlay, Oxford Paperbacks 112, p. 193), I have now decided to return to 'What in this mixture prevails'.

(*b*) The other point (which was mentioned in the second edition of the present book in a footnote on p. 222) is a reference to Charles H. Kahn in connection with the meaning of 'nature' (*phusis*) in B16 : 3 as the 'state of the physical composition' or the 'state of the mixture' of a thing. This sense is discussed by Kahn in his *Anaximander*, 1960, for example on p. 202, in a most informative passage in which Kahn quotes *De Victu*.

Empedocles, but) whole-heartedly to the sensualist epistemology (described so sarcastically by Parmenides) as the only possible theory of human knowledge. Thus truth becomes subjective.

If we assume that Democritus was influenced by both Parmenides and Protagoras, then Democritus's famous dialogue between reason and the senses (Democritus B125) may be described as a dialogue between these two views. For Reason (that is, Eleaticism) attacks the senses: 'Sweet: by convention; bitter: by convention; cold: by convention; colour: by convention. In truth, there are only atoms and the void.'[23] The senses (Protagoras) reply: 'Poor intellect! You who are taking your evidence from us are trying to overthrow us? Our overthrow is your own downfall.'

Epicurus followed in this issue Protagoras rather than Democritus. But the most concise Parmenidean formula, *without* the occurrences of 'erring', seems to be due to St. Thomas Aquinas: 'Nothing is in the intellect that was not before in the senses.'

Two hundred and fifty years after St. Thomas, a kind of restatement of Parmenides's (or Democritus's) position may be found in C. Bovillus (1470–1533), *De intellectu*: 'Nothing is in the senses that was not before in the intellect. Nothing is in the intellect that was not before in the senses. The first is true of angels, the second of humans.' Parmenides said essentially the same: the first is the way of truth, as revealed by the goddess; the second is the way of the deluded opinion of erring mortals.

All this, I propose, is contained in his epistemological fragment, B16.

9. The Presocratics: Unity or Novelty? (1968)

Since writing, late in 1960, the Appendix (now further expanded) to chapter 5, I have read Charles H. Kahn's most admirable book *Anaximander and the Origins of Greek Cosmology* (1960). Kahn rightly emphasizes 'the essential unity' (p. 5) of the early speculation about nature, and points out that the framework of Anaximander's thought dominates the cosmologies of his successors at least down to Plato's *Timaeus*. I find

[23] Democritus was aware of the fact that his atomism was a rationalistic doctrine of a reality behind the appearance, and a development of Eleaticism; and so was Aristotle.

his emphasis important as an antidote to my own emphasis upon the novelty of the successive theories. Yet my thesis that the novelty is *the result of a critical debate* seems to me to cover both points of view: there clearly is both unity *and* novelty.

Perhaps I may add here a point concerning Anaximander's theory of the free suspension of the earth which both Kahn and I find so important. I have suggested that it may well be the result of Anaximander's criticism of Thales. But it seems to me clear that it is also a critical response to a passage in the *Theogony* (720–725). This passage strongly suggests that the earth is equi-distant from those parts of the Universe which surround it: for Tartarus is here said to be exactly as far beneath the earth as heaven (Uranus) is above the earth. (Compare also the *Iliad*, 8, 13–16; *Aeneid*, vi, 577.) This passage also suggests very strongly that *we can draw a diagram* in which, if heaven is conceived as a kind of sphere, the earth would occupy the position assigned to it by Anaximander.[24]

10. An Argument, due to Mark Twain, Against Naïve Empiricism (1989)

My argument on pp. 27 f. against a naive form of empiricism was essentially anticipated by Mark Twain. On his first appointment as a reporter, he tells us, the editor of the newspaper instructed him never to report anything unless he could *verify it or confirm it by personal knowledge*. So he described a social event as follows: 'A woman giving the name of Mrs James Jones, who is reported to be one of the society leaders of the city, is said to have given what purported to be a party yesterday to a number of alleged ladies. The hostess claims to be the wife of a reputed attorney.'

One sees that Mark Twain was quick to realize the silliness of the naive empiricist (verificationist) theory of the sources of our knowledge.

[24] The *Iliad*, 8, 13–16 is quoted by Kahn who, though he refers to the *Theogony*, does not mention *Theogony* 720–5 (perhaps because lines 721 to 725 are missing in some of the MSS, or otherwise suspect?). This may explain why he says (p. 82) about *Theogony* 727 ff., etc.: 'It would be hopeless to draw a diagram to accompany such a description.'

A very interesting discussion of the relationship between Hesiod and Anaximander can be found in Paul Seligman, *The 'Apeiron' of Anaximander*, 1962.

Index of Mottoes

(The various translations throughout the book are mine.)

Index of Names

Note: 'q' means quoted; 'n' means footnote

Index of Subjects

Note: 'n' means footnote